AGING OF CELLS IN AND OUTSIDE THE BODY

BIOLOGY OF AGING AND ITS MODULATION

VOLUME 2

AGING OF CELLS IN AND OUTSIDE THE BODY

Edited by

SUNIL C. KAUL
and
RENU WADHWA

National Institute of Advanced Industrial
Science and Technology
Tsukuba Science City
Ibaraki, Japan

SPRINGER-SCIENCE+BUSINESS MEDIA, B.V.

Library of Congress Cataloging-in-Publication Data

Aging of cells in and outside the body / edited by Sunil C. Kaul and Renu Wadwa
p. cm. – (Biology of aging and its modulation)
ISBN 978-90-481-6314-4 ISBN 978-94-017-0669-8 (eBook)
DOI 10.1007/978-94-017-0669-8
1. Cells–Aging I. Series

2003048941

ISBN 978-90-481-6314-4

Printed on acid-free paper

Contents

About the series "Biology of aging and its modulation"

During the last 40 years, the study of the biological basis of aging has progressed tremendously, and it has now become an independent and respectable field of study and research. Several universities, medical institutes and research centers throughout the world now offer full-fledged courses on biogerontology. The interest of students taking such courses, followed by undertaking research projects for MSc and PhD studies, has also increased significantly. Cosmetic, cosmeceutical and pharmaceutical industry's ever increasing interest in aging research and therapy is also obvious. Moreover, increased financial support by the national and international financial agencies to biogerontological research has given much impetus to its further development.

This five-volume series titled "Biology of Aging and its Modulation" fulfills the demand for books on the biology of aging, which can provide critical and comprehensive overview of the wide range of topics, including the descriptive, conceptual and interventive aspects of biogerontology. The titles of the books in this series and the names of their respective editors are:

1. Aging at the molecular level (Thomas von Zglinicki, UK)
2. Aging of cells in and outside the body (S. Kaul and R. Wadhwa, Japan)
3. Aging of organs and systems (R. Aspinall, UK)
4. Aging of organisms (H. D. Osiewacz, Germany)
5. Modulating aging and longevity (S. Rattan, Denmark)

The target readership is both the undergraduate and graduate students in the universities, medical and nursing colleges, and the post-graduates taking up research projects on different aspects of biogerontology. We hope that these books will be an important series for the college, university and state libraries maintaining a good database in biology, medical and biomedical sciences. Furthermore, these books will also be of much interest to pharmaceutical, cosmaceutical, nutraceutical and health-care industry for an easy access to accurate and reliable information in the field of aging research and intervention.

Suresh I.S. Rattan, Ph.D., D.Sc.
Series Editor and Editor-in-Chief, Biogerontology
Danish Centre for Molecular Gerontology, Department of Molecular Biology,
University of Aarhus, Denmark

Preface

Aging is an intrinsic property and is manifested at all levels of a living organization. Forty years have passed since its first manifestation at the cellular level by Hayflick and co-workers. Thanks to all who aboard the journey of 40 years of thoughts, predictions, experimentations, demonstrations and validations that have eventually established that lifespan of almost all kind of normal somatic cells is counted not by the calendar time but by the number of doublings that they undergo. Volumes of literature are available on the features of aged, but viable and metabolically active, cells. The most consistent of these are their loss of division capacity, unresponsiveness to mitogens, functional decline and failure of homeodynamics. Genetically altered (accelerated aging syndromes) or environmentally challenged cells show these features much earlier in their life as compared to the normal cells.

It has become increasingly clear that living system is not equipped with any special genes that cause aging rather it is an inevitable outcome of imperfect maintenance and repair mechanisms both at the genetic and cellular level. It is a byproduct of gene activities that are essential for keeping the order, integrity and normal function of a living organism early in lifespan. The gene functions during different phases of life transpire in a highly programmed and orderly fashion thus it is not surprising that their byproducts also appear to progress in a programmed way. However, an important component is contributed by the environment and epigenetic factors that in turn control gene activities, and the defense mechanisms that control the accumulation of cellular damage. Established as a dominant trait over immortality (a common feature of cancer cells) and serving as a tumor suppressive mechanism in all living organisms, it experiences a fine tuning and we see that our close genetic relative, mouse that has only 1% difference in genetic composition, undergo differential regulation of aging mechanisms.

With understanding of the most basic aspects and mechanisms regulating aging of cells *in vitro* today, the future challenge lies in the use of most recent advancements of molecular technology to understand further its *in vivo* aspects and intervention.

We are very grateful to all the authors for their interest, enthusiasm and contributions in this book project. We would like to offer our sincerest thanks to the Kluwer Academic Publisher, who made this publication possible.

Sunil C. Kaul
Renu Wadhwa

Aging Outside the Body: Usefulness of the Hayflick System

Suresh I.S. Rattan
Laboratory of Cellular Ageing, Danish Centre for Molecular Gerontology, Department of Molecular Biology, University of Aarhus, DK-8000 Aarhus C, Denmark

Introduction

Until 1961, it was generally believed that the phenomena of aging and death were primarily a characteristic of organisms, whereas the cells isolated from the body did not undergo aging and were potentially immortal. The dramatic shift in paradigm began to occur with the publication of a series of papers by Leonard Hayflick and colleagues [1, 2], followed by hundreds of other publications which helped to establish the new paradigm that normal, diploid and differentiated cells have a limited proliferative potential – an intrinsic limit which is best observed *in vitro*, and is most likely to exist *in vivo*. As recounted by Hayflick in one of his recent interviews [3], Sir Macfarlane Burnett, Nobel Laureate from Australia, coined the phrase "Hayflick limit" for this characteristic of normal cells. Until now, the reality of the Hayflick limit has been demonstrated in several normal cell types including fibroblasts, epithelial cells, endothelial cells, keratinocytes, melanocytes, chondrocytes, muscle satellite cells, articular cartilage cells, glial cells, lymphocytes, trabecular osteoblasts and bone marrow cells, isolated from a variety of tissues and animals.

In a similar vein, the phenomenon of serial sub-cultivation, divided into a Phase I of culture establishment, a Phase II of constant growth and proliferation, and a Phase III of slowing down and eventual growth arrest [1, 2], is often known as the Hayflick phenomenon, and the complete experimental set up as the Hayflick system. However, in recent times, the Hayflick phenomenon has been termed as replicative senescence. What is often forgotten in the contemporary discussion of replicative senescence is that the emergence of the ultimate phenotype of irreversible growth arrest of normal cells is preceded by hundreds of physical, physiological, biochemical and molecular changes [4–6]. Most of those changes are seen to occur progressively as a function of cumulative population doubling level (PDL) achieved by serial sub-

Sunil C. Kaul and Renu Wadhwa (eds.), Aging of Cells In and Outside the Body, 1–8.

cultivation of normal cells, and are independent of the chronological time spent in a non-dividing state. However, the ultimate phenotype of replicative senescence being interpreted as a cancer-preventive mechanism and the tremendous success in elucidating the complex genetic network of cell-cycle-regulating pathways have overshadowed the contribution of the Hayflick system in unravelling the basic process of aging. Therefore, in the following sections, the usefulness and significance of studies on cellular aging *in vitro* in establishing the basis for understanding aging *in vivo* are recapitulated.

Failure of homeodynamics

Homeodynamics (previously termed as homeostasis) is the ability of all living systems to respond to internal and external stress, and to counteract by neutralization and/or by adaptation any disturbances threatening their survival. Incomplete homeodynamics leads to the impairment in functional ability at all levels of organization and increased possibilities of a plethora of diseases and eventual death. Several theories of aging imply directly or indirectly that a progressive failure of homeodynamics is a crucial determinant of the process of aging [5]. For example, inability to protect against free-radical induced damage to macromolecules, inefficient mechanisms of maintaining the fidelity of genetic information transfer, and incomplete elimination of abnormal proteins are some of the main molecular mechanisms of aging, which also represent the failure of homeodynamics [5, 6].

Since appropriate and efficient cellular responsiveness is integral to successful homeodynamics, the Hayflick system has proved to be of great value for elucidating the age-related changes in cellular responses to various physical, chemical and biological factors. It has been shown that the mitogenic and growth-stimulating effects of growth factors, hormones and other agents are reduced significantly during cellular aging. In contrast, the sensitivity of aging cells to toxic agents including antibiotics, phorbol esters, irradiation and other free radical-inducers increases with age [7]. Further studies using this system have shown that the reasons for the age-related changes in the responsiveness of cells are not due to any qualitative and quantitative changes in the cell surface receptors, but are mainly due to defective pathways of signal transduction involving protein kinases and calcium mobilization [8, 9]. Studies on altered responsiveness of lymphocytes during aging have also utilized extensively the *in vitro* system of cellular aging [10, 11].

Another important homeodynamic pathway is the so-called heat shock (HS) response, which is one of the primordial intracellular defence mechanisms against stressful conditions [12, 13]. Exposure of cells and organisms to stresses such as high temperature, caloric restriction, exercise, oxidative and osmotic stress, heavy metals, proteasome inhibitors, amino acids analogues, ethanol, glutathione depletion, calcium ionophores and metabolic poisons induce the cellular stress response leading to the preferential transcription and translation of heat shock proteins (HSP). When a cell encounters a stressor, modifications of the cytoskeleton, cytoplasmic structures, cell surface morphology, cellular redox status, DNA synthesis, protein metabolism and protein stability occur [13]. Stress generates molecular damage, especially abnormally folded proteins, which can aggregate and initiate a sequence of stress response. Stress activates major signalling pathways such as transcription factors HSF1, NFκb and p53, and other pathways regulated by protein kinases [9]. Optimal HS response in terms of HSP synthesis and activity is essential for cell survival. Using the Hayflick system it has been demonstrated that a progressive attenuation of HS response is a critical determinant of the aging process [14].

One of the recent approaches in gerontomodulation is also based in making use of the homeodynamic characteristic discussed above, and it has been hypothesized that if cells and organisms are exposed to brief periods of stress so that their stress response-induced gene expression is upregulated and the related pathways of maintenance and repair are stimulated, one should observe anti-aging and longevity-promoting effects. Such a phenomenon in which stimulatory responses to low doses of otherwise harmful conditions improve health and enhance lifespan is known as hormesis [15]. The Hayflick system has proved to be quite useful in testing the hypothesis of hormesis. For example, anti-aging and replicative lifespan-prolonging effects of low doses of irradiation have been reported for human fibroblasts [16–18]. Using a mild stress regime of repeatedly exposing human fibroblasts to 41°C for 1 h twice a week throughout their replicative lifespan *in vitro*, we have reported several beneficial anti-aging effects. These effects included a reduction in the accumulation of oxidized and glycoxidized proteins, an increase in the levels of various heat shock proteins (hsp70, hsc70 and hsp27), an increase in proteasomal activities, an increase in antioxidative abilities, and increased resistance to ethanol, hydrogen peroxide and UV-A irradiation [19–23]. An important aspect of these studies is the observation that anti-aging and beneficial effects of repeated mild heat shock on human cells were observed without inducing additional cell proliferation. It appears that the progression of cellular aging *in vitro* in terms of accumulation of molecular damage can be slowed down without interfering with the regulatory mechanisms of cell cycle checkpoints.

Instability of the genome

Mortality of organisms is often linked to a progressive instability of the genome in the wake of various physical, chemical and biological factors which continuously challenge it. Both the nuclear and the mitochondrial genomes are prone to such damage. The Hayflick system has proved to be very useful in studying spontaneous and induced DNA damage and repair during aging [24–27]. Several types of DNA damage, including single- and double-strand breaks, oxidatively modified bases, and mutations and deletions been reported to increase during cellular aging [28]. The role of a progressive loss of DNA methylation [29, 30] and of telomeres [31] in aging was also first reported using the Hayflick system. Other studies performed with this system showed that increased instability of the genome of senescent cells made them more prone to viral- and radiation-induced transformation and carcinogeneis [32]. These observations have significance in understanding the relationship between human aging and cancer, since the incidence of cancer is known to increase exponentially with age [33].

With respect to the DNA repair capacity, a positive relation between the ability of fibroblasts to repair UV-induced bulk DNA damage and species lifespan has been reported [34]. However, in the case of serially passaged human fibroblasts and osteoblasts, no change in the capacity to repair UV-induced DNA damage both in bulk DNA and in certain active and inactive gene sequences was observed [35]. It is possible that there are cell type-specific and gene-specific differences in the DNA repair capacity, which can be elucidated using the Hayflick system comprising different cell types.

Misregulation of genetic information transfer

The fidelity of DNA, RNA and protein synthesis is crucial for the normal growth, development and survival of a living system. Therefore, misregulation of genetic information transfer is considered as a crucial molecular determinant of aging. At least three lines of evidence from experiments performed with the Hayflick system have shown that defective genetic information transfer has important role to play during aging. First, cell extracts prepared from serially passaged human fibroblasts showed increased misincorporation of nucleotides and amino acids by senescent cells. Second, an accumulation of altered and abnormal proteins, presumably due to errors in synthesis and/or post-translational modifications, has been reported [36]. Third, an experimental increase in transcriptional and translational errors by either feeding the cells with nucleotide and amino acid analogues, or by treating cells with error-inducing drugs accelerates aging of human cells in culture [5, 37, 38].

The Hayflick system has also been useful in elucidating age-related changes in gene expression, bulk protein synthesis, components of the protein synthetic machinery, pattern of proteins synthesised and the turnover of proteins [5, 37, 38]. Gene-array based analysis of thousands of genes have identified similarities and differences in the aging pattern of normal fibroblasts and those from premature aging syndromess [39–41]. Furthermore, several genes have been identified whose products either act as active regulators of cell cycle-arrest or correlate with replicative senescence. Some of the main genes associated with the induction and maintenance of permanent growth arrest of cells are $p21^{sdi\text{-}1/waf\text{-}1/cip\text{-}1}$, $p16^{INK4}$, and $p19^{ARF}$, and these are often referred to as senescence-specific genes [42–48]. These genes are generally involved in either the activation or the inhibition of protein phosphorylation and dephosphorylation cascade involving various transcription factors and cyclin complexes in association with p53 and Rb genes [49–52].

However, the so-called senescence specific upregulation of genes is the consequence of progressive cellular aging during which molecular damage gets accumulated and becomes the signal for the activation of various genes whose products then act towards arresting and maintaining the senescent cell in a permanent state of growth arrest [53, 54]. Experimental induction of macromolecular damage by free radicals and an excessive loss of telomeres by oxidative stress results in the premature onset of the senescent phenotype by the resultant rapid activation of the so-called senescence-specific genes [55, 56].

Conclusion

It is thus clear that the development and use of the Hayflick system has revolutionised our understanding of the biological basis of aging. First of all, it provided the firm basis for the shift in focus to look for the causes of aging in cellular and intracellular events, rather than in extracellular phenomena. Second, it facilitated making a distinction between aging as a progressive and stochastic increase in molecular disorder and longevity of cells and organisms, which has strict genetic regulation [4, 57]. Unfortunately, too much focus on the end-point replicative senescence has often been the basis of doubts raised against the validity of the Haflick limit *in vivo* pointing to the fact that several cell types in the body may never reach that limit. In all such arguments it is often conveniently ignored that the Hayflick system primarily deals with the question of progressive accumulation of damage and its consequences in changing the cellular physiology of normal, diploid and differentiated cells. Whereas replicative senescence may be a crucial factor for the aging and diseases of some cell types *in vivo,* for example lymphocytes and AIDS [58–60], altered cellular responsiveness, instability of the genome and misregulation of macromolecular synthesis and turnover can have multiple and harmful effects on the surrounding cells, tissues and organs. For example, there is some evidence that senescent fibroblasts promote epithelial cell growth and tumorigenesis in mice [61].

There is an obvious and urgent need of comprehensive studies to answer the question of the occurrence of senescent cells in tissues, and its consequences on the functional integrity of the organs and organisms. The previous forty years of gaining knowledge from the use of the Hayflick system about aging outside the body have provided a solid foundation for understanding and modulating aging inside the body.

References

1. Hayflick L, Moorhead PS (1961). The serial cultivation of human diploid strains. *Exp Cell Res.* 25: 585–621.
2. Hayflick L (1965). The limited *in vitro* lifetime of human diploid cell strains. *Exp Cell Res.* 37: 614–36.
3. Rattan SIS (2000). "Just a fellow who did his job ...", an interview with Leonard Hayflick. *Biogerontology* 1: 79–87.
4. Hayflick L (1994). *How and Why We Age.* New York: Ballantine Books.
5. Holliday R (1995). *Understanding Ageing.* Cambridge: Cambridge University Press.
6. Rattan SIS (1995). Ageing – a biological perspective. *Molec Aspects Med.* 16: 439–508.
7. Rattan SIS, Derventzi A (1991). Altered cellular responsiveness during ageing. *BioEssays* 13: 601–6.
8. De Tata V, Ptasznik A, Cristofalo VJ (1993). Effect of tumor promoter phorbol 12-myristate 13-acetate (PMA) on proliferation of young and senescent WI-38 human diploid fibroblasts. *Exp Cell Res.* 205: 261–9.
9. Finkel T, Holbrook NJ (2000). Oxidants, oxidative stress and the biology of ageing. *Nature* 408: 239–47.

10. Pawelec G, Müller R, Rehbeib A, Hähnel K, Ziegler, BL (1999). Finite lifespans of T cell clones derived from CD34+ human haematopoietic stem cells in *vitro*. *Exp Gerontol.* 34: 69–77.
11. Pawelec G, Wagner W, Adibzadeh M, Engel A (1999). T cell immunosenescence *in vitro* and *in vivo*. *Exp Gerontol.* 34: 419–29.
12. Feder ME, Hofmann GE (1999). Heat-shock proteins, molecular chaperones, and the stress response. *Annu Rev Physiol.* 61: 243–82.
13. Verbeke P, Fonager J, Clark BFC, Rattan SIS (2001). Heat shock response and ageing: mechanisms and applications. *Cell Biol Int.* 25: 845–57.
14. Liu AYC, Choi HS, Lee YK, Chen KY (1991). Molecular events involved in transcriptional activation of heat shock genes become progressively refractory to heat stimulation during aging of human diploid fibroblasts. *J Cell Physiol.* 149: 560–6.
15. Rattan SIS (2001). Applying hormesis in aging research and therapy. *Human Exp Toxicol.* 20: 281–5.
16. Macieira-Coelho A, Diatloff C, Billardon C, Bourgeois CA, Malaise E (1977). Effect of low dose rate ionising radiation on the division potential of cells *in vitro*. III. Human lung fibroblasts. *Exp Cell Res.* 104: 215–21.
17. Holliday R (1991). A re-examination of the effects of ionizing radiation on lifespan and transformation of human diploid fibroblasts. *Mutat Res.* 256: 295–302.
18. Suzuki M, Yang Z, Nakano K, *et al.* (1998). Extension of *in vitro* life-span of gamma-irradiated human embryo cells accompanied by chromosome instability. *J Radiat Res.* 39: 203–13.
19. Rattan SIS (1998). Repeated mild heat shock delays ageing in cultured human skin fibroblasts. *Biochem Mol Biol Int.* 45: 753–9.
20. Verbeke P, Clark BFC, Rattan, SIS (2000). Modulating celluar aging *in vitro*: hormetic effects of repeated mild heat stress on protein oxidation and glycation. *Exp Gerontol.* 35: 787–94.
21. Verbeke P, Clark BFC, Rattan, SIS (2001). Reduced levels of oxidized and glycoxidized proteins in human fibroblasts exposed to repeated mild heat shock during serial passaging *in vitro*. *Free Rad Biol Med.* 31: 1593–602.
22. Verbeke P, Deries M, Clark BFC, Rattan SIS (2002). Hormetic action of mild heat stress decreases the inducibility of prtoein oxidation and glycoxidation in human fibroblasts. *Biogerontology* 3: 105–8.
23. Fonager J, Beedholm R, Clark BFC, Rattan SIS (2002). Mild stress-induced stimulation of heat shock protein synthesis and improved functional ability of human fibroblasts undergoing aging *in vitro*. *Exp Gerontol.* 37: 240–7.
24. Rattan SIS (1989). DNA damage and repair during cellular aging. *Int Rev Cytol.* 116: 47–88.
25. Choi AMK, Pignolo RJ, Rhys CMJ, Cristofalo VJ, Holbrook NJ (1995). Alterations in the molecular response to DNA damage during cellular aging of cultured fibroblasts: reduced AP-1 activation and collagenase expression. *J Cell Physiol.* 164: 65–73.
26. Norwood TH, Gray M (1996). The role of DNA damage in cellular aging: is it time for a reassessment? *Exp Gerontol.* 31: 61–8.
27. Kaneko T, Tahara S, Taguchi T, Kondo H (2001). Accumulation of oxidative damage, 8-oxo-2′-deoxyguanosine, and change of repair systems during *in vitro* cellular aging of cultured human skin fibroblast. *Mutat Res.* 487: 19–30.
28. Wolf FI, Torsello A, Covacci V, *et al.* (2002). Oxidative DNA damage as a marker of aging in WI-38 human fibroblasts. *Exp Gerontol.* 37: 647–56.

29. Wilson VL, Jones PA (1983). DNA methylation decreases in aging but not in immortal cells. *Science* 220: 1055–7.
30. Catania J, Fairweather DS (1991). DNA methylation and cellular ageing. *Mutat Res.* 256: 283–93.
31. Harley CB, Futcher AB, Greider CW (1990). Telomeres shorten during ageing of human fibroblasts. *Nature* 345: 458–60.
32. Huschtscha LI, Holliday R (1983). The limited and unlimited growth of SV40-transformed cells from human diploid MRC-5 fibroblasts. *J Cell Sci.* 63: 77–99.
33. Macieira-Coelho A (2001). Neoplastic disease through the human life span. *Biogerontology* 2: 179–92.
34. Hart R, Setlow RB (1974). Correlation between deoxyribonucleic acid excision-repair and life span in a number of mammalian species. *Proc Natl Acad Sci USA* 71: 2169–73.
35. Christiansen M, Stevnsner T, Bohr VA, Clark BFC, Rattan SIS (2000). Gene-specific DNA repair of pyrimidine dimers does not decline during cellular aging *in vitro*. *Exp Cell Res.* 256: 308–14.
36. Rosenberger RF (1991). Senescence and the accumulation of abnormal proteins. *Mutat Res.* 256: 255–62.
37. Holliday R (1996). The current status of the protein error theory of aging. *Exp Gerontol.* 31: 449–52.
38. Rattan SIS (1996). Synthesis, modifications and turnover of proteins during aging. *Exp Gerontol.* 31: 33–47.
39. Shelton DN, Chang E, Whittier PS, Choi D, Funk WD (1999). Microarray analysis of replicative senescence. *Curr Biol.* 9: 939–45.
40. Ly DH, Lockhart DJ, Lerner RA, Schultz PG (2000). Mitotic misregulation and human aging. *Science* 287: 2486–92.
41. Park WY, Hwang CI, Kang MJ, *et al.* (2001). Gene profile of replicative senescence is different from progeria or elderly donor. *Biochem Biophys Res Commun.* 282: 934–9.
42. Noda A, Ning Y, Venable SF, Pereira-Smith OM, Smith JR (1994). Cloning of senescent cell-derived inhibitors of DNA synthesis using an expression screen. *Exp Cell Res.* 211: 90–8.
43. Wong H, Riabowol K (1996). Differential CDK-inhibitor gene expression in aging human diploid fibroblasts. *Exp Gerontol.* 31: 311–25.
44. Whitaker NJ, Bryan TM, Bonnefin P, *et al.* (1995). Involvement of RB-1, p53, $p16^{INK4}$ and telomerase in immortalisation of human cells. *Oncogene* 11: 971–6.
45. Stein GH, Drullinger LF, Soulard A, Dulic V (1999). Differential roles for cyclin-dependent kinase inhibitors p21 and p16 in the mechanisms of senescence and differentiation in human fibroblasts. *Mol Cell Biol.* 19: 2109–17.
46. Alcorta DA, Xiong Y, Phelps D, Hannon G, Beach D, Barrett JC (1996). Involvement of the cyclin-dependent kinase inhibitor p16 (INK4a) in replicative senescence of normal human fibroblasts. *Proc Natl Acad Sci USA* 93: 13742–7.
47. Kamb A, Gruis NA, Weaver-Feldhaus J, *et al.* (1994). A cell cycle regulator potentially involved in genesis of many tumor types. *Science* 264: 436–40.
48. Jacobs JJL, Kieboom K, Marino S, DePinho RA, van Lohuizen M (1999). The oncogene and Polycom-group gene *bmi-1* regulates cell proliferation and senescence through *ink4a* locus. *Nature* 397: 164–8.
49. Derventzi A, Rattan SIS, Gonos ES (1996). Molecular links between cellular mortality and immortality. *Anticancer Res.* 16: 2901–10.
50. Gao CY, Zelenka PS (1997). Cyclins, cyclin-dependent kinases and differentiaion. *BioEssays* 19: 307–15.

51. Kaul SC, Mitsui Y, Wadhwa R (1998). Molecular insights to cellular mortality and immortalization. *Ind J.Exp Biol.* 36: 345–52.
52. Duncan EL, Wadhwa R, Kaul SC (2000). Senescence and immortalization of human cells. *Biogerontology* 1: 103–21.
53. Reddel RR (1998). A reassessment of the telomere hypothesis of senescence. *BioEssays* 20: 977–84.
54. Faragher RGA, Jones CJ, Kipling D (1998). Telomerase and cellular lifespan: ending the debate? *Nature Biotechnology* 16: 701–2.
55. von Zglinicki T (2002). Oxidative stress shortens telomeres. *Trends Biochem Sci.* 27: 339–44.
56. von Zglinicki T, Bürkle A, Kirkwood TBL (2001). Stress, DNA damage and ageing – an integrative approach. *Exp Gerontol.* 36: 1049–62.
57. Hayflick L (2000). The future of ageing. *Nature* 408: 267–9.
58. Effros RB (1996). Insights on immunological aging derived from the T lymphocyte cellular senescence model. *Exp Gerontol.* 31: 21–7.
59. Effros RB, Pawelec G (1997). Replicative senescence of T lymphocytes: does the Hayflick limit lead to immune exhaustion? *Immunol Today* 18: 450–4.
60. Holliday R (1998). Is AIDS due to the clonal senescence of CD4 lymphocytes? *Exp Gerontol.* 33: 213–15.
61. Krtolica A, Parrinello S, Lockett S, Desprez P-Y, Campisi J (2001). Senescent fibroblasts promote epithelial cell growth and tumorigenesis: a link between cancer and aging. *Proc Natl Acad Sci USA* 98: 12072–7.

Genetics of Cellular Aging

Olivia M. Pereira-Smith, Paul Rookard, Johanna Echigo and James R. Smith

The University of Texas Health Science Center, Sam and Ann Barshop Center for Longevity and Aging Studies, STCBM Bldg., 15355 Lambda Dr., San Antonio, TX 78245-3207, USA

Introduction

Body homeostasis is maintained by positive and negative regulatory signals in the various cells and organs of an animal. Changes in regulation of these signals could lead to loss of cell proliferative response, as observed in the aging immune system, or alternatively to uncontrolled cell division, one step in the progression toward neoplasia. There is good evidence that the likelihood of both these changes increases with age. Therefore, changes in the regulation of cell proliferation are an extremely important component of not only the general aging phenotype, but also the onset of tumorigenicity. The loss of cell proliferative ability of normal animal cells in culture following the accruement of a defined number of population doublings is well documented. This phenomenon has been proposed as a model for aging at the cellular level [1]. Data in support of the model include the observations that a direct correlation exists between species life span and *in vitro* life span [2], but that this correlation is inverse with donor age [3–7]. The frequency of reversal of the senescence phenotype (i.e., immortalization) varies greatly with different animal species. The probability of immortalization (spontaneous and induced) in mouse cells is extremely high, whereas in human cells spontaneous immortalization has never been shown to occur and induced immortalization using agents such as virus or chemical carcinogens is very low (probably $< 10^{-8}$). Because of this greater genomic stability the majority of studies in the field of cellular aging have utilized human cells.

The mechanisms involved in cellular senescence and changes that occur to yield abnormal tumor cells that can proliferate indefinitely (immortal) have become more clear in recent years. In the past, cell fusion studies and molecular approaches gave some idea of the nature of these phenomena and the events that might be occurring as cells senesce or immortalize. However, with our ever-increasing understanding of

Sunil C. Kaul and Renu Wadhwa (eds.), Aging of Cells In and Outside the Body, 9–26.

the genes and proteins involved there is growing evidence for a role of epigenetic in addition to genetic events in these processes. We here present insights gained from studies over the years as well as more recent results.

Short-term analysis of cell fusion products

Among other criteria, senescence in human cells is defined as the point at which the cells become mitogen non-responsive and the population does not double in a set period of time (2–4 weeks depending on the laboratory). Although nonproliferative, the cells do remain viable and metabolically active at this time, and have been maintained in this state for >2 years [8], own unpublished observations). The major loss of function is in the ability to synthesize DNA and divide. One set of studies attempted to understand the reason for this loss of growth factor response in senescent cells by short-term studies of DNA synthesis in the nuclei of heterokaryons or by reconstructing cells.

Studies with heterokaryons

Norwood *et al.* [9] fused normal young with senescent human cells and determined the ability of the nuclei in a heterodikaryon to synthesize DNA, as measured by tritiated-thymidine autoradiography. The cytoplasm of the cells was prelabeled with either tritiated methionine or 14 C thymidine to allow for identification of heterodikaryons by double-layer autoradiography. They found that senescent cells could inhibit DNA synthesis in the nucleus of the otherwise proliferation-competent young cell. The control young homodikaryons were unaffected in DNA-synthetic capability, indicating that the result was not due to the experimental manipulation. Stein's group independently confirmed these observations [10]. Their studies showed further that the block in senescent cells was at initiation of DNA synthesis, since young nuclei that had already entered the S phase of the cell cycle at the time of fusion were not inhibited in progression through the S phase. These results led to the hypothesis that senescent cells produced an inhibitor of DNA synthesis initiation that was diffusible and could affect a young nucleus in the common cytoplasm of the heterodikaryon. Norwood's group demonstrated that this inhibitory activity could be overcome by fusion of senescent human fibroblasts with the immortal cell line HeLa or with immortal simian virus 40 (SV40)-transformed cells [11]. DNA synthesis was reinitiated in senescent nuclei in a large number of the heterodikaryons resulting from these fusions. However, it was not known if the senescent nucleus carried out a complete round of semiconservative replication. These results showed that, although senescent cells would not synthesize DNA under standard circumstances of mitogen stimulation, they were not structurally incapable of carrying out DNA synthesis. Stein and Yanishevsky [12], on the other hand, shortly thereafter found that senescent human cells were able to inhibit DNA synthesis in the nuclei of the immortal cells T98G (human glioblastoma cells) or Rk-13 (rabbit kidney cells). Stein and co-workers [13] then demonstrated that the ability of an immortal cell to overcome the DNA synthesis-inhibitory activity of senescent cells was dependent on the presence of DNA tumor viral genomes (SV40 in the case of immortal S40-transformed cells and papilloma virus DNA in the case of Hela cells), since a carcinogen-derived immortal cell line was unable to override the senescent cell inhibitory activity. These results led to the idea that there are two classes of immortal cells, one that can respond to the inhibitor of DNA synthesis present in senescent cells and another that is able to overcome the inhibitory activity.

In order to determine whether this inhibitory activity was associated with the cytoplasm or the nucleus of the senescent cell, Norwood's group and ours independently proceeded to analyze the DNA-synthetic capability of young cell nuclei in cybrids made from fusion of senescent cytoplasts with whole young cells. We both determined that the senescent cytoplast was as capable of inhibiting DNA synthesis in the young nucleus as was the whole cell [14–15]. In order to explore whether a protein was involved we treated the cytoplasts before fusion with protein synthesis inhibitors, cycloheximide and puromycin. We demonstrated that the inhibitory activity was lost following treatment with either agent, with even a short treatment (as little as 2 h with cycloheximide) resulting in loss of the inhibitory activity. This indicated that the inhibitor of DNA synthesis in senescent cells was a protein(s) or that this activity was mediated by a protein(s). If the cytoplasts were allowed to recover in the absence of cycloheximide, the inhibitory activity was regained within ~3–4 h, indicating that the message coding for this inhibitory protein(s) was long-lived [15].

To determine whether the protein(s) was associated with the cytoplasm or the membrane of the cells, we treated the senescent cytoplasts with 0.125% trypsin in the cold for 1 min. During such a treatment, the trypsin would not be likely to enter the cytoplasm of the cells. We determined that the trypsin treatment eliminated the inhibitory activity and that recovery of the cytoplasts following trypsin treatment permitted a return to activity in approximately the same time frame as was seen in the case of cycloheximide treatment. These results indicated to us that the protein(s) inhibitor of DNA synthesis in senescent cells was most likely located on the outside surface of the membranes [16]. Stein and we have since independently shown that isolated surface membrane preparations from senescent cells and proteins extracted from these membranes are capable of inhibiting DNA synthesis in proliferation-competent young cells when the membranes or proteins are added to cell culture medium in which the young cells are grown [16–17].

Young, logarithmically growing cells do not express a protein(s) inhibitor of DNA synthesis. However, if young cells are made nondividing (quiescent) by growth factor deprivation for at least 2 weeks, or by growth to high density in the presence of growth factors, they express an inhibitor of DNA synthesis [17–18]. The inhibitor of quiescent cells has many properties similar to senescent cells [16]. It is a protein(s), is trypsin-sensitive, and is present in surface membrane-enriched preparations and in the proteins extracted from these membranes. The major difference is that it is not inactivated by cycloheximide treatment for $\leqslant 24$ h, indicating that it is a very stable protein(s) with a long half-life. We were able to demonstrate that the quiescent cell inhibitory activity did involve a protein by trypsin-treating cytoplasts and allowing them to recover from the treatment in the presence and absence of cycloheximide before fusion. Cytoplasts incubated with cycloheximide never recovered inhibitory activity, whereas those not exposed to the protein synthesis inhibitor did. At this time we do not know if the DNA synthesis inhibitors expressed in senescent and quiescent cells are different or the same protein(s) constitutively expressed in senescent cells and mitogen-regulated in quiescent cells.

Studies to attempt to isolate this inhibitor protein(s) have been hampered by the fact that they are membrane associated and that obtaining large numbers of senescent cells for biochemical anlyses is not an easy task. However, the fact that protein translocation of the MRG proteins from nucleus to cytoplasm (see below) is observed in senescent cells may lead to the identification of an associated repressor protein present in the complexes, which may in turn represent this inhibitory activity.

Studies with reconstructed cells

Attempts to determine the role of the nucleus and the cytoplasm in the process of senescence have been made yet no clear result was achieved. Wright and Hayflick [19] treated cytoplasms of whole normal young human fibroblasts with iodoacetate and rotenone to induce irreversible cytoplasmic injury. After fusing these injured cytoplasms to cytoplasms derived from both young and old donor cells, they determined that the cytoplasm was not involved because the cybrids survived for longer than a few days. Interpretation of these results is complicated by the fact that enucleation efficiency is never 100%; the controls showed that injury to the cytoplasm was not lethal to all cells, and it is possible that the iodoacetate and rotenone treatment might have destroyed some senescent-specific inhibitory activity. Equally indecisive are the experiments performed by Muggleton-Harris and Hayflick [20] using micromanipulation where both the old cytoplasm and the old nucleus had an impact on the division potential of the reconstructed cell. In this case, reconstructed cells included fusion of young cytoplasts with young karyoplasts, old cytoplasts with young karyoplasts, and young cytoplasts with old karyoplasts. They found that whenever an old component was involved, the number of reconstructed cells capable of dividing at least six times decreased dramatically as compared with the young cytoplast-young karyoplast control. These results indicated that both nucleus and cytoplasm played a role in the effect on aging of a cell. However, again there was some difficulty in interpreting the results of this study; namely, the technical difficulties involved in micromanipulation, which affects viability of the reconstructed cells, and the fact that the cells were followed through only a limited number of divisions.

Long-term proliferation potential of hybrids

Fusions involving only normal diploid cells

A major problem in attempting long-term studies of the proliferative potential of hybrids from fusions involving normal cells is the generation of mutants that allow for biochemical selection for hybrids. Normal human cell cultures have limited life spans of ~50–70 population doublings (PD). The cultures are heterogenous with respect to the doubling potential of the individual cells in the population (therefore, all cells would not be capable of 50 PD), and 20 PD would be used up in the process of obtaining a million cells of a mutant clone. Littlefield [21] made an attempt to fuse senescent normal human cells with each other and with young cells. In the latter case he used a hypoxanthine, aminopterin, thymidine (HAT) selection system by fusing young Lesch-Nyhan cells with senescent cells and reported that he was unable to obtain hybrids that could grow to any significant extent. He did obtain small clones that survived the selection system and might have been hybrids, but since he could not karyotype the clones, he had no further confirmation that they were hybrids. One of his conclusions was that the failure of young and aged fibroblasts to form large hybrid clones was "surprising, as if aging were dominant in such a cross." In fact, his result and conclusion were correct, as later studies showed. Hoehn and co-workers [22–23] attempted to circumvent this problem by fusing populations of young and old (but not senescent) diploid human cells having different glucose-6-phosphate dehydrogenase (G6PD) isoenzymes. Hybrids were selected by flow cytometery (sorting for a $2n$ DNA content) and confirmed by the presence of a G6PDAB heteropolymer band. The conclusion from this study was that the life span of hybrids from short- and long-lived fusions was intermediate between that of the parental cells fused. The problem with this interpretation is that the data were very limited. Analysis of hybrid life spans was restricted to only rapidly proliferating hybrid clones, and the *in vitro* life span of these hybrid clones was compared with the *in vitro* life span of the mass culture of the parental cell lines. The clonal life spans of the parent cultures were not examined. It has been documented by Martin *et al.* [24] and Smith and Hayflick [25] that there is tremendous heterogeneity in the division potential of the individual cells in a mass culture of human diploid fibroblasts. Young mass cultures are composed of senescent cells, cells capable of very few doublings, as well as cells capable of extensive division potential. Therefore, it is difficult to interpret the data generated by this study.

We were fortunate to obtain, through continuously improving cell culture reagents, a human diploid fibroblast line derived from fetal lung that was capable of achieving $\leqslant 100$ PD *in vitro* [26]. The cell line was normal despite this extremely long *in vitro* life span. With this cell line, we were able to obtain a spontaneous mutant clone resistant to ouabain, which was grown up and used to obtain a hypoxanthine phosphoribosyltransferase-negative (HPRT$^-$) mutation by selection in 6-thioguanine. This cell line had $\sim$45 PD remaining to it at this time and could be used as a universal hybridizer, since fusing it with any wild-type cell would allow for selection in medium containing HAT and ouabain. We also obtained a ouabain-resistant mutant clone from a Lesch-Nyhan (HRPT$^-$) cell line that had $\sim$28 PD remaining to it after growth to a million cells. Isolation of these cell lines allowed us to proceed with studies of the proliferative potential of the hybrids from fusions of young and old normal cells [27].

We decreased the heterogeneity of the parental populations by using in both cases a clone rather than a mass culture of cells. We had previously shown that clonal populations were much more homogeneous with respect to the division potential of the individual cells in the population than were mass-culture populations [28]. In clonal populations, one obtains two clear modes of cells. In the case of a clone early in its *in vitro* life span, one observes a low-population-doubling mode with $\leqslant 10$ PD and a high-population-doubling mode that usually ranges between $\sim$35 and 50 PD. The number of cells in the low-doubling-potential mode increases with increasing *in vitro* life span until one finally observes only this mode in an old clonal culture. Following fusion of clonal populations early and late in their *in vitro* life span, we determined the proliferation potential of the hybrids and compared these with the proliferative potentials of the individual cells in each parent population fused. This analysis demonstrated that the division potential of the hybrids was more like that of the old parental cells than the young, indicating that senescence was dominant as had been shown in the case of both the reconstruction and short-term heterokaryon experiments, as well as Littlefield's fusion results [21]. We were also able to demonstrate that no complementation occurred when we fused clones at the end of their *in vitro* life span with each other to yield hybrids having life spans greater than either of the parents. If, as has been hypothesized, senescence is the result of accumulation of damage through random errors or mutations, one would expect to observe such complementation. The conclusions from fusions involving normal cells are that the senescent phenotype is dominant and that senescence is not the result of random accumulation of damage.

Fusions of normal with immortal cells
In early studies in which hybrids from fusion of normal with immortal human cells were used to assay tumorgenicity, the conclusion had been that the phenotype of immortality was dominant, since it was possible to obtain hybrids that could grow to large cell numbers [29–30]. The interpretation of this result was that the senescent cells lacked some genes necessary for DNA synthesis and cell division and that these were turned on again in a dominant fashion in the immortal cells. This view was held for a long time until a study reported that some hybrids obtained from the fusion of HeLa cells with normal human diploid fibroblasts yielded hybrids that had limited division potential [31]. This was a very exciting but unexpected result and was questioned on the possibility that the particular HeLa cell line they had used was genetically unstable and that some critical immortalizing Hela DNA had been lost in the hybrids that had limited life spans [31]. Bunn and Tarrant [31] also observed that if they maintained these nondoubling hybrid populations for varying periods of time in culture, in some cases they would observe foci of dividing cells appearing at a frequency of ~ 1–2 in 10^5 cells. These would have regained the immortal phenotype and could grow without limit. Muggleton-Harris and DeSimone [32] fused normal cells with immortal SV40-transformed cells by micromanipulation, and reported that the majority of the fusion products (98%) had an extremely limited division potential of ≤ 6 PD. The 2% that resulted in large clones were assumed to be immortal and were not carried to the end of their *in vitro* life span. We decided to complete this experiment using a biochemical selection system for hybrids and fused immortal SV40-transformed cells with normal human cells [33]. What we observed was that the majority of the hybrids (70%) had extremely limited division potential (≤ 7 PD). About 30% of the hybrids proliferated more extensively between 16 and 62 PD, but all eventually ceased division. At the time that the hybrid populations had ceased doubling, the hybrids continued to express the SV40 T antigen, indicating that the viral genome was present and expressed in these cells. We tested the functional ability of the expressed SV40 T antigen to induce DNA synthesis in senescent normal cell

nuclei in heterokaryons and determined that it was intact [34]. We also observed the occurrence of rare immortal variants in these nondoubling populations at frequencies similar to those observed by Bunn and Tarrant [31]. Since we had shown that both small hybrid clones (those having <7 PD) as well as the one that could proliferate as many as 62 PD still expressed SV40 T antigen, there was one other possibility that could explain our result: namely, that integration of the viral genome was different in hybrids that had limited doubling potential. We performed a restriction endonuclease analysis of the immortal SV40-transformed parent used in the fusion, hybrids that ceased doubling, and the immortal variants that arose in these hybrid populations. There were no differences in the Southern blot band patterns of the integrated viral genomes in any of the cells. The conclusion from this set of studies was that the phenotype of immortality was recessive and that cellular senescence was dominant. An explanation for the early results is that in those studies the emphasis was on the study of the tumorigenic phenotype and that probably the most viable, rapidly growing hybrids were selected. These studies did not plate the fusion mixtures at clonal densities such as those of Bunn and Tarrant and ours [31, 33] therefore, immortal variants probably arose frequently in the large populations of cells. The other possibility is that limited-life span hybrids might have been observed in these studies if the actual number of population doublings that the hybrids could achieve before they were used for the tumorigenicity assay had been determined. To determine the generality of the dominance of cellular senescence in hybrids, we then fused normal human cells with a variety of immortal human cells: independently derived SV40-immortalized human cell lines and tumor-derived cell lines, some of which were expressing activated oncogenes. In all fusions the hybrids that we obtained had limited division potential [27]. We also used various normal cell types as fusion partners including T lymphocytes and human umbilical vein endothelial cells [35] and observed the same results. On the basis of these data, we conclude, that immortality occurs as a result of recessive changes in the growth control mechanisms of the normal cell, that there is a genetic basis to the process of cellular senescence.

Fusions of various immortal cell lines with each other

Since the phenotype of immortality was recessive, we were able to exploit this fact to determine the number of ways by which cells could become immortal. This allowed us to identify the number of genes or sets of genes or processes that were involved in cellular senescence that could be modified to yield immortal cells. The approach we took was to fuse different immortal human cell lines with each other. If the immortal parents fused had the same recessive change, we would obtain hybrids that were immortal. If, however, the immortal parents fused had immortalized by different events, complementation of these defects would occur to yield hybrids with limited life span. In an initial study, we fused an immortal SV40-transformed cell line with a small number of immortal human cell lines and were able to identify two complementation groups for indefinite division [35]. We followed this with a more extensive study involving 26 different immortal human cell lines and to date have identified four complementation groups for indefinite division [ref. 36, and unpublished observations].

The analysis involved using a cell line (having mutations that would allow for hybrid selection) that we arbitrarily assigned to complementation group A, which was fused to all of the other test cell lines. If immortal hybrids were obtained, the test cell lines were assigned to group A; if limited-lifespan hybrids were obtained, the test cell lines were assigned to not-group A. We then selected a cell line from not-group A to represent group B and continued fusions with not-group A cell lines. To strengthen the analysis, cell lines that had been assigned to group A were also fused with cell lines representative of group B, C, and D (as they were identified) to demonstrate that immortal hybrids were obtained only in the fusion with the cell line representative of group A and that there was complementation in all other fusions. An additional control was fusion of our double-mutant cell lines representative of each group, with the wild-type cell lines from which they were derived to demonstrate that indeed the hybrids proliferated indefinitely. We have been unable to assign a cell line to more than one complementation group. This result, along with the fact that the mutant cell lines that had been generated following mutagen treatment do not assign to groups other than that of the wild-type parent from which they were derived, indicates that we are working with a limited number of very specific genes that can be modified to yield immortal cells. This is reflected in the fact that the frequency of spontaneous immortalization in human cells is very low. It also indicates that the genes are most likely autosomal and that it takes two events to eliminate both alleles. We included a wide variety of immortal human cell lines in this analysis to determine if we could find any parameter that might correlate with complementation group assignment. Our findings were primarily negative. There was no correlation between cell type, embryonal layer of origin, or type of tumor and complementation group assignment. Expression of activated oncogenes did not affect group assignment. The one strong correlation we did see was that 9 in 10 of the immortal SV40-transformed cell lines we examined (derived from skin and lung fibroblasts, keratinocytes, and amnion cells) all were assigned to the same group, indicating that this virus immortalizes different human cells by the same processes [34–36]. The one exception was a cell line derived from xeroderma pigmentosum (XP) fibroblasts following transfection with an origin-defective SV40 virus. We could not attribute the different assignment of the cell line to the XP nature of the cells or use of origin-defective virus because another independently derived immortalized XP cell line (of the same XP complementation group) and an origin-defective virus-transformed cell line were assigned along with the rest of the SV40 cell lines. It has not been shown why this cell line assigned differently. Additionally, it has been shown that fibroblasts and epithelial cells can be immortalized at similarly high frequencies when transfected with an SV40 early region origin-minus expression plasmid, without the need for a selectable marker. A direct comparison was made of the immortalization frequencies of clones of normal epithelial cells and fibroblasts derived from the same individual. There was found to be no intrinsic difference between the SV40 early region-induced immortalization potentials of human fibroblasts and epithelial cells [37].

However, there are studies in which immortal cells do not always assign to a single complementation group. Roger Reddel and Ken Parkinson's groups have fused five other immortal cell lines with universal hybridiser cell lines for each of the four complementation groups we had generated and found they assigned to more than one complementation group. A possible explanation for this could be these cell lines have lost the function of two or more putative senescence genes [38–39].

Cloning and identification of mortality factor on human chromosome 4 (morf4)

Microcell-mediated chromosome transfer was now used to identify chromosomes encoding senescence related genes specifically altered in immortal cell lines assigned to a given complementation group. The senescence related gene for cells assigned to complementation Group B was found to reside on human chromosome 4 [40], the senescence gene for Group C cell lines on chromosome 1q [41], and the gene for Group D cells on chromosome 7 [42]. In these studies, the stringent genetic criteria used were that a chromosome should induce senescence in multiple cell lines assigned to a group and that the chromosome should have no effect on cell lines assigned to the other groups. By using microcell-mediated transfer, chromosomes 2, 3 6q, 11, 18, and X have been implicated in causing senescence in specific immortal cell lines [43–51]. However the genetic criteria of the studies described above was not followed, raising the possibility that the observed loss of proliferation in the immortal cell lines was due to a dosage effect and not a true genetic event. However, it is possible that additional complementation groups exist, and that there are cells with recessive changes in more than one senescence-related gene or pathway.

Microcell transfer experiments had demonstrated the chromosomal loci of cell senescence-related genes. The challenge now was to clone the genes. Through serendipitous circumstances, we obtained a fragment of human chromosome 4 in a mouse cell background that contained the neomycin resistance gene, and was able to induce senescence in cell lines assigned to group B. Alu polymerase chain reaction (Alu PCR) probes specific to the fragment of human chromosome 4 were used to probe human genomic bacterial artificial chromosome (BAC) filters from Genome systems Inc. and a (BAC) contig to the region, 4q33-34 was obtained [52]. The BACs then were used as probes on high-density cDNA filters from Genome Systems Inc. and six cDNAs obtained. The smallest genomic equivalent of three of the cDNAs was cloned into a promoterless vector and transfected into various immortal human cells. Transfection of the gene MORF-4 (mortality factor on human chromosome 4) into immortal human cell lines assigned to Group B resulted in the loss of proliferation, and there was no effect on the growth of cell lines assigned to other complementation groups. The loss of cell growth was accompanied by other changes, enlarged senescent-like morphology, activation of senescent associated B-galactosidase expression, and a change in the staining pattern of mortalin from that of an immortal to a normal cell. Introduction of a frame shift mutation at the 5′ end of the gene resulted in the loss of senescence inducing ability [53].

Sequencing of MORF 4 revealed that it was an intronless gene inserted into a LINE element and that it was a truncated version of the corresponding cDNA we had identified from the high-density cDNA filters. This suggested that MORF 4 was a member of a gene family and indeed, six additional MORF-related genes (MRG) were then identified. Of these, only two genes, MRG15 and MRGX, on chromosomes 15 and X, were found to be expressed. The other MRG genes are pseudogenes. MORF4, MRG15, and MRGX are highly conserved in the 3′ region of the gene. They all have nuclear localization signals, helix-loop helix and leucine zipper regions, and phosphorylation sites. MRG15 has an additional 5′ region that encodes a chromodomain, MRGX has a unique 5′ region. Our current hypothesis is that MRG15 and MRGX are involved in transcriptional complexes that effect global changes in gene expression and that MORF4 can disrupt these complexes by acting in a dominant negative manner. We are currently in the process of determining the precise mechanism of action of these genes/proteins. By using the yeast two-hybrid system, we have identified a novel gene, PAM14 (**P**rotein **A**ssociated with **MRG, 14** kD), the protein product of which interacts with MRG15. In that same protein complex the retinoblastoma tumor suppressor (Rb) interacts with Mrg15. We have found that MRG15 activates transcription of the B-myb promoter through a nucleoprotein complex that contains Rb and PAM14, and requires the HLH and leucine zipper regions of MRG15 [54]. We have also found that MRG15 is in multiple complexes by sucrose gradient analysis and that one complex involves RB and PAM14, whereas in another is present a histone acetyl transferase (HAT), hMOF. A HAT activity is associated with MRG15 but not MRGX and this is dependent on the presence of the chromodomain [55]. Thus MRG15 appears to be also acting via chromatin modification and acetylation of histones. MRG15 is a highly conserved protein and is present in humans to yeast as well as plants, suggesting an important functional role. MRGX is conserved only through vertebrates [55–56]. We have generated knockout animals of MRGX and MRG15 and observe no phenotype with MRGX null animals but that mice homozygous for MRG15 are embryonic lethal, consistent with the level of conservation across species. Analyses with these animals should reveal the role of these proteins in aging and immortalization. We have preliminary evidence that transfection of MORF4 into cells results in translocation of the MRG proteins from the nucleus to the cytoplasm and that a large fraction of the MRG proteins is in the cytoplasm of senescent cells. Determining the importance and mechanistic basis of this will reveal the role of MORF4 in causing loss of proliferation in immortal cells.

Genes and pathways that lead to senescence

Effect of mRNA isolated from normal cells

The experiments described above suggested that senescent cells produced an inhibitor of DNA synthesis. If this were the case, introduction of mRNA from senescent cells into young cells might inhibit cell proliferation. The most potent mRNA inhibitory activity that has been described thus far is the one isolated from senescent human diploid fibroblasts [57]. Significant inhibition was obtained when as little as 0.01–0.03 mg/ml of poly(A) $^{+}$RNA was injected into young human fibroblasts. This was in contrast to very little inhibition obtained when as much as 5 mg/ml of RNA isolated from young proliferating cells was injected. Inhibitory activity was also obtained with RNA isolated from quiescent human diploid fibroblasts. However, the RNA had to be used at a concentration >1 mg/ml in order to obtain significant inhibitory activity upon microinjection. The activity found in senescent human fibroblasts has some features in common with normal resting cells found in the body. For example, RNA isolated from resting rat liver has been found to inhibit normal human fibroblasts, HeLa cells, and NIH 3T3 cells at concentrations of 0.5–1 mg/ml [58] and mRNA isolated from resting T lymphocytes will inhibit HeLa cells or normal human diploid fibroblast IMR-90 when microinjected at concentrations of $\sim$0.5 mg/ml [59].

Role of $p21^{SDI1/CIP1/WAF1}$ in senescence

The finding that normal senescent cells contained relatively high levels of an mRNA that encoded a potent inhibitor of cell cycle progression prompted us to attempt to clone the cDNA of this gene using an expression cloning technique [60]. This led to the cloning of a gene that we called SDI1. The cell cycle inhibitory protein p21(Sdi1/Cip1/WAF1) was cloned as a growth inhibitor that is over-expressed in senescent cells. p21 associates with and inactivates cyclin E and cyclin D associated kinase activity [61]. Overexpression of p21 results in G1 cell cycle arrest and a senescent phenotype. SDI1 was expressed at high levels in senescent compared to young cells and inhibited DNA synthesis in young cells when introduced by transfection of a mammalian expression vector. It was also shown that it could inhibit a number of tumor derived, immortal cell lines. p21 has emerged as a major factor in the senescence of human fibroblasts (however, in other cell types such as keratinocytes and melanocytes, other Cdk inhibitors may be more important). The fact that it is critical for senescence of human fibroblasts was shown by making fibroblasts deficient in p21 by double homologous recombination [62]. p21 deficient fibroblasts do not undergo a normal senescence process but continue dividing until they reach "crisis." This senescence bypass is achieved in spite of high levels of p16 expression, indicating that the role of p21 in normal fibroblast senescence is essential, and cannot be substituted for nor replaced by any other CdKI [62]. Indeed, SV40-immortalized cells contain very low p21 levels and high p16 levels, supporting the idea of p21 as the main regulator of the block in proliferation that occurs in senescent cells. Recently, Stein [63] suggested that the inability of p16 to induce senescence in the absence of p21 could be explained, at least in part, by the lack of complete CdK4 inhibition by p16. Although SDI1 was first discovered as an inhibitor of cell proliferation derived from senescent human fibroblasts, it was later found that p21 associates with and inactivates cyclin E and cyclin D associated kinase activity [61]. It has become commonly known as $p21^{SDI1/CIP1/WAF1}$ and has been shown to be critical in cell cycle arrest due to DNA damage and its transcription can be induced by p53. However, its level is also controlled by p53 independent mechanisms.

Telomeres and telomerase

An interesting aspect of the phenomenon of replicative senescence is that the cells appear to have a counting mechanism, which keeps track of the number of doublings they have undergone. Telomeres are repeat sequences at chromosome ends. Due to incomplete replication of chromosome ends, telomeres shorten after each cell division. The telomere hypothesis proposes that the shortening serves as the counting mechanism in cells, and when the length of telomeres shortens to a critical length, it is (somehow) sensed as DNA damage that results in growth arrest [64]. One way to maintain telomere length is by the expression of the enzyme telomerase. In support of the telomere hypothesis is the fact that many immortal cell lines, germ cells and stem cells express high levels of telomerase activity, whereas most normal cells do not have detectable telomerase activity [65–67]. The telomere theory of cell aging appears to be confirmed by the observation that expression of the telomerase catalytic subunit hTERT in normal human cells extended their proliferative potential far beyond that of telomerase negative cells [62]. Although p21 is transcriptionally activated by p53 in response to double stranded DNA damage, p21 responds to p53 independent signals as well [68–70]. It has also been reported that the ability of HPV16 E6 protein to down-regulate p21 may be a result of post-transcriptional regulation, again suggesting mechanisms other than p53 for p21 regulation [71]. Although, it is generally assumed that short telomeres cause senescence by the induction of p21 through a p53 dependent pathway, compelling evidence for this mechanism is lacking.

A major challenge for the telomere hypothesis is the complementation group data. Cuthbert *et al.* [72] introduced human chromosome 3 into a human breast carcinoma cell line 21 NT by micro-cell transfer and observed strong repression of telomerase in the hybrids. The vast majority of hybrid clones with repressed telomerase entered permanent growth arrest after 10-18 PDs. Thus, chromosome 3 may encode a repressor of telomerase and the mechanism by which it induces senescence may act through telomere length regulation. However, in studies of complementing hybrids obtained from fusion of different immortal human cell lines, Bryan *et al.* [73] did not observe a correlation between loss of telomerase expression and loss of cell proliferation. We have similar results from studies of whole cell and micro-cell hybrids (Ran and Pereira-Smith, unpublished data). Carman *et al.* [74] have also observed senescence in telomerase expressing Syrian hamster cells. Thus, it is becoming clear that telomerase is not linked to proliferation when analyzing the genetically defined complementation groups. An ALT pathway as an alternative mechanism to maintenance of telomeres has been proposed [75] and described in more detail in the chapter by Roger Reddel in this book.

In addition, there are two other observations that the telomere hypothesis fails to address. One is the mechanism by which p21 is induced by short telomeres. The other is to provide an explanation for the widespread changes in gene expression observed in senescent cells.

DNA methylation levels

Sequential loss of DNA methylation could act as an alternative or additional counting mechanism for cellular aging. A progressive loss of 5-methylcytosine in genomic DNA occurs during serial passage of normal cells in culture. The extent of CpG methylation also decreases during aging of organisms [76–77]. On the other hand, immortal cell lines maintain constant levels of DNA methylation. In addition, it has been shown that transformation triggered by oncogenic Ras or SV40 T antigen elevates cellular DNA methyltransferase (DMeTase) [78] and that pretreatment of cells with DMeTase anti-sense oligonucleotides prevents T antigen-driven cellular transformation. This suggests that increased methylation is required for initiation of cell transformation [79]. Methylation of cytosine residues in CpG islands has been shown to cause silencing of gene expression. Thus, selective or random loss of 5 methylcytosine during repeated sub-cultivation has the potential to cause widespread changes in gene expression and might account for the major changes in gene expression observed to occur when cells become senescent. In addition, cultivation of human cells with the base analogues 5-aza-cytidine or 5-aza-2-deoxycytidine induced demethylation and shortened the *in vitro* life span of the cells [79]. However, the definitive experiment, demonstrating an extension in life span by manipulation of the 5-methylcytosine content of DNA, has not been reported. Thus, there are no conclusive data demonstrating a role for demethylation of DNA in normal cellular aging.

Replicative senescence, in addition to growth arrest, is accompanied by a large number of changes in gene expression. Increased acetylation of histones is associated with a chromatin structure that is more permissive for transcription. The degree of histone acetylation is determined in part by the activities of histone acetyl transferases and histone deacetylases (HDACs). DNA methyltransferase may also play a role in this because it is thought to direct HDACs to specific DNA sites. DNA hypermethylation generally is associated with silencing of genes. Therefore, gene silencing may be the result of the interaction of DNA methyltransferase with methylcytosine and the recruitment of HDACs to the site. The involvement of histone acetylation in cellular senescence of normal human fibroblasts was indicated by treating human fibroblasts with the HDAC inhibitors, sodium butyrate or trichostatin A. This resulted in a marked reduction in proliferative life span and the induction of a phenotype characteristic of senescent cells. The treated cells had an enlarged morphology, were positive for SA-β gal and could reinitiate DNA synthesis in response to SV40 T antigen [80]. The involvement of chromatin structure is further supported by the fact that treatment of normal human cells with topoisomerase inhibitors also resulted in the induction of a senescent-like state. In addition, the finding that the mutated genes underlying syndromes of accelerated aging are helicases [81–82] links chromatin structure modification to senescence. Although, DNA methylation and DNA methyltransferase appear to be implicated in cellular aging, their participation may be indirect through histone acetylation and perhaps other chromatin modifications.

Overall, it is interesting to compare the sequence of the genetic studies of cellular aging with those in the field of tumorigenicity and see how similar they have been. Initial hybrid studies had indicated that tumorigenicity was a dominant phenotype, but later studies from many laboratories demonstrated that this phenotype was recessive. Genetic studies established that senescence is dominant over immortality, which led to the identification of four complementation groups for indefinite division, as well as several chromosomal loci that harbor senescence-related genes. Attempts have been made to identify complementation groups for tumorigenicity but have not yielded clear results because of the complex nature of the tumorigenic phenotype. Moreover, some of these senescence-related genes may act through telomere length regulation, though in the majority of cases there is no correlation between expression of telomerase and genetic complementation. Although it is currently difficult to clone and characterize these genes of interest because it involves positional cloning, the sequencing of the entire human genome makes their future identification more hopeful. And as other players such as chromatin remodeling and epigenetics appear, our understanding of aging and the role of senescence in tumor suppression will be greatly advanced.

References

1. Hayflick L (1965). The limited *in vitro* lifetime of human diploid cell strains. *Exp Cell Res.* 37: 614–36.
2. Rohme D (1981). Evidence for a relationship between longevity of mammalian species and life-spans of normal fibroblasts *in vitro* and erythrocytes *in vivo. Proc Natl Acad Sci USA* 78: 5009–13.
3. Martin GM, Sprague CA, Epstein CJ (1970). Replicative life-span of cultivated human cells: effects of donor age, tissue, and genotype. *Lab Invest.* 23: 86–92.
4. Schneider EL, Mitsui Y (1976). The relationship between *in vitro* cellular aging and *in vivo* human age. *Proc Natl Acad Sci USA* 73: 3584–8.
5. LeGuilly Y, Simon M, Lenoir P, Boure M (1973). Long-term culture of human adult liver cells: morphological changes related to *in vitro* senescence and effect of donor's age on growth potential. *Gerontologia* 19: 303–13.
6. Bierman EL (1978). The effect of donor age on the *in vitro* life span of cultured human arterial smooth muscle cells. *In Vitro* 14: 951–5.
7. Smith JR, Venable S, Roberts TW, Metter EJ, Monticone R, Schneider EL (2002). Relationship between *in vivo* age and *in vitro* aging: assessment of 669 cell cultures derived from members of the Baltimore Longitudinal Study of Aging. *J Gerontol A Biol Sci Med Sci.* 57: B239–46.
8. Matsumura T, Zerrudo Z, Hayflick L (1979). Senescent human diploid cells in culture: survival, DNA synthesis and morphology. *J Gerontol.* 34: 328–34.
9. Norwood TH, Pendergrass WR, Sprague CA, Martin GM (1974). Dominance of the senescent phenotype in heterokaryons between replicative and post-replicative human fibroblast-like cells. *Proc Natl Acad Sci USA* 71: 2231–5.
10. Yanishevsky RM, Stein GH (1980). Ongoing DNA synthesis continues in young human diploid cells (HDC) fused to senescent HDC, but entry into S phase is inhibited. *Exp Cell Res.* 126: 469–72.
11. Norwood TH, Pendergrass WR, Martin GM (1975). Reinitiation of DNA synthesis in senescent human fibroblasts upon fusion with cells of unlimited growth potential. *J Cell Biol.* 64: 551–6.
12. Stein GH, Yanishevsky RM (1979). Entry into S phase is inhibited in two immortal cell lines fused to senescent human diploid cells. *Exp Cell Res.* 120: 155–65.
13. Stein GH, Yanishevsky RM, Gordon L, Beeson M (1982). Carcinogen-transformed human cells are inhibited from entry into S phase by fusion to senescent cells but cells transformed by DNA tumor viruses overcome the inhibition. *Proc Natl Acad Sci USA* 79: 5287–91.
14. Burmer GC, Motulsky H, Zeigler CJ, Norwood TH (1983). Inhibition of DNA synthesis in young cycling human diploid fibroblast-like cells upon fusion to enucleate cytoplasts from senescent cells. *Exp Cell Res.* 145: 79–84.

15. Drescher-Lincoln CK, Smith JR (1984). Inhibition of DNA synthesis in senescent-proliferating human cybrids is mediated by endogenous proteins. *Exp Cell Res.* 153: 208–17.
16. Pereira-Smith OM, Fisher SF, Smith JR (1985). Senescent and quiescent cell inhibitors of DNA synthesis. Membrane- associated proteins. *Exp Cell Res.* 160: 297–306.
17. Stein GH, Atkins L (1986). Membrane-associated inhibitor of DNA synthesis in senescent human diploid fibroblasts: characterization and comparison to quiescent cell inhibitor. *Proc Natl Acad Sci USA* 83: 9030–4.
18. Stein GH, Atkins L, Beeson, M., Gordon, L. (1986). Quiescent human diploid fibroblasts. Common mechanism for inhibition of DNA replication in density-inhibited and serum-deprived cells. *Exp Cell Res.* 162: 255–60.
19. Wright WE, Hayflick L (1975). Contributions of cytoplasmic factors to *in vitro* cellular senescence. *Fed Proc.* 34: 76–9.
20. Muggleton HAL, Hayflick L (1976). Cellular aging studied by the reconstruction of replicating cells from nuclei and cytoplasms isolated from normal human diploid cells. *Exp Cell Res.* 103: 321–30.
21. Littlefield JW (1973). Attempted hybridizations with senescent human fibroblasts. *J Cell Physiol.* 82: 129–32.
22. Bryant EM, Crouch E, Bornstein P, Martin GM, Johnston P, Hoehn H (1978). Regulation of growth and gene activity in euploid hybrids between human neonatal fibroblasts and epithelioid amniotic fluid cells. *Am J Hum Genet.* 30: 392–405.
23. Hoehn H, Bryant EM, Martin GM (1978). The replicative life spans of euploid hybrids derived from short-lived and long-lived human skin fibroblast cultures. *Cytogenet Cell Genet.* 21: 282–95.
24. Martin GM, Sprague CA, Norwood TH, Pendergrass WR (1974). Clonal selection, attenuation and differentiation in an *in vitro* model of hyperplasia. *Am J Pathol.* 74: 137–54.
25. Smith JR, Hayflick L (1974). Variation in the life-span of clones derived from human diploid cell strains. *J Cell Biol.* 62: 48–53.
26. Duthu GS, Braunschweiger KI, Pereira-Smith OM, Norwood TH, Smith JR (1982). A long-lived human diploid fibroblast line for cellular aging studies: applications in cell hybridization. *Mech Ageing Dev.* 20: 243–52.
27. Pereira-Smith OM, Smith JR (1982). Phenotype of low proliferative potential is dominant in hybrids of normal human fibroblasts. *Somatic Cell Genet.* 8: 731–42.
28. Smith JR, Whitney RG (1980). Intraclonal variation in proliferative potential of human diploid fibroblasts: stochastic mechanism for cellular aging. *Science* 207: 82–4.
29. Stanbridge EJ (1976). Suppression of malignancy in human cells. *Nature* 260: 17–20.
30. Croce CM, Koprowski H (1974). Positive control of transformed phenotype in hybrids between SV40-transformed and normal human cells. *Science* 184: 1288–9.
31. Bunn CL, Tarrant GM (1980). Limited lifespan in somatic cell hybrids and cybrids. *Exp Cell Res.* 127: 385–96.
32. Muggleton HAL, DeSimone DW (1980). Replicative potentials of various fusion products between WI-38 and SV40 transformed WI-38 cells and their components. *Somatic Cell Genet.* 6: 689–98.
33. Pereira-Smith OM, Smith JR (1981). Expression of SV40 T antigen in finite life-span hybrids of normal and SV40-transformed fibroblasts. *Somatic Cell Genet.* 7: 411–21.
34. Pereira-Smith OM, Smith JR (1987). Functional simian virus 40 T antigen is expressed in hybrid cells having finite proliferative potential. *Mol Cell Biol.* 7: 1541–4.

35. Pereira-Smith OM, Smith JR (1983). Evidence for the recessive nature of cellular immortality. *Science* 221: 964–6.
36. Pereira-Smith OM, Smith JR (1988). Genetic analysis of indefinite division in human cells: identification of four complementation groups. *Proc Natl Acad Sci USA* 85: 6042–6.
37. Reddel RR, Salghetti SE, Willey JC, *et al.* (1993). Development of tumorigenicity in simian virus 40-immortalized human bronchial epithelial cell lines. *Cancer Res.* 53: 985–91.
38. Duncan EL, Whitaker NJ, Moy EL, Reddel RR (1993). Assignment of SV40-immortalized cells to more than one complementation group for immortalization. *Exp Cell Res.* 205: 337–44.
39. Berry IJ, Burns, JE, Parkinson EK (1994). Assignment of two human epidermal squamous cell carcinomas cell lines to more than one complementation group for the immortal phenotype. *Mol Carcinog.* 9: 134–42.
40. Ning Y, Weber JL, Killary AM, Ledbetter DH, Smith JR, Pereira-Smith OM (1991). Genetic analysis of indefinite division in human cells: evidence for a cell senescence-related gene(s) on human chromosome 4. *Proc Natl Acad Sci USA* 88: 5635–9.
41. Hensler PJ, Annab LA, Barrett JC, Pereira-Smith OM (1994). A gene involved in control of human cellular senescence on human chromosome 1q. *Mol Cell Biol.* 14: 2291–7.
42. Ogata T, Ayusawa D, Namba M, Takahashi E, Oshimura M, Oishi M (1993). Chromosome 7 suppresses indefinite division of nontumorigenic immortalized human fibroblast cell lines KMST-6 and SUSM-1. *Mol Cell Biol.* 13: 6036–43.
43. Sugawara O, Oshimura M, Koi M, Annab LA, Barrett JC (1990). Induction of cellular senescence in immortalized cells by human chromosome 1. *Science* 247: 707–10.
44. Klein CB, Conway K, Wang XW, *et al.* (1991). Senescence of nickel-transformed cells by an X chromosome: possible epigenetic control. *Science* 251: 796–9.
45. Koi M, Johnson LA, Kalikin LM, Little PF, Nakamura Y, Feinberg AP (1993). Tumor cell growth arrest caused by subchromosomal transferable DNA fragments from chromosome 11. *Science* 260: 361–4.
46. Sandhu AK, Hubbard K, Kaur GP, Jha KK, Ozer HL, Athwal RS (1994). Senescence of immortal human fibroblasts by the introduction of normal human chromosome 6. *Proc Natl Acad Sci USA* 91: 5498–502.
47. Sasaki M, Honda T, Yamada H, Wake N, Barrett JC, Oshimura M (1994). Evidence for multiple pathways to cellular senescence. *Cancer Res.* 54: 6090–3.
48. Rimessi P, Gualandi F, Morelli C, *et al.* (1994). Transfer of human chromosome 3 to an ovarian carcinoma cell line identifies three regions on 3p involved in ovarian cancer. *Oncogene* 9: 3467–74.
49. Banga SS, Kim S, Hubbard K, *et al.* (1997). SEN6, a locus for SV40-mediated immortalization of human cells, maps to 6q26-27. *Oncogene* 14: 313–21.
50. Horikawa I, Oshimura M, Barrett JC (1998). Repression of the telomerase catalytic subunit by a gene on human chromosome 3 that induces cellular senescence. *Mol Carcinog.* 22: 65–72.
51. Uejima H, Shinohara T, Nakayama Y, Kugoh H, Oshimura M (1998). Mapping a novel cellular-senescence gene to human chromosome 2q37 by irradiation microcell-mediated chromosome transfer. *Mol Carcinog.* 22: 34–45.
52. Bertram MJ, Berube NG, Swanson XH, Pereira-Smith OM (1999). Assembly of a BAC contig of the complementation group B cell senescence gene candidate region at 4q33-q34.1 and identification of expressed sequences. *Genomics* 56: 353–4.

53. Bertram MJ, Berube NG, Hang-Swanson X, *et al.* (1999). Identification of a gene that reverses the immortal phenotype of a subset of cells, is a member of a novel family of transcription factor-like genes. *Mol Cell Biol.* 19: 1479–85.
54. Leung JK, Berube N, Venable S, Ahmed S, Timchenko N, Pereira-Smith OM (2001). MRG15 activates the B-myb promoter through formation of a nuclear complex with the retinoblastoma protein and the novel protein PAM14. *J Biol Chem.* 276: 39171–8.
55. Pardo PS, Leung JK, Lucchesi JC, Pereira-Smith OM (2002). MRG15 a novel chromodomain protein is present in two distinct multiprotein complexes involved in transcriptional activation. *J Biol Chem.* 277: 50860–6.
56. Bertram MJ, Pereira-Smith OM (2001). Conservation of the MORF4 related gene family: identification of a new chromo domain subfamily and novel protein motif. *Gene* 266: 111–21.
57. Lumpkin CKJ, McClung JK, Pereira-Smith OM, Smith JR (1986). Existence of high abundance antiproliferative mRNA's in senescent human diploid fibroblasts. *Science* 232: 393–5.
58. Pepperkok R, Zanetti M, King R,*et al.* (1988). Automatic microinjection system facilitates detection of growth inhibitory mRNA. *Proc Natl Acad Sci USA* 85: 6748–52.
59. Pepperkok R, Schneider C, Philipson L, Ansorge W (1988). Single cell assay with an automated capillary microinjection system. *Exp Cell Res.* 178: 369–76.
60. Noda A, Ning Y, Venable SF, Pereira-Smith OM, Smith JR (1994). Cloning of senescent cell-derived inhibitors of DNA synthesis using an expression screen. *Exp Cell Res.* 211: 90–8.
61. Harper JW, Adami GR, Wei N, Keyomarsi K, Elledge SJ (1993). The p21 Cdk-interacting protein Cip1 is a potent inhibitor of G1 cyclin-dependent kinases. *Cell* 75: 805–16.
62. Brown JP, Wei W, Sedivy JM (1997). Bypass of senescence after disruption of p21CIP1/WAF1 gene in normal diploid human fibroblasts. *Science* 277: 831–4.
63. Stein GH, Drullinger LF, Soulard A, Dulic V (1999). Differential roles for cyclin-dependent kinase inhibitors p21 and p16 in the mechanisms of senescence and differentiation in human fibroblasts. *Mol Cell Biol.* 19. 2109–17.
64. Harley CB, Sherwood SW (1997). Telomerase, checkpoints and cancer. *Cancer Surv.* 29: 263–84.
65. Kim NW, Piatyszek MA, Prowse KR, *et al.* (1994). Specific association of human telomerase activity with immortal cells and cancer. *Science* 266: 2011–15.
66. Igarashi H, Sakaguchi N (1996). Telomerase activity is induced by the stimulation to antigen receptor in human peripheral lymphocytes. *Biochem Biophys Res Commun.* 219: 649–55.
67. Bryan TM, Englezou A, Gupta J, Bacchetti S, Reddel RR (1995). Telomere elongation in immortal human cells without detectable telomerase activity. *EMBO J.* 14: 4240–8.
68. Johnson M, Dimitrov D, Vojta PJ, *et al.* (1994). Evidence for a p53-independent pathway for upregulation of SDI1/CIP1/WAF1/p21 RNA in human cells. *Mol Carcinog.* 11: 59–64.
69. Halevy O, Novitch BG, Spicer DB, *et al.* (1995). Correlation of terminal cell cycle arrest of skeletal muscle with induction of p21 by MyoD. *Science* 267: 1018–21.
70. Parker SB, Eichele G, Zhang P, *et al.* (1995). p53-independent expression of p21Cip1 in muscle and other terminally differentiating cells. *Science* 267: 1024–7.
71. Burkhart BA, Alcorta DA, Chiao C, Isaacs JS, Barrett JC (1999). Two posttranscriptional pathways that regulate p21(Cip1/Waf1/Sdi1) are identified by HPV16-E6 interaction and correlate with life span and cellular senescence. *Exp Cell Res.* 247: 168–75.

72. Cuthbert AP, Bond J, Trott DA, *et al.* (1999). Telomerase repressor sequences on chromosome 3 and induction of permanent growth arrest in human breast cancer cells. *J Natl Cancer Inst.* 91: 37–45.
73. Bryan TM, Reddel RR (1997). Telomere dynamics and telomerase activity in *in vitro* immortalised human cells. *Eur J Cancer* 33: 767–73.
74. Carman TA, Afshari CA, Barrett JC (1998). Cellular senescence in telomerase-expressing Syrian hamster embryo cells. *Exp Cell Res.* 244: 33–42.
75. Bryan TM, Englezou A, Dalla-Pozza L, Dunham MA, Reddel RR (1997). Evidence for an alternative mechanism for maintaining telomere length in human tumors and tumor-derived cell lines. *Nat Med.* 3: 1271–4.
76. Wilson VL, Jones PA (1983). DNA methylation decreases in aging but not in immortal cells. *Science* 220: 1055–7.
77. Singhal RP, Mays-Hoopes LL, Eichhorn GL (1987). DNA methylation in aging of mice. *Mech Ageing Dev.* 41: 199–210.
78. Slack A, Cervoni N, Pinard M, Szyf M (1999). DNA methyltransferase is a downstream effector of cellular transformation triggered by simian virus 40 large T antigen. *J Biol Chem.* 274: 10105–12.
79. Fairweather DE, Fox M, Margison GP (1987). The *in vitro* lifespan of MRL-5 cells is shortened by 5-azacytidine-induced demethylation. *Exp Cell Res.* 168: 153–9.
80. Ogryzko VV, Hirai TH, Russanova VR, Barbie DA, Howard BH (1996). Human fibroblast commitment to a senescence-like state in response to histone deacetylase inhibitors is cell cycle dependent. *Mol Cell Biol.* 16: 5210–18.
81. Michishita E, Nakabayashi K, Ogino H, Suzuki T, Fujii M, Ayusawa D (1998). DNA topoisomerase inhibitors induce reversible senescence in normal human fibroblasts. *Biochem Biophys Res Commun.* 253: 667–71.
82. Gray MD, Jesch SA, Stein GH (1991). 5-Azacytidine-induced demethylation of DNA to senescent level does not block proliferation of human fibroblasts. *J Cell Physiol.* 149: 477–84.

Aging of Fibroblasts

Dimitris Kletsas
Laboratory for Cell Proliferation & Ageing, Institute of Biology, NCSR "Demokritos," 153 10 Ag. Paraskevi, Athens, Greece

Introduction

The demographic shift towards a "graying" society and the increase in the incidence of age-related impairments and diseases made aging an important biological and social issue. In the effort to unravel the basic mechanisms that regulate the aging process fibroblasts was and still is the most "popular" cellular model of experimentation. This article will try to focus on some of the features of aging human fibroblasts, the mechanisms underlying their phenotype and their role on the failure of tissue homeostasis that characterizes aging.

The "Hayflick Limit"

From the beginning of the 20th century until the early 1960s it was generally believed that normal cells were immortal, i.e., when properly maintained in culture they could proliferate indefinitely. As a consequence, aging was thought to be determined extracellularly and to have little to do with intracellular events [1]. However, in the 1960s, L. Hayflick overthrew this dogma by showing that normal human diploid fibroblasts have a finite replicative capacity – now called "Hayflick limit" – i.e., they can undergo only a limited number of cell divisions, after which they remain metabolically active but are unable to proliferate; a process called *in vitro* aging (or alternatively, *in vitro* senescence) [2, 3]. Subsequent studies in several other cell types, beyond fibroblasts, have proven that *in vitro* aging is a general characteristic of most normal cells [4]. So, a distinction has been introduced between normal mortal cells with finite "lifespans" and immortal cells, usually derived from cloned cancer cells, whose cultures can grow *ad infinitum* [1]. Interestingly, the lifespan of normal cells has nothing to do with calendar time but is determined by the number of doublings that each culture of can undergo. Even more, cells that are removed from the culture

Sunil C. Kaul and Renu Wadhwa (eds.), Aging of Cells In and Outside the Body, 27–46.

and appropriately cryopreserved for long periods, when returned to culture they "remember" to accomplish the number of divisions needed in order to achieve senescence. Thus, there is a clear difference between biological and chronological time in this aging process, suggesting the existence of a "counting mechanism" for the determination of the lifespan of each cell culture.

Morphology and function

According to the initial nomenclature introduced by Hayflick, the lifespan of normal human diploid fibroblasts can be divided in three phases: Phase I corresponds to the primary cultures established from tissue explants by mechanical or enzymatic disaggregation. Subsequently, the cultures enter a long period characterized by intense proliferation at a nearly constant rate (Phase II). Finally, they abruptly enter Phase III, where proliferation slows down and eventually cell division ceases. However, spontaneous or induced cell transformation can lead to unlimited proliferative potential, a process known as immortalization [5] (Figure 1).

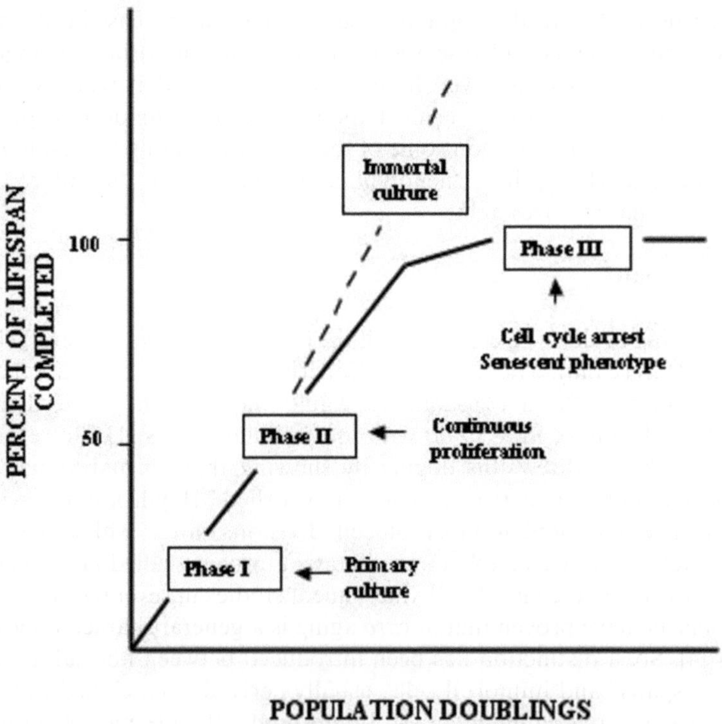

***Figure 1**. Graphical representation of the lifespan of normal diploid fibroblasts.*

Normal young (early-passage) human fibroblasts, grown under conventional culture conditions, exhibit a characteristic elongated, spindle-like morphology with a near-rounded to elliptical nucleus. In contrast, senescent (late-passage) cells are highly enlarged, flattened and irregular in shape. Furthermore, the cytoplasm is filled with an increased number of vacuoles and dense lysosomal autofluorescent granules. Cell nuclei are irregularly shaped and highly lobed. Electron microscopy studies have characterized the ultrastructural alterations that accompany the aging of human fibroblasts in culture. The number of mitochondria remains constant, although increasing proportion of bizarre-shaped organelles with few cristae appear. On the other hand, a significant increase in the number of lysosomes and autophagic vacuoles occurs. The Golgi apparatus is much more prominent, with swollen cisternae and surrounded by numerous vacuoles. The endoplasmatic reticulum usually has an "empty" configuration, while ribosome-free ER can be seen [6].

At the functional level, senescent fibroblasts remain viable and metabolically active. However, their most characteristic alteration is their inability to proliferate, even in the presence of increased amounts of mitogens in their environment. For this reason, the term "replicative senescence" is probably the most appropriate to characterize the phenomenon of *in vitro* aging after serial subculturing. Another important feature of senescent cells concerns the formation of extracellular matrix (ECM). Fibroblasts are the major ECM producers, at least in the connective tissue, and changes in ECM appropriate formation can have detrimental effects on tissue repair and homeostasis. *In vitro* aged fibroblasts show a decreased collagen synthesis and an increased expression of interstitial collagenase (MMP-1) and stromelysin, as well as an increase in the ratio of MMP-1 to its inhibitor (TIMP-1), indicating a switch from a matrix-producing to a matrix-degrading phenotype [7–9]. A more "global" characterization of the senescent-fibroblast gene expression profile has been made by using cDNA microarray technology [10]. Senescent cells express a pro-inflammatory phenotype, marked by the up-regulation of matrix proteases and inflammatory cytokines. Interestingly, this phenotype overlaps substantially with the gene expression patterns observed in activated fibroblasts during the early remodeling phase of the wound healing process [10].

The counting mechanism

The near constant number of population doublings that each fibroblast culture can undergo before becoming senescent gave rise to the idea of the existence of an inherent counting mechanism, a "mitotic clock." It was at the early 1970s when Olovnikov [11] hypothesized that it is the shortening of telomeres after each cell division that can act as a counting mechanism and lead to senescence. Telomeres are DNA-protein structures at the end of chromosomes. They contain many kilobases of repetitive TTAGGG sequences (15–20 kb in human germ cells) and they protect the cells from recognizing the ends of chromosomes as double-strand breaks, and thus inhibit end-to-end chromosome fusion events [12, 13]. In addition, they are also associated with specific proteins, such as TRF2 or TANK1, with poly-ADP ribosylase activity [12, 14]. The latter, is clearly associated with genome protection, as well as mammalian longevity [15]. Olovnikov [16] and Watson [17] have separately described the "end-replication problem," i.e., the inability of DNA polymerase to fully replicate the ends of chromosomes. Indeed, 3′–5′ parental DNA strand is copied in a continuous manner by DNA polymerase, while the 5′–3′ strand is copied discontinuously as series of Okazaki fragments, by using labile RNA primers. After degradation of these primers, the 3′ end of the parental strand is left incompletely copied and is subsequently degraded by nucleases [18]. As a result, a reduction of telomeric DNA should occur in each cell duplication. This hypothesis has been proved experimentally several years later [19], and it has been shown that when telomeres reach an average length of 4–7 kb the cells become senescent. Accordingly, telomere length has been proposed to be the molecular basis of the "mitotic clock."

On the other hand, telomerase, a ribonucleoprotein complex that add telomeric repeat sequences to the 3′ telomeric ends, is not expressed in most somatic cells. It is expressed in germ cells and several cancer cells, leading to the maintenance of a critical size of telomeres and finally to unlimited proliferative capacity, i.e., to immortalization. Even more, the ectopic expression of the catalytic unit of telomerase (hTERT) prevents not only telomere shortening, but also senescence in human fibroblasts [20], further supporting the decisive role of telomeres in senescence and immortalization.

Loss of the proliferative capacity with aging

The most prominent functional alteration of senescent cells is clearly their inability to proliferate. As a consequence, the loss of proliferative capacity of *in vitro* aged cells became one of the central themes in the field of biogerontology. Interestingly, the impressive new findings in this area during the last years have revolutionized our ideas on the aging process but also on other eminent fields, like cell cycle regulation or even carcinogenesis.

Brief description of mitogenic signaling and cell cycle machinery

Cellular proliferation is controlled – positively or negatively – by a group of regulatory molecules, such as growth factors or cytokines. "Classical" growth factors, e.g., platelet-derived growth factor (PDGF) or epidermal growth factor (EGF), exert their action after binding to cell surface transmembrane receptors with tyrosine kinase activity. Ligand binding leads to receptor dimerization and autophosphorylation. The activated receptor phosphorylates a series of cytoplasmic substrates, thus initiating intracellular signaling cascades that convey the mitogenic signal from the cell surface to the nucleus. Probably, the most prominent pathway for the regulation of cell proliferation is the Ras/Raf/MEK/ERK pathway. Activated (phosphorylated) ERK translocates from the cytoplasm to the cell nucleus where it activates transcription factors that lead to the expression of immediate early-genes, such as the members of the c-Jun and c-Fos families, that form the AP-1 transcription complex (Figure 2).

Activation of the transcription machinery leads to the expression of cell cycle regulatory molecules, such as cyclins, which form complexes with the cyclin-

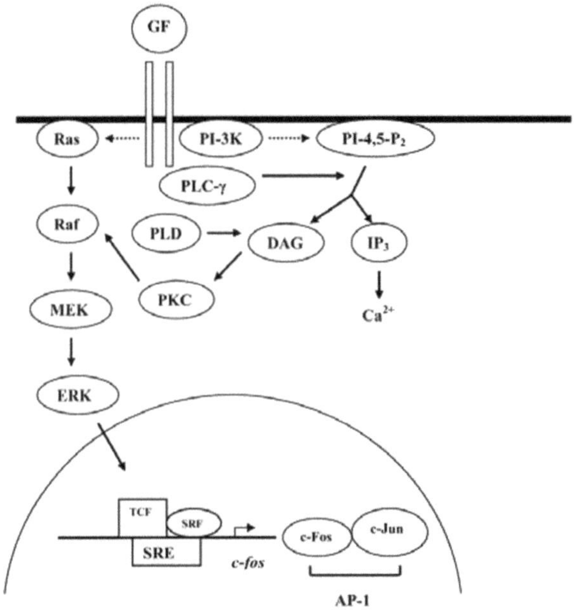

***Figure 2**. Signal transduction pathways from tyrosine-kinase growth factor receptors. (See text for details).*

dependent kinases (CDKs), and eventually phosphorylate the product of the retinoblastoma tumor suppressor gene (pRb) [21]. Hypophosphorylated pRb associates with several transcription factors, most prominent being the E2F proteins. Release of E2F after pRb phosphorylation triggers the initiation of DNA synthesis [22]. This pathway is tightly controlled by a group of CDK inhibitors (CKIs), such as the $p16^{INK4a}$, $p19^{ARF}$ or $p21^{WAF1}$ that prevent pRb phosphorylation and lead to cell cycle arrest [23, 24].

Mitogenic signaling in senescent cells

What alterations occur in senescent cells that render them unable to respond to exogenous growth factors and proliferate? The initial stimulation, i.e., growth factor binding to their specific receptors, seem to remain unaltered during aging [25–27]. Even more, as the density of receptor molecules on the cellular surface remains constant, an increase in total growth factor binding is observed due to the enlargement of senescent cells [25–27]. In addition, no change has been observed in the autophosporylation of EGF- and PDGF-receptors between young and senescent cells [28]. Ligand binding is usually followed by a reduction of plasma membrane receptors due to endocytosis via clathrin-coated pits. This down-regulation is restored much faster in late-passage cells, compared to young fibroblasts [25]. Thus, although divergent reports have also appeared [29], the general consensus is that the senescence-specific growth arrest is not dependent to altered growth factor binding but rather to the down-stream events. Still at the membrane level, it has been recently shown that senescent fibroblasts lack functional caveolae, i.e., membrane invaginations where components of the major signal transduction pathways are localized [30].

Concerning components of the tyrosine kinase receptor-mediated pathways, phospholipase D (PLD) is not activated in senescent cells, probably due to the abundance of its inhibitor, ceramide [31]. This inactivation leads to the defective generation of phosphatidic acid, a potent mitogen [32] and of diacylglycerol, an intermediate step in the activation of protein kinase C (PKC), one of the key components of mitogenic signal transduction machinery. In addition, pharmacological inhibition of the phosphoinositol-3-kinase (PI3K) pathway can lead to a senescent phenotype [33]. Calcium, an important component of the signal transduction machinery is also affected: the capacitative Ca^{2+} entry after depletion of intracellular stores is severely diminished in senescent fibroblasts [123]. On the other hand, the Ras/Raf/MEK/ERK pathway is activated in senescent cells in response to growth factors [34]. However, the phosporylated ERK fail to translocate inside the nucleus [34, 35], possibly leading to the loss of the transcription factor $TCF^{Elk\text{-}1}$-activation. In parallel, the serum response factor (SRF) has found to be phosphorylated in senescent fibroblasts, and thus unable to bind to the serum response element (SRE) of several gene promoters. As a consequence, although several immediate-early genes (such as *c-jun*, *junB* or *c-myc*) are adequately expressed in senescent cells [36, 37], the induction of others, e.g., *c-fos*, *erg-1*, *Id-1H* and *Id-2H*, is severely diminished [38, 39]. Furthermore, AP-1 composition is altered [40] and the DNA binding activity of this transcriptional complex is substantially decreased as fibroblasts age [41].

Altered cell cycle regulation in senescent cells

The most striking and decisive alteration in cell cycle regulation in senescent cells is their inability to fully phosphorylate the pRb protein after growth-factor stimulation. pRb protein phosphorylation, and the subsequent activation of the E2F transcription factors, is a critical step in the transition of cells from the G_1 to the S phase of the cell cycle. This is achieved by the E2F-mediated induction of a series of genes that are implicated in DNA synthesis. However, in senescent fibroblasts a "global" change has been observed in gene expression in the G_1/S boundary. In particular, a severe decrease is observed in the mRNA production of a group of "G_1/S" genes, such as thymidine kinase, thymidylate synthetase, dihydrofolate reductase, ribonucleotide reductase, proliferating cell nuclear antigen and several histone subtypes [42]. As mentioned above, pRb is phosphorylated by the action of cyclin/CDK complexes, such as cyclin D/CDK4–6 or cyclin E/CDK2. In senescent cells, the cyclin-dependent kinase inhibitor (CKI) $p16^{INK4a}$ is overexpressed. This protein inhibits the function of the cyclin D/CDK4–6 complexes and thus pRb phosphorylation, leading to E2F silencing and cell-cycle arrest. In accordance to this, inactivation of the $p16^{INK4a}$/pRb pathway has been shown to lead to an extension of the lifespan or even immortalization [24]. Interestingly, the same tumor suppressor locus (*INK4A/*

ARF) in the short arm of chromosome 9, encodes, by alternative splicing, another CKI, i.e., $p19^{ARF}$ [43]. $p19^{ARF}$ (alternatively called $p14^{ARF}$ when referring to humans) inhibits the oncoprotein MDM2, thus stabilizing the tumor suppressor p53. p53 is an important tumor suppressor protein, found to be inactivated in approximately 50% of human cancers. It exerts its action on cell cycle by inducing the expression of the CDI $p21^{WAF1}$, that inhibits the cyclin E/CDK2-mediated pRb phosphorylation. $p19^{ARF}$ is overexpressed in senescent murine fibroblasts and seems to be important in the aging process in this cell type [44] (Figure 3).

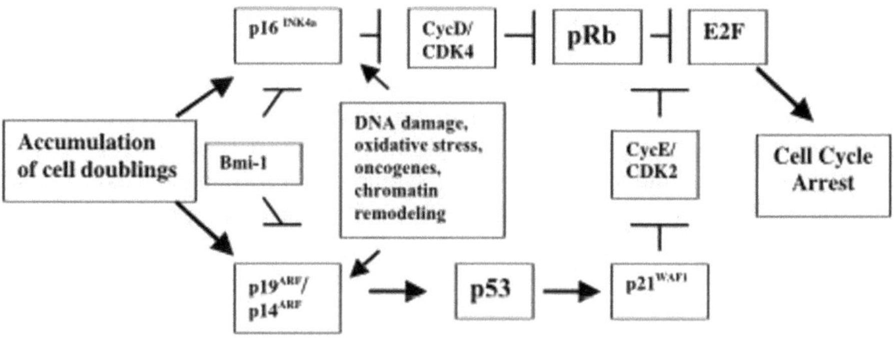

Figure 3. *Cell cycle regulation in senescent fibroblasts. (See text for details).*

As a consequence, p53 and $p21^{WAF1}$ are also upregulated in senescent cells [45, 46]. However, beyond its activation by CKIs, an alternative hypothesis for p53 over-expression in senescent cells, based on the well known ability of this protein to act as a "sensor" of DNA damage and to respond by promoting cell cycle arrest or apoptosis, could be that telomere shortening in senescent cells can be "recognized" as a sort of DNA damage and lead to p53 induction.

It is generally believed that the upregulation of both $p21^{WAF1}$ and $p16^{INK4a}$ is necessary for the induction and the maintenance of the senescent phenotype. However, it has been shown that at the very last population doublings the kinetics of the expression of these two proteins are not identical. After the end of the cellular lifespan a decline of $p21^{WAF1}$ occurs, while $p16^{INK4a}$ expression continues to rise [47]. So, it seems that $p21^{WAF1}$ is both necessary and sufficient for cell cycle arrest. On the other hand, $p16^{INK4a}$ – that synergises with $p21^{WAF1}$ on growth arrest – initiates also a "differentiation" programme that ensures the irreversibility of the senescent cell cycle arrest [47].

Tumor suppressors and immortalization

From the above, it is clear that the role of tumor suppressor proteins p53 and pRb is not confined solely to cancer but includes also replicative senescence. This is further substantiated by the ability of viral proteins to induce cellular immortalization. Probably, the most popular model is that based on the expression of the simian virus (SV) 40 L T antigen. This antigen drives normal fibroblasts to crisis and the surviving cells can give clones with unlimited proliferative potential. This effect has been associated with the ability of SV40 L T antigen to bind and inactivate both pRb and p53 [48]. Interestingly, immortalization can also be achieved by a combined infection of the cells with E6 and E7 proteins of the papiloma virus that can inactivate p53 and pRb, respectively. Several laboratories have further studied this process by conditionally immortalize murine or human cells with the thermolabile tsA58 T antigen. When these immortalized cells are placed in a non-permissive temperature, where T antigen is no longer expressed, the subsequent activation of p53 and pRb leads to the acquisition of the morphological and functional characteristics of the cells that have undergone replicative senescence [49–51]. However, beyond viral proteins, cellular proteins also that can bind to and abrogate the function of p53, like the hsp70 family member mortalin (mot-2), can also extend human fibroblasts' lifespan [48, 52].

Stress and senescence

One of the earliest theories on aging, the "free radical theory" [53], suggests that the continuous oxidative damage that occurs during lifespan is a critical factor for the aging process. Endogenous oxygen radicals and hydrogen peroxide are continuously produced mainly by the mitochondria as a by-product of metabolism. These reactive oxygen species (ROS) can efficiently react with DNA and cause the formation of 8-oxo-2′-deoxyguanosine (oxo ^{8}dG) and other mutagenic adducts [54]. Furthermore, cellular senescence is also characterized by the increased levels of "abnormal" proteins, mostly formed by post-translational modifications [5], such as oxidation and glycation [55]. A reduction of the lysosomal-mediated protein degradation is observed [56], leading to the appearance of lipofuscin – yellow, autofluorescent vesicles – in the lysosomes of senescent cells. Furthermore, downregulation of proteasome, the large multicatalytic proteolytic complex that selectively degrades damaged and ubiquitinated proteins, has been recently suggested to occur in senescent cells [57]. Finally, the production of the molecular chaperones heat shock proteins (hsps), that represent another cellular tool for protection against protein misfolding and aggregation, is decreased with aging [58]. As a result, damaged proteins accumulate in senescent cells due to defective proteolysis, leading to impaired cellular functions.

According to the above, in addition to serial subculturing, cellular senescence can be provoked by the application of an exogenous oxidative stress. Indeed, early-passage cells when exposed to sublethal oxidative stress, such as H_2O_2 [59, 60] or tert-butylhydroperoxide [61] can enter a permanent arrested state. This phenomenon is now called "stress-induced premature senescence" (SIPS), as the stressed cells resemble those that have undergone *in vitro* senescence after serial subculturing. In particular, they have a similar morphology and analogous regulation of growth arrest (i.e., overexpression of $p21^{WAF1}$ and $p16^{INK4a}$, pRb hypophosphorylation and E2F1 inactivation) and gene expression (such as *c-fos* down-regulation, overexpression of collagenase-1, fibronectin and type 1 procollagen) [62]. Furthermore, SIPS is induced when cells are cultured under hyperoxic conditions (oxygen partial pressure 40% compared to conventional culture condition of 20%). On the other hand, cultures growth in hypoxic conditions (3% oxygen tension) or in the presence of antioxidants or free radical scavengers exhibit a significant increased in their lifespan [62].

The mechanism underlying stress-induced premature senescence in human fibroblasts includes the direct impairment of multiple cellular functions and the accumulation of non-functional proteins. Interestingly, it has also been associated with the activation of the secretion of transforming growth factor-beta (TGF-β), an important regulator of tissue homeostasis [63]. TGF-β is secreted as a response to treatment with H_2O_2 and inhibition of its action by neutralizing antibodies against its receptors reverses the SIPS induced by this oxidative stress [64]. This could probably be attributed to the ability of TGF-β to activate a NADH oxidase and secondarily induce H_2O_2 production [65]. In line with the above, fibroblasts from a group of donors with elevated glucocorticoid levels that exhibit increased *in vitro* lifespan secrete lower levels of active TGF-β, indicating the inverse relationship between TGF-β secretion and cellular longevity [66].

Premature senescence by non-oxidative stress

Normal human cells can also enter a senescence-like state after certain types of DNA damage [67]. Furthermore, alterations in the epigenetic maintenance mechanisms can result to senescence too [68]. Absence of Bmi1, a component of the Polycomb Group complexes, lead to premature senescence through the upregulation of cyclin-dependent inhibitors $p16^{INK4a}$ and $p19^{ARF}$ [69]. In addition, chromatin remodeling can have the same effect: DNA demethylation agents [70] and inhibitors of histone deacetylases can also induce premature senescence [71].

However, the most spectacular and probably unexpected induction of premature senescence is that provoked by oncogene overexpression. Ras protein is a central point where several extracellular mitogenic stimuli converge and then, through the activation of several intracellular signal transduction cascades (such as the Ras/Raf/Mek/Erk pathway) cell cycle progression is stimulated. In addition, hyperactive Ras is found in approximately 20% of human tumors and, furthermore, expression of an oncogenic form of this protein into murine immortal cell lines results in cellular transformation. In contrast, when incorporated into normal (mortal) cells, after a short period of increased proliferation, the cells enter into a senescence-like permanent cell-cycle arrest, accompanied by overexpression of $p16^{INK4a}$ and $p19^{ARF}$ and by activation of p53 and pRb [72–74]. In contrast to the classical replicative senescence, this type of premature senescence cannot be prevented by the expression of active telomerase [75]. It must be mentioned, that overexpression of c-Myc – another oncoprotein involved in cell proliferation and also mutated in several human cancers – can drive normal human fibroblasts not to senescence, but to increased sensitivity for apoptosis [76]. So, it seems that cellular senescence and apoptosis represent two alternative "fail-safe" anti-tumor mechanisms, in order to ensure the maintenance of homeostasis, after oncogene deregulation.

What decides if a normal cell subjected to an "oncogenic stress" will activate a programme towards senescence or apoptosis? An answer to this can be given by the cellular response to the overexpression of E2F1. This transcription factor was initially suggested to be an oncogene, while more recent data indicate also a tumor suppressor role [77, 78]. Inappropriate E2F1 expression in normal human fibroblasts may result in $p14^{ARF}$- and p53-dependent senescence, while in p53-compromised cells can lead to lifespan extension and in pRb-compromised cells to apoptosis [77]. The above clearly indicate that it is the overall cellular context that directs the ultimate response to this oncogenic stress.

The different types of stress, i.e., "oxidant" or "oncogenic," that can induce premature senescence, seem not to work in isolation. For example, oncogenic Ras under normoxic culture conditions (20% oxygen) can increase mitochondrial reactive oxygen species and activate the senescence programme. In a hypoxic environment (1% oxygen) Ras is unable to increase the levels of the CDK inhibitor $p21^{WAF1}$ and induce senescence, indicating that Ras-induced senescence could be mediated, at least in part, by oxidative stress [79]. In the same context, forced expression of $p21^{WAF1}$ lead to increase of intracellular levels of ROS and cellular senescence, and this process can be blocked by N-acetyl-L-cysteine, a ROS inhibitor, indicating also that ROS overexpression is a necessary step towards permanent growth arrest [80].

Finally, the observation that some types of premature senescence, such as after oncogene overexpression, do not require telomere shortening, raises now a terminology issue. For years the terms "*in vitro* senescence" and "replicative senescence" were synonymous. It is now proposed that the permanent arrest that does not include telomere shortening or an equivalent counting mechanism to be referred as "stress" [68, 81], or alternatively to use "cellular senescence" to embrace all processes leading to this type of permanent arrest and "proliferative senescence" for those resulting from telomere shortening [82]. However, it is clear that the study of all these phenotypes will provide complementary information for understanding in depth the mechanisms underlying the senescent state.

Life-sustaining stress

In contrast to the various stresses that affect cellular functions leading prematurely to a senescent phenotype, it was shown that a mild, sublethal stress, by activating a moderate stress response, can counteract the negative effects of aging [83]. Several mild stresses (irradiation, temperature shock, hypergravity or exercise) can increase longevity in several model systems. At the cellular level, several mild heat shock treatments although do not lead to lifespan extension, delayed the appearance of aging features in cultured human fibroblasts [84]. Considering stress at the organismic level, caloric-restriction that represents the most efficient and reliable method to increase longevity, at least in rodents [83], also delays the loss of fibroblasts' proliferative capacity with aging [85]. Caloric-restricted animals display several physiological alterations, such as elevated levels of stress-inducible compounds, e.g., glucocorticoids. Interestingly, fibroblasts from patients with elevated glucocorticoid levels in the absence of any exogenous stress (Cushing's syndrome) also exhibit a considerable increase of their *in vitro* lifespan [66], indicating the close relationship between stress response and longevity.

Senescence and death

The empirical association between organismal senescence and death often leads to a similar connection between cellular senescence and an increased propensity to cell death. In particular, apoptotic cell death gained much consideration due to its "programmed" nature, in analogy to the near constant number of cell doublings required for *in vitro* cellular senescence. However, in contrast to this idea, initial studies have shown that senescent human fibroblasts are more resistant, compared to early-passage cells, to apoptotic cell death provoked by serum withdrawal, probably due to the increased levels of the anti-apoptotic protein Bcl-2 [86, 87]. The idea that senescent fibroblasts are specialized "survivors" [88] became a dogma in biogerontology for several years, leading also to the hypothesis of a progressive accumulation of senescent cells in the tissues of the elderly [9]. However, it has been recently shown that okadaic acid, TNF-α or Fas are able to drive senescent fibroblasts to apoptosis [89, 90]. Still some differences exist in the mechanism underlying apoptotic induction between young and senescent fibroblasts. For examples, in Fas-mediated apoptosis of senescent cells one of the hallmarks of this type of cell death, i.e., DNA fragmentation, is absent, and it has been hypothesized that this is due to reduced levels of endonucleases or their activators [89]. In addition, while young fibroblasts were able to undergo p53-dependent and p53-independent apoptosis in response to DNA damage, senescent cells are unable to die in a p53-dependent manner. This seems to occur due to their inability to stabilize p53 in response to genotoxic stress [91]. Finally, telomerase can inhibit hydroxyl radical-induced apoptosis in telomerase-negative normal human fibroblasts [92]. So, maintenance of telomere length seems to be crucial for the resistance to both replicative senescence and apoptosis.

Aging fibroblasts and tissue homeostasis

Aging of cells in culture is now an established model for studying various aspects of cellular physiology. However, the question "which is the physiological significance of these alterations in organismic aging and tissue homeostasis, in general?" still remains to be answered.

In vitro senescence and aging *in vivo*

After the initial description of the *in vitro* aging process, the analogy between this phenomenon and aging *in vivo* has been thoroughly investigated. Initial studies have shown that there is a negative correlation between the age of an individual and the proliferative lifespan of his fibroblasts *in vitro* [93, 94]. In support to this, a good correlation has been reported between species lifespan and proliferative capacity in culture, for a variety of animals, from mice (maximum assumed lifespan 2 years) to humans (maximum assumed lifespan 100 years) [95, 96]. However, the most direct evidence that aging *in vivo* and *in vitro* are closely related, comes from cultures established from donors with accelerated aging (progeria), such as patients suffering from Werner syndrome. These cells exhibit significantly shorter proliferative lifespan in culture compared to normal donors [97]. The above have supported the idea that cellular aging *in vivo* and *in vitro* are strongly related phenomena, probably based on common mechanisms. However, in 1998 Cristofalo and colleagues [98] criticized this concept by showing no significant correlation between the proliferative potential of fibroblast cell strains and the age of the donor [98]. They have also reported that multiple cell strains developed from the same individual at different ages failed to reveal significant trends between proliferative capacity and donor age. Finally, recently the analysis of a large number of cultures from healthy donors showed a trend (approaching statistical significance) towards low proliferative potential as donors aged [99]. All these reports indicate also the mosaic nature of cellular aging within each body, as well as the importance of the health status of the donor and the conditions of the biopsy.

However, one must consider that senescent cells in a tissue could be more difficult to "adapt" in a primary culture, compared to "younger" cells, and thus lead to an underestimation of their presence and physiological consequences. So, the need for a "biomarker" for identifying aged cells in a tissue *in situ* was necessary. In the mid 1990's it has been shown that *in vitro* senescent cells – in contrast to quiescent or immortal cells – specifically express β-galactosidase, histochemically detectable at pH 6.0 (called senescence associated β-galactosidase or SA β-gal staining). Furthermore, an age-dependent increase of this marker in skin samples has been reported, providing the evidence that senescent cells may exist and accumulate with age inside the tissue *in vivo* [100]. However, more recent data have questioned these findings. In particular, SA β-gal-positive cells were found in quiescent immortal cells or even in human cancer cells chemically stimulated to differentiate [101]. In addition, it has been found in human cells that this marker is valid for distinguishing senescent cells only under specific culture conditions [102]. A more recent analysis of tissue sections from donors of different ages has shown no age-associated increase in SA β-gal staining, suggesting that this indicator is a valuable marker solely for replicative senescent cells [102]. It must be noted however that due to their limited proliferation in the tissue, fibroblasts is probably not the best cell type to prove the existence of senescent cells *in vivo*. Other cell types, such as epithelial or smooth muscle cell, when derived from aged tissues exhibit much shorter lifespans, increased SA β-gal staining and decreased telomere lengths [82]. In support to the above, fibroblasts from chronic wounds (ulcers), that are continuously stimulated to proliferate in the tissue, become prematurely senescent when cultured *in vitro* [103, 104].

Aged fibroblasts, normal tissue turnover and wound repair

The cells throughout their lifespan in the tissue *in vivo* are continuously exposed to several noxes, such as oxidative stress, UV light, etc., known to affect their functions and also shorten their *in vitro* lifespan. Controversial data appear in the literature concerning the functional characteristics of the *in vivo* aged fibroblasts, probably due to inherent viability among donors or to the "history" of each tissue, e.g., differential exposure to UV light. In general, it seems that *in vivo* aged fibroblasts, although they proliferate normally in response to isolated growth factors and also secrete equal amounts of collagen as young cells, they exhibit also some functional differences, such as impaired migratory capacity [105–107]. The most striking age-related change is the up-regulation of matrix metalloproteases secretion, a common alteration in fibroblasts aged *in vivo* and *in vitro* [108, 109]. These alterations can severely influence, in long terms, the normal tissue turnover and lead to the characteristic structural and functional changes in the aged skin, i.e., decreased dermal thickness, decline in collagen content, alterations in glycosaminoglycan profile, loss of elasticity and finally impaired wound repair [110]. However, beyond alterations in fibroblast function, the critical role of systemic factors, such as the altered hormonal profile or nutritional deficiencies in the elderly, as well as their specific interplay with aged cells, must not be overlooked in the explanation of age-related alterations in tissue homeostasis [106]. On the other hand, fibroblasts from long-lived caloric-restricted mice have preserved their proliferative, biosynthetic and contractile capacity, important parameters of tissue repair. In addition, these animals, after a short-term refeeding, demonstrate enhanced capacity for wound repair in comparison to *ad libitum* fed mice [111]. Furthermore, dermal fibroblasts from patients with elevated glucocorticoid levels exhibit *in vitro*, a lifespan expansion (see above), as well as an anabolic phenotype, characterized by increased proliferative capacity and ability to secrete ECM components (our unpublished observations).

Replicative senescence and carcinogenesis

It has been suggested that fibroblast replicative senescence, due to the permanent cell cycle arrest, is a "fail safe" anti-tumor mechanism. However, advancing age is the major risk factor for cancer development. Are these two facts contradictory? Here, it must be mentioned that in humans most age-related cancers are of epithelial origin (carcinomas), while sarcomas and lymphomas prevail in the paediatric population [112]. Although, the development of carcinomas is predominately the result of mutation accumulation or epigenetic gene silencing [112, 113], there is increasing evidence for the crucial role of stromal-epithelial interactions in the growth, survival and differentiation of neoplastic epithelial cells. Indeed, normal microenvironment can effectively suppress carcinoma development [114], while on the other hand, ectopic expression of matrix degrading proteinases can cause malignant conversion of epithelial cells [115]. Fibroblasts that represent the major cell type of stroma, when activated by growth factors or by the vicinity to tumors can promote carcinogenesis [116, 117]. According to the above, it has been hypothesized that age-related alterations of human fibroblasts, such as matrix metalloprotease hypersecretion, can lead to a "permissive" environment for the growth of pro-neoplastic epithelial cells [9]. Recent evidence has provided strong support for this hypothesis. By using co-cultures of fibroblast and epithelial cells, it has been shown that senescent fibroblasts (and not presenescent ones) stimulate the growth of preneoplastic (immortalized but not tumorigenic) and neoplastic (tumor-derived) epithelial cells. In contrast, normal epithelial cells have no growth advantage under the same culture conditions [118, 119]. The inducer of fibroblast senescence seems to be irrelevant for this phenomenon, as fibroblasts that became senescent after serial subculturing, oxidative damage or forced expression of oncogenic Ras all have the same effect on the growth of epithelial cells [119]. Furthermore, this fibroblast-derived promotion of epithelial cell growth seem to be due both the secretion of soluble factors and the deposition of insoluble extracellular matrix components [118]. In addition, when co-injected into immunocompromised mice, senescent fibroblasts have synergized with preneoplastic epithelial cells and gave rise to large and aggressive tumors [118], indicating that senescent fibroblasts can promote tumorigenesis *in vivo*, thus crucially affecting tissue homeostasis.

The idea that replicative senescence represents an anti-tumor mechanism in young organisms and promotes cancer in aged ones is consistent with the evolutionary theory of "antagonistic pleiotropy." According to this, a trait that has been selected in order to enhance the fitness early in life, could became deleterious in aged individuals [120]. In this context it is interesting to mention novel data on the role of the tumor suppressor protein p53. It is well know that this protein is non-functional in approximately 50% of human tumors, while is activated in senescent cells. Recently, a strain of genetically altered mice carrying a truncated form of p53 that increases wild-type p53 activity has been developed [121]. These mice, as expected, exhibit enhanced resistance to spontaneous tumors. However, they also display an early onset of phenotypes associated with normal aging, such as osteoporosis, atrophy of muscle and skin, reduced body mass, depletion of hematopoetic stem cells, reduced stress resistance and impaired wound healing. Finally, their median and maximum lifespan is approximately 20% shorter, compared with wild-type littermates. These data clearly indicate the connection between cellular and organismic aging, and also the antagonistic pleiotropy that characterizes the action of important senescence-related genes, such as p53.

The study of fibroblast aging during the last decades has provided an amazing bulk of information that has revolutionized our ideas not only on the aging process, but also on cellular physiology and even carcinogenesis. It has also provided the scientific basis for the development of new therapeutic approaches against age-related diseases, such as the use of anti-oxidants or, more recently, telomerase inhibitors [122]. In addition, it may indicate life-style changes for confronting age-related impairments, towards prolonged survival and, most important, successful aging.

Acknowledgments

I would like to thank all the members of the laboratory for their hard work and stimulating discussions and specially Harris Pratsinis for the critical reading of this manuscript.

References

1. Hayflick L (1998). How and why we age. *Exp Gerontol.* 33: 639–53.
2. Hayflick L, Moorhead P (1961). The serial cultivation of human diploid cell strains. *Exp Cell Res.* 25: 585–621.

3. Hayflick L (1965). The limited *in vitro* lifetime of human diploid cell strains. *Exp Cell Res.* 37: 614–36.
4. Smith JR, Lincoln DW II (1984). Aging of cells in culture. *Int Rev Cytol.* 89: 151–77.
5. Rattan SIS (1995). Ageing – A biological perspective. *Mol Aspects Med.* 16: 439–508.
6. Lipetz J, Cristofalo VJ (1972). Ultrastructural changes accompanying the aging of human diploid cells in culture. *J Ultrastruct Res.* 39: 43–56.
7. Takeda K, Gosiewska A, Peterkofsky B (1992). Similar, but not identical, modulation of expression of extracellular matrix components during *in vivo* and *in vitro* aging of human skin fibroblasts. *J Cell Physiol.* 153: 450–9.
8. Millis AJ, Hoyle M, McCue HM, Martini H (1992). Differential expression of metalloproteinase and tissue inhibitor of metalloproteinase genes in aged human fibroblasts. *Exp Cell Res.* 201: 373–9.
9. Campisi J (1997). Aging and cancer: The double edged sword of replicative senescence. *J Am Geriatr Soc.* 45: 482–8.
10. Shelton DN, Chang E, Whittier PS, Choi D, Funk WD (1999). Microarray analysis of replicative senescence. *Curr Biol.* 9: 939–45.
11. Olovnikov AM (1996). Telomeres, telomerase, and aging: origin of a theory. *Exp Gerontol.* 31: 443–8.
12. Campisi J, Kim S-h, Lim C-S, Rubio M (2001). Cellular senescence, cancer and aging: the telomere connection. *Exp Gerontol.* 36: 1619–37.
13. Reddel RR (2000). The role of senescence and immortalization in carcinogenesis. *Carcinogenesis* 21: 477–84.
14. Cook BD, Dynek JN, Chang W, Shostak G, Smith S (2002). Role for the related Poly (ADP-Ribose) Polymerases Tankyrase 1 and 2 at human telomeres. *Mol Cell Biol.* 22: 332–42.
15. Bürkle A (2000). Poly (ADP-ribosyl)ation: a post translational protein modification linked with genomic protection and mammalian longevity. *Biogerontology* 1: 41–6.
16. Olovnikov AM (1971). Principles of marginotomy in template synthesis of polynucleotides. *Dokl Akad Nauk (S.S.S.R.)* 201: 1496–9.
17. Watson JD (1972). Origin of concatameric T4 DNA. *Nature* 345: 456–8.
18. Harley CB (1991). Telomere loss: mitotic clock or genetic time bomb? *Mutat Res.* 256: 271–82.
19. Harley CB, Futcher AB, Greiner CW (1990). Telomeres shorten during ageing of human fibroblasts. *Nature* 345: 458–60.
20. Bodnar AG, Ouellette M, Frolkis M, *et al.* (1998). Extension of life-span by introduction of telomerase into normal human cells. *Science* 279: 349–52.
21. Sher C, Roberts JM (1999). CDK inhibitors: Positive and negative regulators of G1–phase progression. *Genes Dev.* 13: 1501–12.
22. Kaelin WG Jr. (1999). Functions of the retinoblastoma protein. *Bioessays* 21: 950–8.
23. Ruas M, Peters G (1998). The $p16^{INK4a/CDKN2A}$ tumor suppressor and its relatives. *Biochim Biophys Acta* 14: 115–77.
24. Bringold M, Serrano M (2000). Tumor suppressors and oncogenes in cellular senescence. *Exp Gerontol.* 35: 317–29.
25. Psarras S, Kletsas D, Stathakos D (1994). Restoration of down-regulated PDGF receptors by TGF-β in human embryonic fibroblasts. Enhanced response during cellular *in vitro* aging. *FEBS Lett.* 339: 84–8.
26. Philips PD, Kuhle E, Cristofalo VJ (1983). [^{125}I] EGF binding ability is stable throughout the replicative life-span of WI-38 cells. *J Cell Physiol.* 114: 311–16.

27. Philips PD, Pignolo RJ, Cristofalo VJ (1987). Insulin-like growth factor-I: specific binding to high and low affinity sites and mitogenic action throughout the replicative life-span of WI-38 cells. *J Cell Physiol.* 133: 135–43.
28. Gerhard GS, Phillips PD, Cristofalo VJ (1991). EGF- and PDGF-phosphorylation in young and senescent WI-38 cells. *Exp Cell Res.* 193: 87–92.
29. Shiraha H, Gupta K, Drabik K, Wells A (2000). Aging fibroblasts present reduced Epidermal Growth Factor (EGF) responsiveness due to preferential loss of EGF receptors. *J Biol Chem.* 275: 19343–51.
30. Wheaton K, Sampsel K, Boisvert F-M, Davy A, Robbins S, Riabowol K (2001). Loss of functional caveolae during senescence oh human fibroblasts. *J Cell Physiol.* 187: 226–35.
31. Venable ME, Blobe GC, Obeid LM (1994). Identification of a defect in the phospholipase D/diacylglycerol pathway in cellular senescence. *J Biol Chem.* 269: 26040–4.
32. Exton JH (1994). Phosphatilylocholine breakdown and signal transduction. *Biochim Biophys Acta* 1212: 26–42.
33. Tresini M, Mawal-Dewan M, Cristofalo VJ, Sell C (1998). A phosphatidylinositol 3-kinase inhibitor induces a senescent-like growth arrest in human diploid fibroblasts. *Cancer Res.* 58: 1–4.
34. Lim IK, Hong KW, Kwak IH, Yoon G, Park SC (2000). Cytoplasmic retention of p-Erk1/2 and nuclear accumulation of actin proteins during cellular senescence in human diploid fibroblasts. *Mech Ageing Dev.* 119: 113–29.
35. Tresini M, Lorenzini A, Frisoni L, Allen RG, Cristofalo VJ (2001). Lack of Elk-1 phosphorylation and dysregulation of the extracellular regulated kinase signaling pathway in senescent human fibroblasts. *Exp Cell Res.* 269: 287–300.
36. Stein GH, Dulic V (1995). Origins of G1 arrest in senescent human fibroblasts. *Bioessays* 17: 537–43.
37. Kletsas D, Stathakos D, Sorrentino V, Philipson L (1995). The growth inhibitory block of TGF-β is located close to the G1/S border of the cell cycle. *Exp Cell Res.* 217: 477–83.
38. Seshadri T, Campisi J (1990). Repression of c-fos transcription and an altered genetic program in senescent human fibroblasts. *Science* 247: 205–9.
39. Hara E, Yamaguchi T, Nojima H, *et al.* (1994). Id-related genes encoding helix-loop-helix proteins are required for G1 progression and are repressed in senescent human fibroblasts. *J Biol Chem.* 269: 2139–45.
40. Sheerin A, Thompson KS, Goyns MH (2001). Altered composition and DNA binding activity of the AP-1 transcription factor during the ageing of human fibroblasts. *Mech Ageing Dev.* 122: 1813–24.
41. Atadja PW, Stringer KF, Riabowol K (1994). Loss of serum response element-binding activity and hyperphosphorylation of serum response factor during cellular aging. *Mol Cell Biol.* 119: 4991–9.
42. Pang JH, Chen KY (1994). Global change of gene expression at late G1/S boundary may occur in human IMR-90 diploid fibroblasts during senescence. *J Cell Physiol.* 160: 531–8.
43. Sharpless NE, DePinho RA (1999). The *INK4A/ARF* locus and its two gene products. *Curr Opin Genet Dev.* 9: 22–30.
44. Kamijo T, Zindy F, Roussel MF, *et al.* (1997). Tumor suppression at the mouse INK4a locus mediated by the alternative reading frame product $p19^{ARF}$. *Cell* 91: 649–59.
45. Bond J, Haughton M, Blaydes J, Gire V, Wynford-Thomas D, Wyllie F (1996). Evidence that transcription activation of p53 plays a direct role in the induction of cellular senescence. *Oncogene* 13: 2097–104.

46. Noda A, Ning Y, Venable SF, Pereira-Smith OM, Smith JR (1994). Cloning of senescent cell-derived inhibitors of DNA synthesis using an expression screen. *Exp Cell Res.* 211: 90–8.
47. Stein GH, Drullinger LF, Soulard A, Dulic V (1999). Differential roles of cyclin-dependent kinase inhibitors p21 and p16 in the mechanisms of senescence and differentiation in human fibroblasts. *Mol Cell Biol.* 19: 2109–17.
48. Duncan EL, Wadhwa R, Kaul SC (2000). Senescence and immortalization of human cells. *Biogerontology* 1: 103–21.
49. Gonos ES, Burns JS, Mazars GR, *et al.* (1996). Rat embryo fibroblasts immortalized with simian virus 40 large T antigen undergo senescence upon its inactivation. *Mol Cell Biol.* 16: 5127–38.
50. Hsieh JL, Kletsas D, Glunn G, Hughes A, Schachter M, Demoliou-Mason C (2000). p53, P21 WAF1/CIP1 and MDM2 involvement in the proliferation and apoptosis in an *in vitro* model of conditionaly immortalized human vascular muscle cells. *Arterioscler Thromb Vasc Biol.* 20: 973–81.
51. Pratsinis H, Demoliou-Mason C, Hughes A, Kletsas D (2000). A novel *in vitro* model of conditionally immortalized human vascular smooth muscle cells. A tool for aging studies. *Ann NY Acad Sci.* 908: 321–3.
52. Wadhwa R, Takano S, Robert M, Yoshida A, Reddell RR, Mitsui Y, Kaul SC (1998). Inactivation of tumor suppressor p53 by mot-2, an hsp70 family member. *J Biol Chem.* 273: 29586–91.
53. Harman D (1956). Aging: a theory based on free radical and radiation chemistry. *J Gerontol.* 11: 298–300.
54. Chen Q, Fischer A, Reagan JD, Yan L-J, Ames BN (1995). Oxidative DNA damage and senescence of human fibrioblasts. *Proc Natl Acad Sci USA* 92: 4337–41.
55. Verbeke P, Clark BFC, Rattan SIS (2000). Modulating cellular aging *in vitro*: Hormetic effects of repeated mild heat stress on protein oxidation and glycation. *Exp Gerontol.* 35: 787–94.
56. Cuervo AM, Dice JF (1998). How do intracellular proteolytic systems change with age? *Front Biochim.* 1: 25–43.
57. Szweda PA, Friguet B, Szweda LI (2002). Proteolysis, free radicals and aging. *Free Rad Biol Med.* 33: 29–36.
58. Feder ME, Hofmann GE (1999). Heat-shock proteins, molecular chaperones, and the stress response: evolutionary and ecological physiology. *Ann Rev Physiol.* 61: 243–82.
59. Chen Q, Ames BN (1994). Senescence-like growth arrest induced by hydrogen peroxide in human diploid fibroblasts F65 cells. *Proc Natl Acad Sci USA* 91: 4130–4.
60. von Zglinicki T (2000). The role of oxidative stress in telomere length regulation and replicative senescence. *Ann NY Acad Sci.* 908: 99–110.
61. Toussaint O, Houbion A, Remacle J (1992). Aging of a multi-step process characterized by a lowering of energy production leading the cell to a sequence of defined stages II. Testing of some predictions on aging human fibroblasts in culture. *Mech Ageing Dev.* 65: 65–83.
62. Toussaint O, Medrano EE, von Zglinicki T (2000). Cellular and molecular mechanisms of stress-induced premature senescence (SIPS) of hyman diploid fibroblasts and melanocytes. *Exp Gerontol.* 35: 927–45.
63. Roberts AB, Sporn MB (1996). Transforming growth factor-β. In: Clark RAF, ed. *The Molecular and Cellular Biology of Wound Repair*. New York: Plenum Press, pp. 275–308.

64. Frippiat C, Chen QM, Zdanov S, Magalaes JP, Remacle J, Toussaint O (2001). Sublethal H_2O_2 stress triggers a release of TGF-β1 which induced biomarkers of cellular senescence of human diploid fibroblasts. *J Biol Chem.* 267: 2531–7.
65. Thannickal VJ, Fanburg BL (1995). Activation of an H_2O_2-generating NADH oxidase in human lung fibroblasts by transforming growth factor beta 1. *J Biol Chem.* 270: 30334–8.
66. Pratsinis H, Tsagarakis S, Zervolea Ir, *et al.* (2002). Chronic *in vivo* exposure to glucocorticoids prolongs cellular lifespan: the case of Cushing's syndrome-patients' fibroblasts. *Exp Gerontol.* 37: 1237–45.
67. DiLeonardo A, Linke SP, Clarkin K, Wahl GM (1994). DNA damage triggers a prolonged p53-dependent G1 arrest and long-term induction of Cip1 in normal human fibroblasts. *Genes Dev.* 8: 2540–51.
68. Seranno M, Blasco MA (2001). Putting the stress on senescence. *Curr Opin Cell Biol.* 13: 748–53.
69. Jacobs JJ, Kieboom K, Marino S, DePinho RA, van Lohuizen M (1999). The oncogene and Polycomb-group gene bmi-1 regulates cell proliferation and senescence through the ink4a locus. *Nature* 397: 164–8.
70. Holliday R (1986). Strong effects of 5-azacytidine on the *in vitro* lifespan of human diploid fibroblasts. *Exp Cell Res.* 166: 543–52.
71. Ogryzko VV, Hirai TH, Russanova VR, Barbie DA, Howard BH (1996). Human fibroblast commitment to a senescence-like state in response to histone deacetylase inhibitors is cell cycle dependent. *Mol Cell Biol.* 16: 5210–18.
72. Serrano M, Lin AW, McGurrach ME, Beach D, Lowe SW (1997). Oncogenic ras provokes premature cell senescence associated with accumulation of p53 and $p16^{INK4a}$. *Cell* 88: 593–602.
73. Zhu J, Woods D, McMahon M, Bishop JM (1998). Senescence of human fibroblasts induced by oncogenic Raf. *Genes Dev.* 12: 2997–3007.
74. Lin AW, Barradas M, Stone JC, van Aelst L, Serrano M, Lowe SW (1998). Premature senescence involving p53 and p16 activated in response to constitutive MEK/MAPK mitogenic signaling. *Genes Dev.* 12: 3008–19.
75. Wei S, Wei W, Sedivy JM (1999). Expression of catalytically active telomerase does not prevent premature senescence caused by overexpression of oncogenic Ha-Ras in normal human fibroblasts. *Cancer Res.* 59: 1539–43.
76. Zindy F, Eischen CM, Randle DH, *et al.* (1998). Myc signaling via the ARF tymor suppressor regulates p53-dependent apoptosis and immortalization. *Genes Dev.* 12: 2424–33.
77. Dimri GP, Itahana K, Acosta M, Campisi J (2000). Regulation of a senescence checkpoint response by the E2F1 transcription factor and $p14^{ARF}$ tumor suppressor. *Mol Cell Biol.* 20: 273–85.
78. Gorgoulis VG, Zacharatos P, Mariatos G, *et al.* (2002). Transcription factor E2F-1 acts as a growth promoting factor and is associated with adverse prognosis in Non-Small Cell Lung Carcinomas (NSCLCs). *J Pathol.* 198:142–56.
79. Lee AC, Fenster BE, Ito H, *et al.* (1999). Ras proteins induce senescence by altering the intracellular levels of reactive oxygen species. *J Biol Chem.* 274: 7936–40.
80. Macip S, Igarashi M, Fang L, *et al.* (2002). Inhibition of p21-mediated ROS accumulation can rescue p21-induced senescence. *EMBO J.* 21: 2180–8.
81. Wright WE, Shay JW (2001). Cellular senescence as a tumor-protection mechanism: the essential role of counting. *Curr Opin Genet Dev.* 11: 98–103.

82. Hornsby PJ (2002). Cellular senescence and tissue aging *in vivo*. *J Gerontol.* 57A: B251–6.
83. Minois N (2000). Longevity and aging: beneficial effects of exposure to mild stress. *Biogerontology* 1: 15–29.
84. Rattan SIS (1998). Repeated mild heat shock delays ageing in cultured human skin fibroblasts. *Biochem Mol Biol Int.* 45: 753–9.
85. Pendergraas WR, Li Y, Jiang D, Fei RG, Wolf NS (1995). Caloric restriction: conservation of cellular replicative capacity *in vitro* accompanies life-span extension in mice. *Exp Cell Res.* 217: 309–16.
86. Wang E (1995). Senescent human fibroblasts resist programmed cell death, and failure to suppress bcl-2 is involved. *Cancer Res.* 55: 2284–92.
87. Warner HR, Hodes RJ, Pocinki K (1997). What does cell death has to do with aging? *J Am Geriatr Soc.* 45: 1140–6.
88. Wang E, Lee MJ, Pandey S (1994). Control of fibroblasts senescence and activation of programmed cell death. *J Cell Biochem.* 54: 432–9.
89. Tepper CG, Seldin MF, Mudryj M (2000). Fas-mediated apoptosis of proliferating, transiently growth-arrested, and senescent normal human fibroblasts. *Exp Cell Res.* 260: 9–19.
90. DeJesus V, Rios I, Davis C, Chen Y, Calhoun D, Zakeri Z, Hubbard K (2002). Induction of apoptosis in human replicative senescence. *Exp Cell Res.* 274: 92–9.
91. Seluanov A, Gorbunova V, Falcovitz A, *et al.* (2001). Change of the death pathway in senescent human fibroblasts in response to DNA damage is caused by an inability to stabilize p53. *Mol Cell Biol.* 21: 1552–64.
92. Ren J-G, Xia H-L, Tian Y-M, Just T, Cai G-P, Dai Y-R (2001). Expression of telomerase inhibits hydroxyl radical-induced apoptosis in normal telomerase negative human lung fibroblasts. *FEBS Lett.* 488: 133–8.
93. Martin GM, Sprague CA, Epstein CJ (1970). Replicative lifespan of cultured human cells: effects of donor age, tissue, and genotype. *Lab Invest.* 23: 86–92.
94. Schneider EL, Mitsui Y (1976). The relationship between *in vitro* cellular aging and *in vivo* human age. *Proc Natl Acad Sci USA* 73: 3584–8.
95. Rohme D (1981). Evidence for a relationship between longevity of mammalian species and life spans of normal fibroblasts *in vitro* and erythrocytes *in vivo*. *Proc Natl Acad Sci USA* 78: 5009–13.
96. Shall S (1996). Ageing of cells *in vitro*. In: Rattan SIS, Toussaint O, eds. *Molecular Gerontology: Research Status and Strategies*. New York: Plenum Press, 75–85.
97. Oshima J, Campisi J., Tannock CA, Martin GM (1995). Regulation of c-fos expression in senescing Werner syndrome fibroblasts differs from that observed in senescing fibroblasts from normal donors. *J Cell Physiol.* 162: 277–83.
98. Cristofalo V, Allen RG, Pignolo RJ, Martin BG, Beck JC (1998). Relationship between donor age and replicative life span of human cells in culture: a re-evaluation. *Proc Natl Acad Sci USA* 95: 10614–19.
99. Smith JR, Venable S, Roberts TW, Metter EJ, Monticone R, Schneider EL (2002). Relationship between *in vivo* age and *in vitro* aging: assessment of 669 cell cultures derived from members of the Baltimore Longitudinal Study of Aging. *J Gerontol A Biol Sci Med Sci.* 57: B239–46.
100. Dimri GP, Lee X, Basile G, *et al.* (1995). A biomarker that identifies senescent human cells in culture and in aging skin *in vivo*. *Proc Natl Acad Sci USA* 92: 9363–7.
101. Yegorov YE, Akimov SS, Hass R, Zelenin AV, Prudovsky IA (1998). Endogenous β-galactosidase activity in continuously nonproliferating cells. *Exp Cell Res.* 243: 207–11.

102. Severino J, Allen RG, Balin S, Balin A, Cristofalo VJ (2000). Is β-galactosidase staining a marker of senescence *in vitro* and *in vivo*? *Exp Cell Res.* 257: 162–71.
103. Raffetto JD, Mendez MV, Phillips TJ, Park HY, Menzoian JO (1999). The effect of passage number on fibroblast cellular senescence in patients with chronic venus insufficiency with and without ulcer. *Am J Surg.* 178: 107–12.
104. Vande Berg JS, Rudolph R, Hollan C, Haywood-Reid PL (1998). Fibroblast senescence in pressure ulcers. *Wound Repair Regen.* 6: 38–49.
105. Freedland M, Karmiol S, Rodriguez J, Normolle D, Smith D Jr, Garner W (1995). Fibroblast responses to cytokine are maintained during aging. *Ann Plast Sur.* 35: 290–6.
106. Kletsas D, Pratsinis H, Zervolea Ir, *et al.* (2000). Fibroblast responses to exogenous and autocrine growth factors relevant to tissue repair. The effect of ageing. *Ann NY Acad Sci.* 908: 155–66.
107. Mogford JE, Rawil N, Chen A, Gies D, Xia Y, Mustoe TA (2002). Effect of age and hypoxia on TGFbeta1 receptor expression and signal transduction in human dermal fibroblasts: impact on cell migration. *J Cell Physiol.* 190: 259–65.
108. Ly DH, Lockhart DJ, Lerner RA, Schultz PG (2000). Mitotic misregulation and human aging. *Science* 287: 2486–92.
109. Reed MJ, Ferara NS, Vernon RB (2001). Impaired migration, integrin function and actin cytoskeletal organization in dermal fibroblasts from a subset of aged human donors. *Mech Ageing Dev.* 122: 1203–20.
110. Ashcroft GS, Horan MA and Ferguson MW (1995). The effects of ageing on cutaneous wound healing in mammals. *J Anat.* 187: 1–26.
111. Reed MJ, Penn PE, Li Y, *et al.* (1996). Enhanced cell proliferation and biosynthesis mediate improved wound repair in refed, caloric-restricted mice. *Mech Ageing Dev.* 89: 21–43.
112. DePinho RA (2000). The age of cancer. *Nature* 408: 248–54.
113. Baylin SB, Herman JG (2000). DNA hypermethylation in tumorigenesis: epigenetics joints genetics. *Trends Genet.* 16: 168–74.
114. Park CC, Bissell MJ, Barcellos-Hoff MH (2000). The influence of the microenvironment on the malignant phenotype. *Mol Med Today* 6: 324–9.
115. Sternlicht MD, Lochter A, Sympson CJ, *et al.* (1999). The stromal proteinase MMP3/ stromelysin-1 promotes mammary carcinogenesis. *Cell* 98: 137–46.
116. Skobe M, Fusenig NE (1998). Tumorigenic conversion of immortal human keratinocytes through stromal activation. *Proc Natl Acad Sci USA* 95: 1050–5.
117. Olumi AF, Grossfeld GD, Hayward SW, Carroll PR, Tlsty TD, Cunha GR (1999). Carcinoma-associated fibroblasts direct tumor progression of initiated human prostatic epithelium. *Cancer Res.* 59: 5002–11.
118. Krtolica A, Parrinello S, Lockett S, Desprez P-Y, Campisi J (2001). Senescent fibroblasts promote epithelial cell growth and tumorigenesis: A link between cancer and aging. *Proc Natl Acad Sci USA* 98: 12072–7.
119. Krtolica A, Campisi J (2002). Cancer and aging: a model for the stroma promoting effects of the aging stroma. *Int J Biochem Cell Biol.* 34: 1401–14.
120. Kirkwood TB, Austad SN (2000). Why do you age? *Nature* 408: 233–8.
121. Tyner SD, Venkatachalan S, Choi J, *et al.* (2002). p53 mutant mice that display early ageing-associated phenotypes. *Nature* 415: 45–53.
122. Granger MP, Wright WE, Shay JW (2001). Telomerase in cancer and aging. *Crit Rev Oncol/Hematol.* 41: 29–40.
123. Papazafiri P, Kletsas D (2003). Developmental and age-related alterations of calcium homeostasis in human fibroblasts. *Exp Gerontol.* 38: 307–11.

Aging of Endothelial Cells

Tsutomu Kumazaki
Suzugamine Women's College, 4-6-18 Inokuchi, Nishi-ku, Hiroshima, 733-8623, Japan

Introduction

The endothelium acts as a barrier between blood and vascular smooth muscle cells by virtue of its position within the blood vessel wall. Therefore, the functional integrity of the endothelium monolayer, which is composed of endothelial cells (ECs), is essential to the prevention of vascular leakage and the formation of atherosclerotic lesions [1, 2], which are often found in the human vasculature of the elderly [3, 4]. Aging is an independent risk factor for atherosclerotic cardiovascular disease, leading to progressive impairment of organ function [3, 5, 6]. Changes in the vascular system, including structural and functional alterations of the arterial wall and an increase in blood pressure, frequently are observed during aging [7, 8]. As an important source of vasoactive substances, such as endothelin-1 (ET-1) and nitric oxide (NO), vascular ECs play a key role in the control of vascular tone. Because EC functions change during aging, EC senescence is an important pathogenetic factor in age-related vascular dysfunction [9].

Progressive loss of proliferative potential by somatic cells is considered the most consistent manifestation of cellular senescence, and *in vitro* senescence of human ECs is also characterized by the loss of proliferative potential and an increase in cell size. ECs often have been used as a model for studies of cellular aging, and these studies have revealed that EC aging is accompanied by the alteration of several cellular parameters, including protein synthesis, surface structure, and gene expression. Moreover, an increasing body of evidence indicates that cellular aging also occurs *in vivo* and that the proportion of senescent cells in tissue increases with age [10]. Many of the changes occurring during *in vitro* replicative senescence reflect alterations associated with *in vivo* aging, and a close relationship between *in vitro* and *in vivo* senescence of ECs has now been established. Thus, EC senescence likely plays a key role in age-associated vascular disease. In this chapter, aging-associated changes of ECs, are summarized and discussed.

Sunil C. Kaul and Renu Wadhwa (eds.), Aging of Cells In and Outside the Body, 47–65.

Endothelin

Originally discovered in a culture medium of ECs [11], ET-1 is synthesized and secreted by ECs. ET-1 is a 21-amino acid peptide with strong vasoconstrictor and mitogenic activities [12], and it has been causatively implicated in the structural and functional abnormalities of arterial hypertension, atherosclerosis, and glomerulosclerosis [13–18]. Interestingly expression of ET-1 mRNA is undetectable in the aorta of young donors, but becomes detectable in that of older donors [19]. The level of ET-1 mRNA detected in human aortic ECs (HAECs) is roughly two times higher in 76-year-old donors than in younger donors [20]. *In vitro* aging human umbilical vein ECs (HUVECs) show parallel results [20]. The synthesis of ET-1 peptide is also elevated in HAECs from older donors (over 50 years old) as well as in aged HUVECs (over 51 population doubling levels (PDLs)). These results indicate that ECs age similarly *in vivo* and in culture, by exhausting cell division potential. ET-1 is a possible factor in hypertension, which is often seen in elderly individuals. The observation of age-associated increases in circulating levels of ET-1 and in ET converting enzyme (ECE) activity in the rat [4] supports this hypothesis. On the other hand, the sensitivity of vascular smooth muscle cells to ET-1 decreases with age [21]. Therefore, the higher rate of production of ET-1 by older individuals may reflect a feedback mechanism of the cardiovascular system to compensate for lowered sensitivity. Interestingly, ET-1 synthesis has also been demonstrated in vascular smooth muscle cells [19, 22], renal mesangial [23] and tubulointerstitial cells [24], cardiac myocytes [25], and macrophages [26], although the functional significance of ET-1 production by these cells is not yet clear.

NO acts to relax blood vessels, and it also inhibits ET-1 synthesis *in vitro* [27] and *in vivo* [28]. NO is synthesized by three distinct isoforms of NO synthases (NOSs) that are present in the vascular wall [29]. Senescent HUVECs in culture lack the ability to produce NO, and senescent ECs therefore may be unable to modulate vascular tone *in vivo*. Decreases in the inducible isoform of NOS (iNOS) during *in vitro* aging of HUVECs [2] and in the aorta of older rats [30] have been reported. In fact, basal and stimulated bioactivity of endothelium-derived NO is reduced in aging rats [4, 31] suggesting reduced vascular relaxation and thereby higher blood pressure in older animals. Expression of iNOS is regulated by cytokines in many tissues [32], although it is constitutively expressed in certain organs [33]. It has been shown that impairment of endothelium-dependent vasomotion and atherosclerosis are associated with local activation and/or expression of iNOS [34–36]. On the other hand, there is a report showing that expression of iNOS is elevated in aging rat aorta [37] suggesting suppression of ET-1 synthesis. However, increased expression of ET-1 is observed. Thus, more experiments are necessary to reach a conclusion on the role of iNOS in age-related vascular dysfunction.

Regulation of endothelin synthesis

One of the characteristics of cellular aging is a decrease in the percentage of cells at S phase of the cell cycle because of growth arrest. Most cell specific functions are expressed in non-S (mostly G_1/G_0) phase of confluent culture and most markers of cellular aging are closely related to growth arrest. In particular, S-phase specific gene expression decreases in senescent culture. However, it has been shown that growing cells secrete more ET peptide than do confluent and stationary cells [20]. Higher expression of ET-1 mRNA in S-phase cells than in non-S-phase cells was observed in all of the PDLs examined, with increasing expression in older cells [20], indicating two distinct mechanisms in the up-regulation of ET-1 expression.

Fibronectin

Fibronectin (FN) is an extracellular matrix (ECM) protein and is a major component of the ECM. It binds to integrin, a receptor on the cell surface and therefore is an important factor in cell adhesion and determination of position within the ECM. Interestingly, HAECs from a young donor contain about 4 times more FN mRNA than do young ECs from umbilical vein [38, 39]. The reason for this difference is theorized as follows: Vascular ECs are continuously exposed to hydrodynamic forces and may be carried away from the vascular wall by the blood stream. Because the aorta experiences stronger shear forces than do veins, aortic ECs may require more FN than might HUVECs to remain attached to the vascular wall. If ECs are detached and carried away from the vascular wall by the blood stream, the resulting open space must be filled by division of the surrounding cells. Otherwise, smooth muscle cells, having lost their protective barrier, are exposed directly to the blood, which may result in unusual growth of smooth muscle cells and ultimately to arteriosclerosis. Because senescent cells are large, to remain attached to the substrate they probably require a greater number of adhesive molecules such as FN. With aging, increases in both FN mRNA expression and cell area are observed among HAECs from donors of varying age. A 2.7-fold increase is observed in comparison of HAECs from 5- and 76-year old donors. An increase as great as 2.1 fold is observed during *in vitro* aging of HUVECs along with an increase in cell area [38, 39]. Interestingly, the increasing level of FN expression and cell area with age are almost parallel between *in vitro* aging HUVECs and *in vivo* aged HAECs.

Telomere length

As mentioned above, cellular senescence is accompanied by cessation of cell division with changes in cell function, morphology, and gene expression. These changes in cell phenotype may contribute to age-associated diseases, including atherosclerosis [40]. Accumulating evidence has suggested the critical role of telomere length and telomerase in cellular senescence *in vitro* [41]. Telomere shortening occurs during *in vitro* aging of ECs and other cell types, and telomere length has been shown to be a mitotic clock that counts the divisions cells undergo. By introduction of telomerase catalytic component (hTERT), we have established a HUVEC cell line, that does not exhibit telomere shortening and therefore can proliferate indefinitely. These cells preserve a phenotype of young ECs but do not show any malignant characteristics. Others have demonstrated that hTERT introduction extends the life span of human vascular smooth muscle cells [42]. These studies suggest that telomere stabilization is important for the long-term viability of vascular cells. In fact, progressive telomere shortening in human arteries has been observed in regions susceptible to atherosclerosis [43]. Moreover, telomere length has been reported to correlate inversely with pulse pressure and atherosclerotic grade in humans [44, 45]. These observations strongly suggest that telomere shortening and EC senescence *in vivo* may contribute to the pathogenesis of age-associated vascular disorders.

Interleukin 1α

Cellular senescence *in vitro* is accompanied by the failure of ECs to respond to exogenous growth factors. An increase in the cytokine interleukin-1α (IL-1α is observed in senescent fibroblasts [46], although others have observed no difference in the amount of IL-1α mRNA between young and senescent fibroblasts [47]. The conflicting observations may be derived from the difference in cell strain used, because post-transcriptional regulation of IL-1α has been shown [48]. In HUVECs, senescent cells (of several strains) contain high levels of IL-1α mRNA [47, 48]. IL-1α is well characterized as a mitogen for human fibroblasts [49–52], but it is an inhibitor of proliferation for ECs [53, 54]. In the latter case, IL-1α may function as regulator of immediate-early transcriptional events induced during the formation of the capillary-like, tubular EC phenotype *in vitro* [55, 56]. IL-1α is also a potent modulator of EC surface properties and function, and suppresses gap junction activity of ECs [57].

Treatment of human EC populations with an antisense oligodeoxynucleotide to human IL-1α mRNA reduces IL-1α production and prevents cell senescence, eventually extending the proliferative life span of the cells *in vitro* [47]. Removal of the IL-1α antisense oligodeoxynucleotide results in the quick generation of the senescent phenotype and loss of proliferative potential. These data suggest that human EC senescence *in vitro* is a dynamic process regulated by the potential intracellular activity of IL-1α and involves telomere shortening. However, under conditions of suppressed IL-1α-synthesis, the rate of telomere shortening in ECs might be reduced by about half and life span therefore doubled. Even though aging is slowed by antisense treatment, telomere length of ECs eventually reaches a point at which it signals the cell to stop dividing, and thus cells cannot escape senescence.

Other factors

HUVECs and HAECs have common properties, as illustrated above, but it is widely accepted that ECs from different tissues, such as microvascular and large-vessel ECs, have different properties because they serve different functions *in vivo* [58, 59]. Therefore, the senescence process might differ among different types of ECs. A marked increase in cell area is observed with aging in both bovine aortic (BAECs) and microvascular endothelial cells (BMECs) [60], as observed commonly in HUVECs and HAECs. However, a marked difference in structural alterations between BMECs and BAECs during senescence is also observed [60].

Because overexpression of plasminogen activator inhibitor-1 (PAI-1), an ECM protein, in transgenic mice leads to thrombotic disease, an elevated PAI level *in vivo* is a major risk factor for myocardial infarction and deep vein thrombosis [61, 62]. This may be made worse by EC senescence [63]. PAI-2 is overexpressed in senescent fibroblasts [46]. The level of induction of PAI-1 by serum stimulation is increased in senescent fibroblasts [64]. Furthermore, it has been shown that senescent fibroblasts upregulate urokinase-type plasminogen activator (u-PA) and PAI-1 more than 50-fold relative to growing controls [65]. In senescent HUVECs, an increase in PAI-1 has been shown to be common in fibroblasts [65], and its increase is also observed in senescent BMECs and BAECs [60]. However, no changes are observed in the expression of u-PA and u-PA receptors [60].

One study of BMECs and BAECs has shown no changes in levels of thrombospondine-1, vinculin, α_V integrin, or even FN [60] although an age-associated increase of FN is observed universally in ECs (Table 1) as well as in fibroblasts [20, 38, 64]. However, at the cellular level in BMECs (but not in BAECs), an increase in vinculin and a decrease in integrin was observed [60], indicating changes in redistribution of vinculin and integrin in senescent BMECs. Although FN is organized in short, thin fibrils in young BMECs, it is detected in the form of long thick fibers in senescent BMECs. FN isoforms are produced from mRNAs which are alternatively-spliced at ED-A, ED-B and IIICS sites. Interestingly, the FN isoform containing the ED-A domain, but not that of ED-B, is observed in both senescent BMECs and BAECs. This change might be derived from changes in FN redistribution, because antibodies that specifically recognize senescent fibroblasts in cell staining have been obtained, and they can recognize FNs from both young and senescent cells in Western blot analysis [66]. In relation to this observation, a slight decrease in the ED-B-containing isoform is suggested in senescent fibroblasts, because a slight increase in the number of cells that produce only ED-B minus mRNA is observed, but no change is observed in the ED-A domain [67].

Thrombospondine-1 blocks the proliferation, migration, and morphogenesis of ECs [68], and PAI-1 also has inhibitory activity on cell migration in various cell types [69]. As mentioned above, no changes in the level of thrombospondine-1 are observed in senescent cells, but an increase in PAI-1 without change in its ECM distribution is shown [60]. An age-associated decrease in cell migratory activity induced by basic fibroblast growth factor (bFGF) is observed in both BMECs and BAECs, and this decrease may be explained by the increase in PAI-1. In addition, senescence also causes cessation of omental microvascular EC responses to epidermal growth factor (EGF), which displays angiogenic activity toward those cells [70]. Therefore, EC aging is accompanied by a general loss of angiogenic potential, and this implies a reduced neovascularization activity in elderly individuals, which would have important consequences in those physiopathological processes requiring angiogenesis, such as wound healing and inflammation. In addition, a decrease in angiogenic potential might account for the typical indolence of tumor growth and invasion in aged patients [71], because it is widely accepted that these processes are strictly dependent on the formation of new capillaries [72].

Table 1. *Changes in endothelial cells with aging*

Gene name or object	Level	Change	Cell type/organ	Remarks	Ref. no.
IL-1α	mRNA	↑	HUVEC		47
		↑	HUVEC	Three strains	48
		→	HUVEC	Three other strains	48
IL-1α	Protein	↑	HUVEC	All strains; no increase observed in fibroblasts	48
		↑	HUVEC		88
		↑	HUVEC		89
ET-1	mRNA	↑	HUVEC		20
		↑	HAEC		20
		↑	Rat aorta		37
		↑	Rat kidney		37
ET-1 secretion	peptide	↑	HUVEC		20
		↑	HUVEC		90
		↑	Rat aorta	Aorta < kidney < carotid	37
		↑	Rat kidney		37
		↑	Rat carotid		37
FGF2	mRNA	↓	BAEC		91
eNOS	mRNA	→	Rat lung		30
		↓	Rat aorta		30
		↑	Rat aorta		37
eNOS	Protein	↓	HUVEC		2
eNOS phoshporylation		↓	HUVEC		2

Table 1. (continued)

Gene name or object	Level	Change	Cell type/organ	Remarks	Ref. no.
iNOS	mRNA	↑	Rat aorta		37
iNOS	Protein	↑	Rat lung		30
		↓	Rat aorta		30
Neuronal NOS	mRNA	→	Rat		30
Induction of cGMP formation by NO synthesized by HUVEC		↓ (3-fold)	HUVEC		92
Fibronectin	mRNA	↑	HUVEC		38
		↑	HAEC (*in vivo* aged)		38
Fibronectin	Protein	↑	HUVEC		88
		→	BMEC	Changed at cell level	60
Laminin	mRNA	↑	BAEC		91
MMP1 (collagenase)	Protein	↓ (0.5-fold)	HMEC		93
Collagenase	Protein	↓	HMEC (*in vivo* aged)		93
TIMP-1	Protein	↑ (2.8-fold)	HMEC		93
		↑	HMEC (*in vivo* aged)		93
TIMP-2	Protein	↑	HUVEC		10
PAI-1	mRNA	↑	HUVEC	Not inducible by TPA or IL-1α	89
		↑	HUVEC		48
		↑	HUVEC		10

Table 1. *(continued)*

Gene name or object	Level	Change	Cell type/organ	Remarks	Ref. no.
PAI -1	Protein	↑	HUVEC		89
		↑	BMEC		60
		↑	BAEC		60
Prostacyclin secretion	Protein	↑	HUVEC		90
		↓	HUVEC	Decrease at middle age	94
Thromboxane A_2 secretion	Protein	↑	HUVEC		90
u-PA	Protein	→	BMEC		60
		→	BAEC		60
u-PAR	Protein	→	BMEC		60
		→	BAEC		60
Thrombospondin	Protein	→	BMEC		60
		→	BAEC		60
Vinculin	Protein	→	BMEC	Increased at cell level No change in BAEC	60
α_v integrin	Protein	→	BMEC	Decreased at cell level No change in BAEC	60
F-actin	Protein	→	BMEC		60
		→	BAEC		60
Cyclo-oxygenase 2		↑	HUVEC		47
		↑			48

Table 1. *(continued)*

Gene name or object	Level	Change	Cell type/organ	Remarks	Ref. no.
Cyclin A	mRNA	↓	HUVEC		95
Cyclin A	Protein	↓	HUVEC		95
Cyclin D1	mRNA	↑	HUVEC		95
Cyclin D1	Protein	↑	HUVEC		95
Cyclin E	Protein	↑	HUVEC		95
p16	Protein	↑	HUVEC		95
		↑	HDMEC		96
p21	Protein	↑	HUVEC		2
		↑	HUVEC		95
		↑	HDMEC	Peaked at old, but not at senescence	96
p27	Protein	↑	HUVEC		95
		↑	HDMEC	Peaked at old, but not at senescence	96
p53	Protein	↑ (3.5-fold)	Bovine, corneal, EC		97
pRB	Protein	↓	HUVEC		95
pRB phosphorylation		↓	HUVEC		95
Akt	Proteom	↓	HUVEC		2
Akt phosphorylation		↓	HUVEC		2
ICAM-1	mRNA	↑	HUVEC		98
		→	HAEC		99
G-CSF	mRNA	→	HUVEC		100
G-CSF secretion		↑ (10-fold)	HUVEC		100

***Table 1**. (continued)*

Gene name or object	Level	Change	Cell type/organ	Remarks	Ref. no.
Epo	mRNA	→	HUVEC		100
Inducibility of M-CSF mRNA by IL-1α	mRNA	↓	HUVEC	No induction in old cells	100
Inducibility of G-CSF imRNA by IL-1α	mRNA	→	HUVEC	Inducible in both young and old	100
Inducibility of Epo mRNA by IL-1α	mRNA	→	HUVEC	No induction in both young and old	100
Inducibility of G-CSF secretion by IL-1α	Protein	↓	HUVEC	More than 800-fold induction in young, but only 12-fold in old	100
Inducibility of E-selectin mRNA by IL-1α	mRNA	→	HUVEC	Inducible in both young and old	100
ACE	mRNA	↓	Human adult EC	(3/4 lines studied)	101
ACE		↓	Rat lung		30
ACE activity		↓	Rat plasma		30
		↓	Rat lung		30
		↑	Rat aorta		30
vWF		↓	Human adult EC		101
		↑	Human lung artery	Child vs. adult	102
		↑	Human lung vein		102
CD34		↑	Human lung artery	Child vs. adult	102
		↑	Human lung vein		102
CD31			HUVEC	Detectable in old cells	103
			Human lung artery	Detectable in adult	102
			Human lung vein	Detectable in adult	102

Table 1. *(continued)*

Gene name or object	Level	Change	Cell type/organ	Remarks	Ref. no.
U937 adhesion		↑	HUVEC		98
SA-β-gal		↑	HUVEC		103
		↑	HUVEC		10
		↑	Bovine, corneal, EC		97
Telomere length		↓	*in vivo* aged		43
		↓	HDMEC		96
		↓	Bovine, corneal, EC		97
Proliferation		↓	BAEC		104
Capillary growth by FGF2		↓	Mouse, *in vivo*		105
Bcl-2	Protein	↑	HUVEC		95
Bax	Protein	↑	HUVEC		95
Apoptosis		↑	HUVEC	Inducibility	2
		↑	HUVEC		95
Gap junction-mediated intercellular communication		↓	HUVEC		106
Migration		↓ (0.6-fold)	HMEC	Increases up to 1.4-fold of young cells after TIMP-1 inhibition in senescent cells	93
		↓	BAEC		104
Migration induced by PDGF		↓	RCEC		107
Unidirectional endothelialization		↓	BAEC		104

Table 1. *(continued)*

Gene name or object	Level	Change	Cell type/organ	Remarks	Ref. no.
Trisomy of chromosome 11		↑	Human adult EC (*in vitro* aged)	No trisomy in young cells, but 100% in senescent cells 11/12 examined	101
Trisomy of chromosome 11		↑	Human adult EC (*in vivo* aged)		101
Morphological change			Human aorta	Detected	108
Large EC		↑	Human aorta	Multinucleated variant EC	108
Circulating EC		↑	Rat		30
LDL receptor		↑	Human aorta		108
Cholesterol concentration		↑	Rat		109
Phospholipid concentration		→	Rat		109
Lipid peroxide levels		↑	Rat		109
Plasma membrane fluidity of aortic EC		↓	Rat		109
Contraction of coronary artery ring by ET-1		↓ (0.2-fold)	Dog	9 yr vs. more than 9 yr (same as in rat)	110

ACE: angiotensin converting enzyme; BAEC: bovine aortic endothelial cell; Epo: erythropoietin; ET-1: endothelin 1; FGF2: fibroblast growth factor 2; G-CSF: granulocyte colony-stimulating factor; HAEC: human aortic endothelial cell; HDMEC: human dermal microvascular endothelial cell; HMEC: human microvascular endothelial cell; 5-HT: 5-hydroxytryptamine; HUVEC: human umbilical vein endothelial cell; ICAM-1: intercellular adhesion molecule 1; M-CSF: macrophage colony-stimulating factor; MMP: matrix-degrading metalloproteases; PAI: plasminogen activator inhibitor; PDGF: platelet-derived growth factor; RCEC: rat capillary endothelial cell; TIMP: tissue inhibitor of metalloprotease; u-PA: urokinase-type plasminogen activator; u-PAR: plasminogen activator receptor; vWF: von Willebrand factor.

↑, increase; ↓, decrease; →, no change.

Apoptosis

Apoptosis of ECs may critically disturb the integrity of the endothelial monolayer and thereby contribute to vascular injury and atherosclerosis [73]. Indeed, all classical pro-atherosclerotic factors such as oxidized low density lipoprotein (LDL), proinflammatory cytokines, and reactive oxygen species induce EC apoptosis [74, 75]. ECs die easily by apoptosis in culture under conditions of growth factor deprivation or insufficient adherence. However, ECs *in vivo*, under conditions providing sufficient growth factors seem to be more resistant to apoptosis. ECs are protected against apoptosis by antiapoptotic proteins such as FLIP, which specifically interferes with Fas-induced apoptotic pathways [76]. In addition, NO synthesis by endothelium inhibits apoptosis induced by various apoptotic stimuli [31, 75, 77, 78], because NO interferes with the apoptosis signal-transduction pathway through several mechanisms [79–81]. Because EC apoptosis, similar among various cell types, is executed via activation of the cysteine protease family, the caspases [77, 82], the execution of apoptosis can be blocked by inhibition of caspases via S-nitrosylation of the essential cysteine residue [77, 83–85]. However, advanced age leads to impaired NO synthesis by endothelium, thereby facilitating apoptotic death of ECs, and finally leading to endothelial dysfunction [3, 31, 86], a key pathogenic factor in atherosclerotic disease progression [3, 87].

In summary, cell aging produces a number of changes in ECs (Table 1) as well as in other types of cells. These changes, especially in ECs, probably occur *in vivo* and are important factors in age-associated vascular disease.

References

1. Ross R (1995). Cell biology of atherosclerosis. *Annu Rev Physiol.* 57: 791–804.
2. Hoffmann J, Haendeler J, Aicher A, *et al.* (2001). Aging enhances the sensitivity of endothelial cells toward apoptotic stimuli: important role of nitric oxide. *Circ Res.* 89: 709–15.
3. Zeiher AM, Drexler H, Saurbier B, Just H (1993). Endothelium-mediated coronary blood flow modulation in humans. Effects of age, atherosclerosis, hypercholesterolemia, and hypertension. *J Clin Invest.* 92: 652–62.
4. Barton M, Cosentino F, Brandes R, Moreau P, Shaw S, Luscher T (1997). Anatomic heterogeneity of vascular aging: role of nitric oxide and endothelin. *Hypertension* 30: 817–24.
5. Auerbach O, Hammond EC, Garfinkel L (1968). Thickening of walls of arterioles and small arteries in relation to age and smoking habits. *N Engl J Med.* 278: 980–4.
6. Kappel B, Olsen S (1980). Cortical interstitial tissue and sclerosed glomeruli in the normal human kidney, related to age and sex. A quantitative study. *Virchows Arch A Pathol Anat Histol.* 387: 271–7.
7. Dzau VJ, Gibbons GH, Morishita R, Pratt RE (1994). New perspectives in hypertension research. Potentials of vascular biology. *Hypertension* 23: 1132–40.
8. Safar ME, Frohlich ED (1995). The arterial system in hypertension. A prospective view. *Hypertension* 26: 10–14.
9. Dohi Y, Kojima M, Sato K, Luscher TF (1995). Age-related changes in vascular smooth muscle and endothelium. *Drugs Aging* 7: 278–91.

10. Faragher RGA, Kipling D (1998). How might replicative senescence contribute to human ageing? *Bioessays* 20: 985–91.
11. Yanagisawa M, Kurihara H, Kimura S, *et al.* (1988). A novel potent vasoconstrictor peptide produced by vascular endothelial cells. *Nature* 332: 411–15.
12. Miyauchi T, Masaki T (1999). Pathophysiology of endothelin in the cardiovascular system. *Annu Rev Physiol.* 61: 391–415.
13. Benigni A, Perico N, Gaspari F, *et al.* (1991). Increased renal endothelin production in rats with reduced renal mass. *Am J Physiol.* 260: F331–9.
14. Lerman A, Edwards BS, Hallett JW, Heublein DM, Sandberg SM, Burnett JC Jr (1991). Circulating and tissue endothelin immunoreactivity in advanced atherosclerosis. *N Engl J Med.* 325: 997–1001.
15. Schiffrin EL (1995). Endothelin: potential role in hypertension and vascular hypertrophy. *Hypertension* 25: 1135–43.
16. Hocher B, Thone-Reinecke C, Rohmeiss P, *et al.* (1997). Endothelin-1 transgenic mice develop glomerulosclerosis, interstitial fibrosis, and renal cysts but not hypertension. *J Clin Invest.* 99: 1380–9.
17. Barton M, Haudenschild CC, d'Uscio LV, Shaw S, Munter K, Luscher TF (1998). Endothelin ETA receptor blockade restores NO-mediated endothelial function and inhibits atherosclerosis in apolipoprotein E-deficient mice. *Proc Natl Acad Sci USA* 95: 14367–72.
18. Barton M, d'Uscio L, Shaw S, Meyer P, Moreau P, Luscher TF (1998). ET(A) receptor blockade prevents increased tissue endothelin-1, vascular hypertrophy, and endothelial dysfunction in salt-sensitive hypertension. *Hypertension* 31: 499–504.
19. Tokunaga O, Fan J, Watanabe T, Kobayashi M, Kumazaki T, Mitsui Y (1992). Endothelin: immunohistologic localization in aorta and biosynthesis by cultured human aortic endothelial cells. *Lab Invest.* 67: 210–17.
20. Kumazaki T, Fujii T, Kobayashi M, Mitsui Y (1994). Aging- and growth-dependent modulation of endothelin-1 gene expression in human vascular endothelial cells. *Exp Cell Res.* 211: 6–11.
21. Dohi Y, Luscher TF (1990). Aging differentially affects direct and indirect actions of endothelin-1 in perfused mesenteric arteries of the rat. *Br J Pharmacol.* 100: 889–93.
22. Hahn AW, Resink TJ, Scott-Burden T, Powell J, Dohi Y, Buhler FR (1990). Stimulation of endothelin mRNA and secretion in rat vascular smooth muscle cells: a novel autocrine function. *Cell Regul.* 1: 649–59.
23. Zoja C, Orisio S, Perico N, *et al.* (1991). Constitutive expression of endothelin gene in cultured human mesangial cells and its modulation by transforming growth factor-beta, thrombin, and a thromboxane A2 analogue. *Lab Invest.* 64: 16–20.
24. Kohan DE (1996). Endothelins: renal tubule synthesis and actions. *Clin Exp Pharmacol Physiol.* 23: 337–44.
25. Suzuki T, Kumazaki T, Mitsui Y (1993). Endothelin-1 is produced and secreted by neonatal rat cardiac myocytes *in vitro*. *Biochem Biophys Res Commun.* 191: 823–30.
26. Ehrenreich H, Anderson RW, Fox CH, *et al.*(1990). Endothelins, peptides with potent vasoactive properties, are produced by human macrophages. *J Exp Med.* 172: 1741–8.
27. Boulanger C, Luscher TF (1990). Release of endothelin from the porcine aorta. Inhibition by endothelium-derived nitric oxide. *J Clin Invest.* 85: 587–90.
28. Barton M, Lattmann T, d'Uscio LV, Luscher TF, Shaw S (2000). Inverse regulation of endothelin-1 and nitric oxide metabolites in tissue with aging: implications for the age-dependent increase of cardiorenal disease. *J Cardiovasc Pharmacol.* 36: S153–6.

29. Forstermann U, Boissel JP, Kleinert H (1998). Expressional control of the 'constitutive' isoforms of nitric oxide synthase (NOS I and NOS III). *FASEB J.* 12: 773–90.
30. Challah M, Nadaud S, Philippe M, *et al.* (1997). Circulating and cellular markers of endothelial dysfunction with aging in rats. *Am J Physiol.* 273: H1941–8.
31. Tschudi MR, Barton M, Bersinger NA, *et al.* (1996). Effect of age on kinetics of nitric oxide release in rat aorta and pulmonary artery. *J Clin Invest.* 98: 899–905.
32. Forstermann U, Closs EI, Pollock JS, *et al.* (1994). Nitric oxide synthase isozymes. Characterization, purification, molecular cloning, and functions. *Hypertension* 23: 1121–31.
33. Mohaupt MG, Elzie JL, Ahn KY, Clapp WL, Wilcox CS, Kone BC (1994). Differential expression and induction of mRNAs encoding two inducible nitric oxide synthases in rat kidney. *Kidney Int.* 46: 653–65.
34. Verbeuren TJ, Bonhomme E, Laubie M, Simonet S (1993). Evidence for induction of nonendothelial NO synthase in aortas of cholesterol-fed rabbits. *J Cardiovasc Pharmacol.* 21: 841–5.
35. Kessler P, Bauersachs J, Busse R, Schini-Kerth VB (1997). Inhibition of inducible nitric oxide synthase restores endothelium-dependent relaxations in proinflammatory mediator-induced blood vessels. *Arterioscler Thromb Vasc Biol.* 17: 1746–55.
36. Behr-Roussel D, Rupin A, Sansilvestri-Morel P, Fabiani JN, Verbeuren TJ (2000). Histochemical evidence for inducible nitric oxide synthase in advanced but nonruptured human atherosclerotic carotid arteries. *Histochem J.* 32: 41–51.
37. Goettsch W, Lattmann T, Amann K, *et al.* (2001). Increased expression of endothelin-1 and inducible nitric oxide synthase isoform II in aging arteries *in vivo*: Implications for atherosclerosis. *Biochem Biophys Res Commun.* 280: 908–13.
38. Kumazaki T, Kobayashi M, Mitsui Y (1993). Enhanced expression of fibronectin during *in vivo* cellular aging of human vascular endothelial cells and skin fibroblasts. *Exp Cell Res.* 205, 396–402.
39. Kumazaki T, Wadhwa R, Kaul SC, Mitsui Y (1997). Expression of endothelin, fibronectin, and mortalin as aging and mortality markers. *Exp Gerontol.* 32: 95–103.
40. Minamino T, Miyauchi H, Yoshida T, Ishida Y, Yoshida H, Komuro I (2002). Endothelial cell senescence in human atherosclerosis. Role of telomere in endothelial dysfunction. *Circulation* 105: 1541–4.
41. Shay JW, Wright WE (2001). Telomeres, and telomerase: implication for cancer and aging. *Radiat Res.* 155: 188–93.
42. Minamino T, Mitsialis SA, Kourembanas S (2001). Hypoxia extends the life span of vascular smooth muscle cells through telomerase activation. *Mol Cell Biol.* 21: 3336–42.
43. Chang E, Harley CB (1995). Telomere length and replicative aging in human vascular tissues. *Proc Natl Acad Sci USA* 92: 11190–4.
44. Jeanclos E, Schork NJ, Kyvik KO, Kimura M, Skurnick JH, Aviv A (2000). Telomere length inversely correlates with pulse pressure and is highly familial. *Hypertension* 36: 195–200.
45. Okuda K, Khan MY, Skurnick J, Kimura M, Aviv H (2000). Telomere attrition of the human abdominal aorta: relationships with age and atherosclerosis. *Atherosclerosis* 152: 391–8.
46. Kumar S, Millis AJT, Baglioni C (1992). Expression of interleukin 1-inducible genes and production of interleukin 1 by aging human fibroblasts. *Proc Natl Acad Sci USA* 89: 4683–7.

47. Maier JAM, Voulalas P, Roeder D, Maciag T (1990). Extension of the life-span of human endothelial cells by an interleukin-1 alpha antisense oligomer. *Science* 249: 1570–4.
48. Garfinkel S, Brown S, Wessendorf JH, Maciag T (1994). Post-transcriptional regulation of interleukin 1 alpha in various strains of young and senescent human umbilical vein endothelial cells. *Proc Natl Acad Sci USA* 91: 1559–63.
49. Libby P, Ordovas JM, Biringi LK, Auger KR, Dinarello CA (1986). Inducible interleukin-1 gene expression in human vascular smooth muscle cells. *J Clin Invest.* 78: 1432–8.
50. Stern DM, Bank I, Nawroth PP, *et al.* (1985). Self-regulation of procoagulant events on the endothelial cell surface. *J Exp Med.* 162: 1223–35.
51. Malone DG, Pierce JH, Falko JP, Metcalfe DD (1988). Production of granulocyte-macrophage colony-stimulating factor by primary cultures of unstimulated rat microvascular endothelial cells. *Blood* 71: 684–9.
52. Mantovani A, Dejana E (1989). Cytokines as communication signals between leukocytes and endothelial cells. *Immunol Today* 10: 370–5.
53. Montesano R, Mossaz A, Ryser JE, Orci L, Vassalli P (1984). Leukocyte interleukins induce cultured endothelial cells to produce a highly organized, glycosaminoglycan-rich pericellular matrix. *J Cell Biol.* 99: 1706–15.
54. Montesano R, Orci L, Vassalli P (1985). Human endothelial cell cultures: phenotypic modulation by leukocyte interleukins. *J Cell Physiol.* 122: 424–34.
55. Hla T, Maciag T (1990). Isolation of immediate-early differentiation mRNAs by enzymatic amplification of subtracted cDNA from human endothelial cells. *Biochem Biophys Res Commun.* 167: 637–43.
56. Hla T, Maciag T (1990). An abundant transcript induced in differentiating human endothelial cells encodes a polypeptide with structural similarities to G-protein-coupled receptors. *J Biol Chem.* 265: 9308–13.
57. Hu VW, Xie HQ (1994). Interleukin-1 alpha suppresses gap junction-mediated intercellular communication in human endothelial cells. *Exp Cell Res.* 213:218–23.
58. Madri JA, Kocher O, Merwin JR, Bell L, Yannariello-Brown J (1989). The interactions of vascular cells with solid phase (matrix) and soluble factors. *J Cardiovasc Pharmacol.* 14: S70—5.
59. Brindle NP (1993). Growth factors in endothelial regeneration. *Cardiovasc Res.* 27: 1162–72.
60. Cavallaro U, Castelli V, Monte UD, Soria MR (2000). Phenotypic alterations in senescent large-vessel and microvascular endothelial cells. *Mol Cell Biol Res Commun.* 4: 117–21.
61. Auwerx J, Bouillon R, Collen D, Geboers J (1988). Induction of tissue-type plasminogen activator, plasminogen activator inhibitor in diabetes mellitus. *Arteriosclerosis* 8: 68–72.
62. Erickson LA, Fici GJ, Lund JE, Boyle TP, Polites G, Marrotti KR (1990). Development of venous occlusions in mice transgenic for plasminogen activator inhibitor. *Nature* 346: 74–6.
63. Zhang JC, Fabry A, Paucz L,Wojta J, Binder BR. (1996). Human fibroblasts down-regulate plasminogen activator inhibitor type-1 in cultured human macrovascular and microvascular endothelial cells. *Blood* 88: 3880–6.
64. Murano S, Thweatt R, Shmookler Reis RJ, Jones RA, Moerman EJ, Goldstein S (1991). Diverse gene sequences are overexpressed in werner syndrome fibroblasts undergoing premature replicative senescence. *Mol Cell Biol.* 11: 3905–14.

65. West MD, Shay JW, Wright WE, Linskens MHK (1996). Altered expression of plasminogen activator and plasminogen activator inhibitor during cellular senescence. *Exp Gerontol.* 31: 175–93.
66. Porter MB, Pereira-Smith OM, Smith JR (1990). Novel monoclonal antibodies identify antigenic determinants unique to cellular senescence. *J Cell Physiol.* 142: 425–33.
67. Kumazaki T, Mitsui Y, Hamada K, Sumida H, Nishiyama M (1999). Detection of alternative splicing of fibronectin mRNA in a single cell. *J Cell Sci.* 112: 1449–53.
68. Dawson DW, Pearce SF, Zhong R, Silverstein RL, Frazier WA, Bouck NP (1997). CD36 mediates the *in vitro* inhibitory effects of thrombospondin-1 on endothelial cells. *J Cell Biol.* 138: 707–17.
69. Cajot JF, Bamat J, Bergonzelli GE, *et al.* (1990). Plasminogen-activator inhibitor type 1 is a potent natural inhibitor of extracellular matrix degradation by fibrosarcoma and colon carcinoma cells. *Proc Natl Acad Sci USA* 87: 6939–43.
70. Matsuda T, Okamura K, Sato Y, *et al.* (1992). Decreased response to epidermal growth factor during cellular senescence in cultured human microvascular endothelial cells. *J Cell Physiol.* 150: 510–16.
71. Ershler WB (1987). The change in aggressiveness of neoplasms with age. *Geriatrics* 42: 99–103.
72. Folkman J (1992). The role of angiogenesis in tumor growth. *Semin Cancer Biol.* 3: 65–71.
73. Dimmeler S, Zeiher AM (2000). Endothelial cell apoptosis in angiogenesis and vessel regression. *Circ Res.* 87: 434–9.
74. Dimmeler S, Haendeler J, Galle J, Zeiher AM (1997). Oxidized low density lipoprotein induces apoptosis of human endothelial cells by activation of CPP32-like proteases: a mechanistic clue to the response to injury hypothesis. *Circulation* 95: 1760 –3.
75. Dimmeler S, Rippmann V, Weiland U, Haendeler J, Zeiher AM (1997). Angiotensin II induces apoptosis of human endothelial cells: protective effect of nitric oxide. *Circ Res.* 81: 970–6.
76. Sata M, Walsh K (1998). Endothelial cell apoptosis induced by oxidized LDL is associated with the down-regulation of the cellular caspase inhibitor FLIP. *J Biol Chem.* 273: 33103–6.
77. Dimmeler S, Haendeler J, Nehls M, Zeiher AM (1997). Suppression of apoptosis by nitric oxide via inhibition of ICE-like and CPP32-like proteases. *J Exp Med.* 185: 601–8.
78. Haendeler J, Zeiher AM, Dimmeler S (1999). Nitric oxide and apoptosis. *Vitamin Horm.* 57: 49–77
79. Vasa M, Breitschopf K, Zeiher AM, Dimmeler S (2000). Nitric oxide activates telomerase and delays endothelial cell senescence. *Circ Res.* 87: 540–2.
80. Nicotera P, Brune B, Bagetta G (1997). Nitric oxide: inducer or suppressor of apoptosis? *Trends Pharmacol Sci.* 18: 189–90.
81. Liu L, Stamler JS (1999). NO: an inhibitor of cell death. *Cell Death Differ.* 6:937–42.
82. Nagata S (1997). Apoptosis by death factor. *Cell* 88: 355–65.
83. Tenneti L, D'Emilia DM, Lipton SA. (1997). Suppression of neuronal apoptosis by S-nitrosylation of caspases. *Neurosci Lett.* 236: 139–42.
84. Li J, Billiar TR, Talanian RV, Kim YM (1997). Nitric oxide reversibly inhibits seven members of the caspase family via S-nitrosylation. *Biochem Biophys Res Commun.* 240: 419–24.
85. Mannick JB, Hausladen A, Liu L, *et al.* (1999). Fas-induced caspase denitrosylation. *Science* 284: 651–4.

86. Lucher TF, Noll G (1995). The pathogenesis of cardiovascular disease: role of the endothelium as a target and mediator. *Atherosclerosis* 118: S81–90.
87. Schachinger V, Britten MB, Zeiher AM (2000). Prognostic impact of coronary vasodilator dysfunction on adverse long-term outcome of coronary heart disease. *Circulation* 101: 1899–906.
88. Pagani F, Zagato L, Maier JA, Ragnotti G, Coviello DA, Vergani C (1993). Expression and alternative splicing of fibronectin mRNA in human diploid endothelial cells during aging *in vitro*. *Biochim Biophys Acta* 1173: 172–8.
89. Comi P, Chiaramonte R, Maier JAM (1995). Senescence-dependent regulation of type 1 plasminogen activator inhibitor in human vascular endothelial cells. *Exp Cell Res*. 219: 304–8.
90. Sato I, Kaji K, Morita I, Nagao M, Murota S (1993). Augmentation of endothelin-1, prostacyclin and thromboxane A2 secretion associated with *in vitro* ageing in cultured human umbilical vein endothelial cells. *Mech Ageing Dev*. 71: 73–84.
91. Augustin-Voss HG, Voss AK, Pauli BU (1993). Senescence of aortic endothelial cells in culture: effects of basic fibroblast growth factor expression on cell phenotype, migration, and proliferation. *J Cell Physiol*. 157: 279–88.
92. Sato I, Morita I, Kaji K, Ikeda M, Nagao M, Murota S (1993). Reduction of nitric oxide producing activity associated with *in vitro* aging in cultured human umbilical vein endothelial cell. *Biochem Biophys Res Commun*. 195: 1070–6.
93. Reed MJ, Corsa AC, Kudravi SA, McCormick RS, Arthur WT (2000). A deficit in collagenase activity contributes to impaired migration of aged microvascular endothelial cells. *J Cell Biochem*. 77: 116–26.
94. Hasegawa N, Yamamoto K (1993). A step in the process of prostacyclin production whose decline leads to the age-related decrease in production by human umbilical vein endothelial cells in culture. *Mech Ageing Dev*. 69: 167–78.
95. Wagner M, Hampel B, Bernhard D, Hala M, Zwerschke W, Jansen-Durr P. (2001). Replicative senescence of human endothelial cells *in vitro* involves G1 arrest, polyploidization and senescence-associated apoptosis. *Exp Gerontol*. 36: 1327–47.
96. Tang J, Gordon GM, Nickoloff BJ, Foreman KE (2002). The helix-loop-helix protein Id-1 delays onset of replicative senescence in human endothelial cells. *Lab Invest*. 82: 1073–9.
97. Whikehart DR, Register SJ, Chang Q, Montgomery B (2000). Relationship of telomeres and p53 in aging bovine corneal endothelial cell cultures. *Invest Ophthalmol Vis Sci*. 41: 1070–5.
98. Maier JA, Statuto M, Ragnotti G (1993). Senescence stimulates U937-endothelial cell interactions. *Exp Cell Res*. 208: 270–4.
99. Everett JP, Shipley GD, Mauck KA, Wagner CR, Morris TE, Hosenpud JD (1994). Phenotypic variations in resting and activated levels of ICAM-1 expression by cultured human aortic endothelial cells. *Transplantation* 58: 946–50.
100. Fujii T, Kumazaki T, Nagasawa T, Kobayashi M, Abe T, Mitsui Y (1996). Modulation of hemopoietic factor production in relation to endothelial cell aging by interleukin-1 induction. *Exp Cell Res*. 226: 356–62.
101. Johnson TE, Umbenhauer DR, Hill R, *et al.* (1992). Karyotypic and phenotypic changes during *in vitro* aging of human endothelial cells. *J Cell Physiol*. 150: 17–27.
102. Muller AM, Skrzynski C, Nesslinger M, Skipka G, Muller KM (2002). Correlation of age with *in vivo* expression of endothelial markers. *Exp Gerontol*. 37: 713–19.

103. van der Loo B, Fenton MJ, Erusalimsky JD (1998). Cytochemical detection of a senescence-associated beta-galactosidase in endothelial, smooth muscle cells from human and rabbit blood vessels. *Exp Cell Res.* 241: 309–15.
104. Niu S, Matsuda T (1992). Endothelial cell senescence inhibits unidirectional endothelialization *in vitro. Cell Transplant.* 1: 355–64.
105. Swift ME, Kleinman HK, DiPietro LA (1999). Impaired wound repair and delayed angiogenesis in aged mice. *Lab Invest.* 79: 1479–87.
106. Xie HQ, Hu VW (1994). Modulation of gap junctions in senescent endothelial cells. *Exp Cell Res.* 214: 172–6.
107. Phillips GD, Stone AM (1994). PDGF-BB induced chemotaxis is impaired in aged capillary endothelial cells. *Mech Ageing Dev.* 73: 189–96.
108. Tokunaga O, Satoh T, Yamasaki F, Wu L (1998). Multinucleated variant endothelial cells (MVECs) in human aorta: chromosomal aneuploidy and elevated uptake of LDL. *Semin Thromb Hemost.* 24: 279–84.
109. Quinlan KL, Song IS, Bunnett NW, *et al.* (1998). Neuropeptide regulation of human dermal microvascular endothelial cell ICAM-1 expression and function. *Am J Physiol.* 275:C1580–90.
110. Donoso MV, Fournier A, Peschke H, Faundez H, Domenech R, Huidobro-Toro JP (1994). Aging differentially modifies arterial sensitivity to endothelin-1 and 5-hydroxytryptamine: studies in dog coronary arteries and rat arterial mesenteric bed. *Peptides.* 15: 1489–95.

In Vitro Senescence of Human Osteoblasts

Moustapha Kassem, Karin Stenderup, Jeannette Justesen and Marie Kveiborg
University Department of Endocrinology, University Hospital of Odense, DK-5000 Odense C, Denmark

Introduction

Human aging is associated with bone loss leading to bone fragility and increased risk for fractures, a disease known as osteoporosis. Osteoporosis is one of the most prevalent and serious diseases affecting the elderly population and constitutes a major public health problem. The cellular and molecular causes of age-related bone loss are current intensive topic of investigation with the aim of identifying new approaches to abolish its negative effects on the skeleton. The aim of this chapter is to give a review on the current understanding of the contribution of aging of the osteoblasts (the bone forming cells) to the phenomenon of age-related bone loss.

The concept of bone remodelling

Bone as a tissue, is composed of bone matrix and bone cells. Bone matrix is built up of type I collagen (90%) and the remaining 10% is composed of a large number of non-collagenous proteins (e.g., osteocalcin, osteonectin, bone sialoproteins and various proteoglycans). Non-collagenus proteins participate in the process of matrix maturation, mineralization and may regulate the functional activity of bone cells. Two main types of bone cells have been identified. Osteoblasts (bone forming cells) and osteoclasts (bone resorbing cells) that together with their precursor cells and associated cells (e.g., endothelial cells, nerve cells) are organized in specialized units called bone multicellular units (BMU) (analogous to the organization of kidney cells into nephrons) [1]. The main function of the BMU in the adult skeleton is to mediate a bone replacement mechanism called "bone remodelling" aiming at the maintenance of the integrity of the skeleton by removing old bone of high mineral density and high prevalence of fatigue microfractures and replacing it with young bone of low mineral

Sunil C. Kaul and Renu Wadhwa (eds.), Aging of Cells In and Outside the Body, 67–84.

density and better mechanical properties [1–2]. A process that supports the dual function of the skeleton as a biomechanical organ as well as an important contributor to the divalent ion homeostasis. Bone remodelling takes place in a special compartment covered by a layer of bone lining cells termed bone remodelling compartment [3]. Bone remodelling comprises a specific sequence of cellular events (Figure 1).

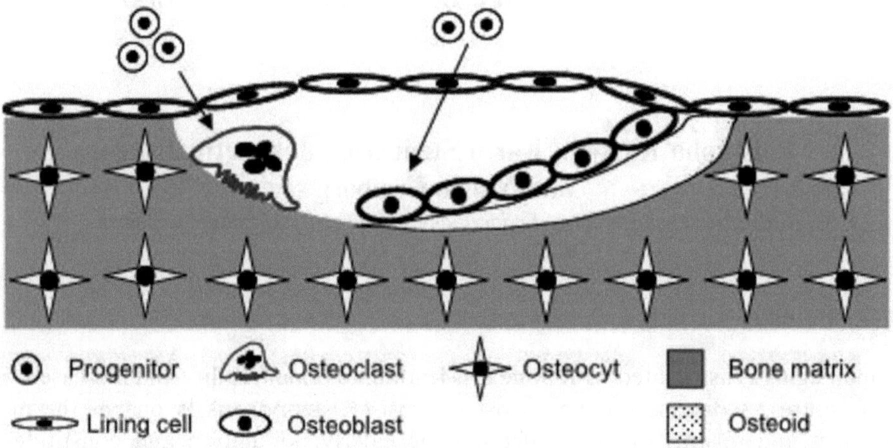

Figure 1. An illustration of bone remodeling sequence. Osteoclast precursors fuse to form multinucleated osteoclasts that resorb bone. Bone resorption is followed by bone formation where osteoblast progenitors are recruited to bone remodeling sites and they differentiate into mature osteoblast that produced osteoid (unmineralized matrix) that mineralizes. Under normal circumstances the amount of bone removed by osteoclasts equals the amount of bone formed by osteoblasts. Bone remodeling takes place in a specialized compartment covered by bone lining cells called bone remodeling compartment [adapted from ref. 3].

It starts with recruitment of osteoclast precursors, their fusion to form osteoclasts and the migration of osteoclasts to bone surfaces initiated for bone remodelling. The osteoclasts remove an amount of bone and create a resorption lacuna. Bone resorption is followed temporally in the same anatomic location by bone formation. Similar to bone resorption, bone formation comprises several cellular events including recruitment of osteoblasts from stem cells and precursor cells in the bone marrow and their migration to the bottom of resorption lacuna. This is followed by matrix production filling the resorptive lacuna with newly formed matrix that mineralizes and forms lamellar bone. Usually the amount of bone removed by the osteoclasts is equal to the amount of bone formed by the osteoblasts and a stable bone mass is maintained. However, this is not the case during aging and different skeletal diseases leading to bone loss.

Cellular basis of age-related bone loss

After a pubertal growth period, the skeletal mass reaches its peak (hence termed peak bone mass) in the third decade of lifespan and afterwards bone loss is observed. Based on bone mass measurements in large cohorts of healthy populations, two patterns of age-related bone loss could be identified [4] (Figure 2).

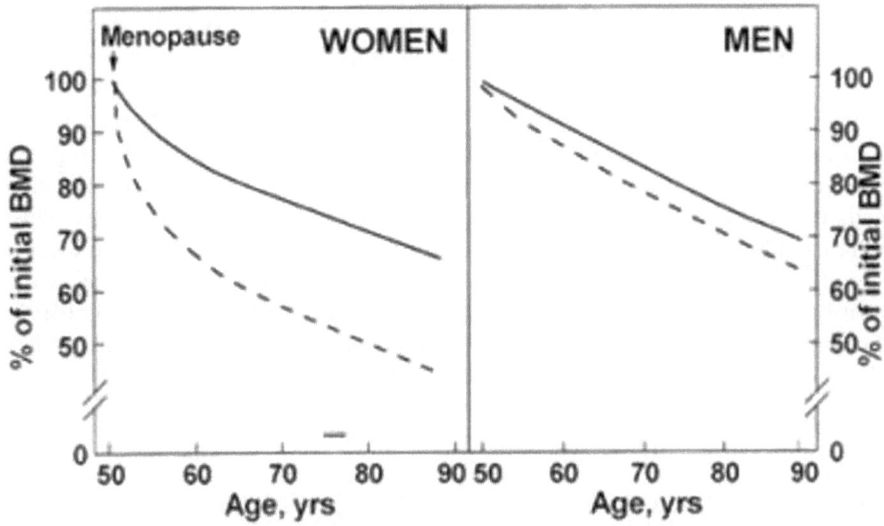

Figure 2. Schematic representation of changes in bone mass over life in cancellous (broken line) and cortical (solid line) bone in women (left panel) and men (right panel) from age 50 onward. In men only one phase of continuous bone loss is observed but in women two phase are recognized: a perimenopausal accelerated phase of bone loss and a late slow phase. Note also that the accelerated phase, but not the slow phase, involves disproportionate loss of cancellous bone (from Riggs et al. 1998, J Bone Miner Res. 13: 763).

A continuous slow phase of age-related bone loss that is observed in both men and women occurring at a constant rate of 1% per year. In perimenopausal women, a rapid phase of bone loss is observed leading to bone loss of 5–10% per year. Histological analysis of iliac crest bone biopsies revealed that the main mechanism of perimenopausal bone loss is increased osteoclastic activity and bone resorption [5]. On the other hand, the continuous age-related bone loss observed in aging men and women is due to an impaired bone formation [2, 6, 7].

Origin of osteoblasts and regulation of bone formation

Bone formation is dependent on the coordinated action of a group of osteoblasts that are recruited at the bone formation sites during bone remodelling. *In vivo*, mature osteoblasts are polarized, cuboidal cells sitting on bone surfaces engaged in active bone formation. Mature osteoblasts are differentiated from osteogenic precursor and stem cells present in the bone marrow. Based on the pioneer work of Friedenstein and co-workers [8, 9], it has been recognized that the non-hematopoietic compartment of bone marrow (known as bone marrow stroma) contains a group of fibroblast-like stem cells with osteogenic differentiation potential (referred to as bone marrow stromal cells, BMSC, mesenchymal stem cells MSC, skeletal stem cells, SSC). *In vivo* transplantation of BMSC in syngenic or immunodeficient mice resulted in the formation of a mixture of tissues including bone, cartilage and hematopoietic-supporting stroma [9, 10]. *In vitro*, individual BMSC are clonogenic forming clones of cells termed colony forming unit-fibroblasts (CFU-f) and they exhibit extensive proliferation potential [9–12]. However, definite stem cell self-renewal characteristics have not been yet described. The *in vivo* identity and localization of BMSC are not known but may lie in close proximity to blood vessels and sinusoids [13, 14]. Few surface markers are available to identify BMSC and one of the most useful is Stro-1 antibody that identifies a glycoprotein component on all the clonogenic cells with osteogenic potential in the bone marrow and thus can be used for the isolation of BMSC [15]. Under appropriate conditions, BMSC are able to differentiate into osteoblasts and other cell types including adipocytes, chondrocytes, hematopoietic-supporting stromal cells. Little is known about the hierarchical relationship between these different differentiation pathways [16]. However, some recent experiments have shown that an extensive plasticity exists between these different cell types. For examples, clonogenic BMSC can differentiate into either osteoblasts or adipcoytes [17]. Also, subcutaneous adipocytes can revert to bone forming cells under appropriate culture conditions (J. Justesen, M.Kassem, unpublished data).

One of the most consistent histomorphometric findings in bone biopsies obtained from elderly persons is the presence of decreased mean wall thickness (MWT) in both trabecular and cortical bone indicating decreased osteoblastic bone forming capacity during bone remodelling [6, 7, 18]. MWT is dependent on both the number of recruited osteoblasts at the beginning of bone formation phase and the activity of individual osteoblast. Recruitment of adequate number of osteoblasts is dependent on the availability of stem and precursor cells and their proper response to growth, differentiation and chemotactic signals in the bone microenvironment. Matrix production and mineralization functions of mature osteoblasts are dependent on the proper response of osteoblasts to hormones, growth factors and cytokines as well as the availability of nutrients and ions necessary for accomplishing these processes. Age-related impairment of osteoblast functions can thus be the result of any mechanism interfering with these processes.

Hypotheses of age-related impairment of bone formation

Based on understanding of the cellular events of bone formation phase of bone remodelling, several investigators have examined potential mechanisms causing age-related osteoblast failure using *in vitro* osteoblast culture systems. We have grouped these studies according to the working hypotheses tested into (Figure 3):

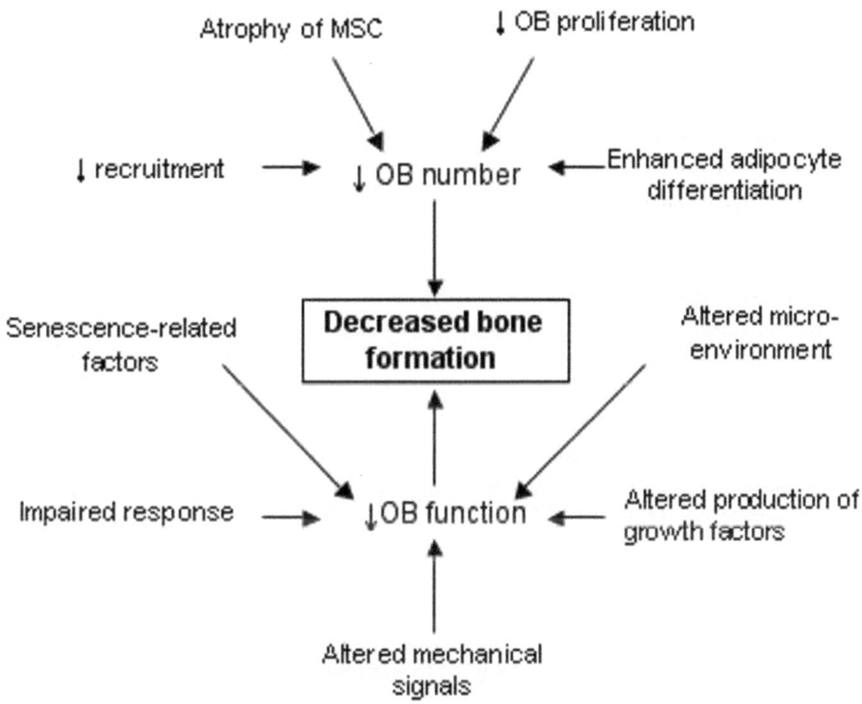

***Figure 3**. Hypotheses proposed to explain age-related decreased bone formation.*

- Decreased osteogenic stem cell population size.
- Decreased number of differentiated osteoblasts due to enhanced adipocyte differentiation in bone marrow.
- Decreased osteoblast maximal life span
- Decreased number of differentiated osteoblastic cells
- Decreased osteoblastic bone formation capacity
- Impaired responsiveness to hormone and growth factors signals.

- Changes in bone microenvironment composition.
- Impaired osteoblast function as a manifestation of the general homeostatic failure of the aged organism.

Large numbers of studies have been performed on murine cell culture models and there exist several discrepancies between results obtained in these culture models and those obtained from human cell cultures. These discrepancies will be pointed out and discussed. Nevertheless, we think that results obtained from human osteoblast cultures are more representative for normal human physiology.

Age-related changes in the osteogenic stem cell population size

Stem cells are needed for the continuous supply of mature, functionally competent cells required for normal tissue turnover and regeneration. An age-related decrease in the number or function of stem cells can lead to tissue deterioration [19]. Several investigators have utilized the *in vitro* characteristic of the osteogenic stem cells in bone marrow: adherence to plastic, clonogenicity (formation of CFU-f) and positive staining for alkaline phophatase (AP) to examine age-related changes in the number of osteoblastic stem cell population. Based on dilution curves, the number of CFU-f formed in cultures is directly proportional to the number of cells explanted (and thus presumably the number of stem cells *in vivo*) [20]. In these studies, BMSC were cultured in serum containing media and the number of CFU-f was counted either manually or by image analysis computer programs. A CFU-f colony was defined as a group of more than 16 cells (several investigators used cut-off value of 50 cells). Some investigators utilized positive AP staining to classify a CFU-f as an osteogenic colony. However, recent *in vivo* clonal analysis of human CFU-f did not reveal a direct correlation between *in vitro* expression of osteogenic markers including AP and their *in vivo* bone formation potential [21].

Three studies that isolated MSC from rats at different ages demonstrated that the total number of CFU-f [22, 23] or AP^+ CFU-f [24] decreased with donor age. While Egrise *et al.* [23] reported that aged rats (21 months) had 30% lowered total number of CFU-f compared with young rats (4 months old), the number of AP^+ CFU-f colonies was similar in cultures of old and young rats.

In aged mice, Khan *et al.* [25] reported a decreased number of CFU-f and AP^+ CFU-f compared to young mice. Other investigators found variable results using similar methodology. Xu *et al.* [26] reported that the total number of CFU-f increased with age in mice between 3 and 15 months. Brockbank *et al.* [27] found that the reported age-related decrease in CFU-f number was due to an age-related increase in bone marrow cellularity and after correction for this there was no decrease in the CFU-f population size.

Another assay that was employed to determine the number of osteogenic precursor cells was to count the number of bone nodules (osteoblastic cell aggregates) formed in osteoblast cultures established from calvarias or BMSC after staining for mineralized matrix with von Kossa stain [28]. Cell dilution studies showed that the number of bone nodules formed is proportional to the number of osteogenic precursor cells explanted [28]. This assay was utilized by Tsuji *et al.* [29] who reported that cultures established from old rats formed significantly lowered number of bone nodules and AP^+ cells compared with cultures of young rat cells. However, the size distribution of bone nodules and the population doubling rate of cells did not change with donor age suggesting that the number of precursor cells and not their proliferative potential is impaired with age [29].

Counting the number of CFU-f as an *in vitro* indicator of the number of osteogenic precursor cells *in vivo* has also been employed in a number of human studies and the reported results are not consistent. Nisahida *et al.* [30] found an age-related decrease in the number of AP^+ CFU-f in bone marrow cultures obtained from 49 females (age 4–88). However, the major decline in the number of AP^+ CFU-f occurred between 4 to 20 years and there were little age-related changes afterwards. Muschler *et al.* [31] found no age-related changes in number of AP^+ CFU-f in a group of 57 patients aged 13-83. However, the number of AP^+ CFU-f decreased with aging in women but not in men. This decrease was based on two outliers samples from donors aged 13 and 14. D'Ippotito *et al.* [32] found in a group of 41 donors, an age-related decrease in the number of AP^+ CFU-f, but the major decline in CFU-f number occurred up to the age of 40 and no change in the number of AP^+CFU-f was observed afterwards. The authors observed also a statistically significant decrease in the CFU-f colony size between young and old donors [32]. Mets *et al.* [33] reported the effect of donor age on the number of CFU-f established from bone marrow cultures obtained from the sternum in 28 individuals aged 17–87. The authors observed the presence of two morphologically distinct types of colonies: CFU-f colonies with fibroblast-like morphology that declined in number with increased donor age and colonies with large cells of epithelial morphology that increased in number with increased donor age. However, the major decline of CFU-f number occurred up to the age of 30 and no age-related decline in the number of CFU-f was observed afterwards [33].

In contrast to these studies, Oreffo *et al.* found no age-related change in the total CFU-f number, size or number of AP^+ CFU-f in 16 normal donors (age 14–48), in a 57 patients with osteoarthritis (age 28–87) or 26 patients with osteoporosis (age 69–97) [34]. However, the number of AP^+ CFU-f in patients with osteoporosis was significantly decreased compared to patients with osteoarthritis but the two populations were not age-matched.

The discrepancies in the results reported in the human studies may be related to differences in the methods utilized by various investigators. The source of bone marrow samples varied between studies. Some studies have used bone marrow obtained during surgical procedures for treatment of osteoarthritis or other operations or from post-mortem samples. These donors may not represent the normal aging population. Also, bone marrow was obtained from different sites (spine, proximal knee, iliac crest) which may influence bone marrow composition and the occurrence of CFU-f. Furthermore, the technique for establishing CFU-f cultures was different. Some authors used bone marrow aspirates and other used "flushing" technique to remove bone marrow cells from the bone fragments obtained during orthopaedic procedures.

In order to overcome some of these limitations, our group [35] studied 38 normal healthy volunteers (23 young and 15 old individuals) and 13 patients with osteoporosis. The CFU-f population was established from Stro-1$^+$ bone marrow cell population. We did not find any age-related effects on the total number of CFU-f, number of AP$^+$ CFU-f or CFU-f colony size distribution (Table 1). In conclusion, human studies show that with aging there is maintenance of CFU-f cell population size in the bone marrow and that the observed decline in the number of CFU-f in early adulthood may represent changes in the skeletal dynamics from a modelling mode characteristic of skeletal growth and consolidation to a remodelling dynamics characteristic of the adult skeleton. This may also explain why experiments employing rodents showed a decline in the CFU-f number as they continue to grow throughout their lifespan.

***Table 1**. Effect of age on stem cell population size in bone marrow*

	Young	Old	*p*-value
Number (*n*)	23	28	
Age (years)	22–44	58–83	
Stro-1$^+$ cells (%)	27 ± 7	24 ± 1	n.s.
CFU-F number / culture	87 ± 12	113 ± 12	n.s.
CFU-F size (mm^2)	7.2 ± 0.5	6.9 ± 0.3	n.s.
CFU-F size distribution			
$1\ mm^2 < x < 5\ mm^2$ (%)	38 ± 5	38 ± 4	n.s.
$5\ mm^2 < x < 10\ mm^2$ (%)	38 ± 3	39 ± 3	n.s.
$> 10\ mm^2$ (%)	24 ± 4	23 ± 2	n.s.
ALP positive CFU-Fs (%)	66 ± 6	68 ± 4	n.s.
ALP positive cells (%)	34 ± 5	40 ± 3	n.s.

Low-density mononuclear cells were enriched in bone marrow stromal cells employing Stro-1 antibody. Stro-1$^+$ cells were seeded (1×10^5 cells/cm^2) into chamber slides. After 13 days cells were stained for alkaline phosphatase (ALP). CFU-F = colony-forming unit fibroblast, n.s. = not significant.

Age-related enhancement of adipocyte differentiation in bone marrow

Marrow stromal cells can differentiate into multiple lineages including osteoblasts and adipocytes. Thus enhanced adipocyte differentiation from BMSC may lead to reduction in the number of stem cells available for osteoblastic cell differentiation and bone formation. The hypothesis of the presence of an inverse relationship between adipocyte and osteoblast differentiation is based on a number of observations obtained from histomorphometric and *in vitro* culture studies. Several investigators have demonstrated that with age the observed decrease in trabecular bone volume (TBV) is associated with increased bone marrow adipocytic volume [36–38]. Also, in cultures of BMSC it has been reported that osteoblast differentiation is inversely correlated to adipocyte cell differentiation [39, 40]. Furthermore, in a mouse model of accelerated senescence (SAM-6 mouse) decreased bone formation was associated with enhanced adipogenesis *in vivo* and *in vitro* [41]. All these findings led to the suggestion that inhibiting adipocytic differentiation in bone marrow may be a novel approach to increase bone mass [42].

We examined the effect of donor age on adipocyte forming capacity of BMSC. BMSC cultures were established from 34 young and 20 elderly healthy persons and the cells were induced to form adipocytes. We found that the adipocyte-forming capacity of BMSC did not change with donor age [43]. Similarly, we did not find age-related changes in the expression of mRNA levels of adipocyte and osteoblast differentiation [43]. In order to test the causal relationship between osteoblastogenesis and adipogensis in bone marrow, we examined the effects of induction of adipogensis on bone formation in a mouse model treated for 10 months with troglitazone (a PPAR-γ agonist). Histomorphometric analysis of bone marrow cellular composition revealed the presence of an enhanced adipogensis in bone marrow but no change in bone formation [44] suggesting that bone formation and adipocyte formation can be regulated independently and that the increase in bone marrow adipocyte volume with age is not related necessarily to decreased bone formation

Age-related changes in osteoblast life span

While the previous experiments reporting the growth dynamics of CFU-f employed short term cultures, they may not be able to detect subtle growth defects. Hayflick model for replicative senescence has been employed in biogerontology research to unravel mechanisms of age-related cellular defects [45]. Using this model, several investigators have reported an inverse relationship between the donor age and maximal proliferative potential of the cells *in vitro* [46]. We have characterized a Hayflick model for replicative senescence of human osteoblasts [47, 48]. During continuous culture *in vitro*, human osteoblasts exhibited typical senescence-related phenotype including senescent-associated decrease in osteoblast marker production (AP, osteoclacin, collagen type I), decrease mean telomere fragment length and increase in the number of senescence-associated β-galactosidase (SA β-gal) [49, 50]. We further examined the effect of donor age on the maximal proliferative potential of BMSC. An age-related decline in the maximal life span from 41 ± 10 population doublings (PD) in young donors to 24 ± 11 PD in old donors was observed. Our results thus suggest that human aging is associated with reduced maximal proliferation potential of BMSC. However, the proliferation potential of aged BMSC is still very high and thus the contribution of the observed decreased maximal proliferative potential to age-related decreased bone formation *in vivo* is not clear.

Age-related changes in the number of differentiated osteoblastic cells

Bone formation requires the presence of an adequate number of mature osteoblastic cells and thus impaired osteoblast differentiation can lead to impaired bone formation. In a study by Roholl *et al.* [51] in aged rat model, various populations of osteoblastic cells (preosteoblasts, osteoblasts) and their relationship to bone formation surfaces were examined. The authors found that with increasing donor age, there is an accumulation of pre-osteoblastic cells and decreased number of mature osteoblasts suggesting that impaired osteoblast differentiation is a potential mechanism for age-related impaired bone formation. To address this possibility, Long *et al.* [52] isolated a population of osteoblastic cells from human bone marrow employing a two-step physicochemical separation (gradient centrifugation to isolate low density mononuclear cells and adherence to plastic for a short period to select for non-adherent cells). This population of cells expressed osteocalcin and osteonectin characteristic of mature osteoblasts and examined the changes in their frequency with donor-age. The authors found that old donor cells expressed high levels of osteocalcin and osteonectin compared to cells obtained from young individuals which do not support the hypothesis of impaired osteoblast differentiation. Also, Katzburg *et al.* isolated osteoblastic cells from bone biopsies taken at the time of corrective orthopaedic surgery from a group of young and old men and women and the authors did not find age-related differences between the groups in the basal levels of AP or osteocalcin [53]. Thus, impaired differentiation of mature osteoblasts from precursor cells was not confirmed in human studies.

Age-related changes in osteoblastic bone formation capacity

The main function of osteoblasts is to form bone. The effect of age on bone forming ability of the osteoblasts was tested in a number of studies. Some investigators employed osteoblast expression of biochemical markers of bone formation (AP, osteocalcin, osteonectin etc.) as surrogate markers for bone forming ability. Fedarko *et al.* [54] examined the effect of donor age on production of a number of extracellular matrix proteins in osteoblast cultures obtained from trabecular bone explants and found an age-dependent decrease in production of collagen and decorin. In a similar approach, Koshihara *et al.* [55] employed a model system for periosteal osteoblasts to compare collagen production in cultures of young and old donors. The study included few study subjects and the young and old donors were not matched for gender. The study however showed that osteoblastic collagen production increased with donor age that may be caused by increased collagenase inhibitory activity in aged donors. Also as mentioned above, Katzburg *et al.* [53] did not find any differences in the basal levels of alkaline phosphatase or osteocalcin production in osteoblast cultures from young and old men.

The effect of donor age on bone forming capacity has been examined more directly using *in vivo* bone formation assay. In this assay BMSC mixed with hydroxyapatite/tricalcium phosphate (HA/TCP) as a carrier are implanted subcutaneously in syngenic animals. Both cells from young and old donors were tested in both young and old recipient animals. BMSC derived from aged rats exhibited decreased bone formation capacity compared with MSC from young donors [56]. Similarly, Quarto *et al.* [24] reported decrease in the amount of bone formed by BMSC from aged rats irrespective of the age of recipient. Interestingly, in these experiments, treatment of BMSC from aged rats with dexamethasone abolished the differences in bone forming capacity between young and old rats [24]. Huibregtse *et al.* [57] employing a similar assay, reported that rabbit BMSC exhibited an age-related decreased in the chondro-osteogenic potential. We have employed similar methodology to test the difference between BMSC obtained form young and old human donors. After implantation in immune-deficient mice, we found no difference in the amount of bone formed between the two age groups (K Stenderup, J Justesen, M Kassem, unpublished results).

Age-related changes in responsiveness to calcitropic hormones and growth factors

Osteoblast cell proliferation and function are dependent on adequate response to growth factors and hormones known to control bone remodelling and bone formation. Impaired response to growth factors can lead to decrease the number and/or the function of osteoblastic cells and consequently bone formation. Pfeilsh-cifter *et al.* [58] examined the effect of donor age on the responses of cultured human osteoblasts to a number of important growth factors. Osteoblasts from aged donors exhibited a decrease in proliferative responses to growth hormone and platelet-derived growth factors compared with young donor cells. Similarly, osteoblast cultures obtained from young persons exhibited a better response to estradiol [59] and insulin-like growth factor (IGF)-I [60] compared with cells obtained from older donors. Our group has examined the effect of serum and calcitriol on the expression of osteoblastic markers in BMSC of young and old donors and found no significant differences in any of the markers examined between the two age groups [43].

Age-related changes in bone microenvironment

Osteoblastic cells produce a large number of growth factors and cytokines in the local microenvironment. Several of these bone active molecules are sequestrated in bone matrix and exert their main effects locally as a controller of osteoblast activity during bone remodelling. Bone growth factors may also function as "coupling factors" linking bone resorption with bone formation as they are released in bone micro-environment by the bone degrading activities of the osteoclasts and can stimulate bone formation by the osteoblasts [61]. Based on these considerations several investigators measured the effect of donor age on the amount of growth factors present in the bone matrix directly or indirectly through bone inductive effects of demineralized bone matrix powder.

IGF-I and IGF-II are important regulator of osteoblastic cell proliferation and differentiation. Seck *et al.* [62] examined the relationship between donor age, cortical bone content of IGF-I and IGF-II as well as cortical bone remodeling in a large number of donors at different ages. Bone samples were obtained from the proximal femur neck. The authors found an age-related decrease in bone matrix concentration of IGF-I in both men and women and of IGF-II in men only. However, there was no significant correlation between bone content of IGF-I, IGF-II and any parameter of bone remodeling. From the same group, production of transforming growth factor (TGF)-beta did not change with donor age [63].

Demineralized bone matrix (DBM) powder was implanted subcutaneously in animals of different ages and the amount of bone formed was quantitated using histomorphometric or biochemical techniques. Irving *et al.* [64] found that bone induction capacity of bone matrix is impaired in old animals. Bone powder obtained from tibiae and femurs of 6-month-old rats and implanted subcutaneously in young (6 weeks), adult (6 months), and old (2 years) rat recipients. Bone was formed after 14 days in young rats, after 15 days in adult rats and after 25 days in old rats. Similarly, the bone induction of powder obtained from bones of 4-month-old rats was dependent on recipient age with gradual decline in rats aged 1 months old to 16 months old [65]. Also, decreased in rates of bone formation and amount of bone obtained after subcutaneous implantation of DBM powder were observed in old rats (4 months old) compared with young rats (4 weeks old) [66]. The decreased bone formation found in rats was confirmed in a study employing rabbits. Rabbits aged 3, 6, 18, and 28 months received intra-muscular implantation of DBM powder from 6 month-old male rabbit donors [67]. The amount of new bone formed was negatively correlated with age of the recipient. Similar experiments were performed using human DBM powder from young (8 donors, 18–46 years) and old (9 donor, 62–90 years) that were implanted subcutaneously in mice [68]. The amount of bone formed by old donor DBM powder was decreased compared with young donors but the difference was not statistically significant. These studies suggest that age-related decreased bone formation is due to both impaired osteoblast functions as well as changes in the bone microenvironment of aged organism.

Age-related osteoblast failure is part of the general homeostatic failure of the aged organism

Age-related bone loss is one manifestation of a generalized degeneration occurring in the aged organism. Thus, age-related deterioration of osteoblast functions can be conceived as part of the general deterioration observed in various organs in the aging body. Osteoporotic bone loss and fractures are one of the cardinal features of the aging phenotype as well as premature aging syndromes (e.g., Werner syndrome). Thus, studying aging models provides insight in the pathological mechanisms affecting osteoblast functions. In a recent study by De Boer *et al.* [69] of mice deficient in Xpd, a gene important for repairing a range of genetic lesions by nucleotide excision repair (NER). The mouse exhibited all the phenotype characteristics of premature aging including osteoporosis. Also, in a mouse model with activated mutation in p53 gene increased resistance to neoplasia was associated with premature aging including reduced longevity, osteoporosis and generalized organ atrophy [70]. Telomere shortening has been proposed as one pathological mechanism leading to age-related proliferative failure of cells *in vitro* and possibly *in vivo* [50]. We have demonstrated the presence of a progressive shortening in telomere length with osteoblast senescence *in vitro* that was associated with decreased expression of osteoblastic functional markers e.g., osteocalcin and collagen type I [48, 71]. In order to test the role of telomere maintenance in osteoblast biology, we created a stably-transduced human BMSC lines with human telomerase reverse transcriptase (hTERT). The cells exhibit in addition to a prolonged life span, enhanced bone formation capacity [72] (Figure 4). Similar findings were reported by Shi *et al.* [73]. Understanding the molecular pathways leading to maintenance of telomere length and improved osteoblast functions may provide new approaches for abolishing age-related effect on bone.

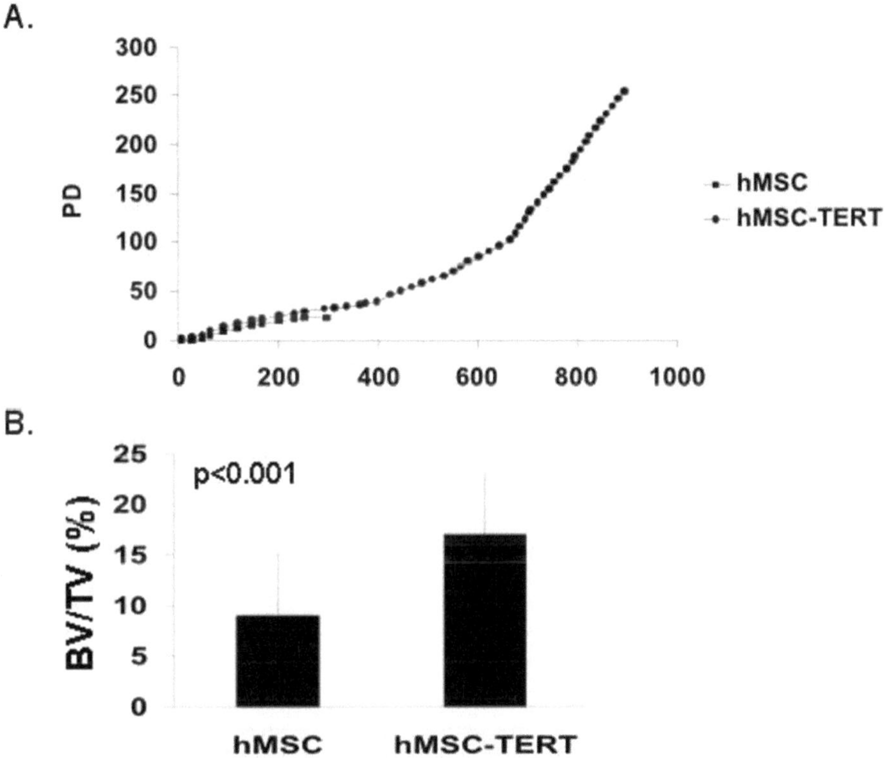

***Figure 4**. Over expression of human telomerase reverse transcriptase gene increases the proliferative lifespan of human marrow stromal cells (A) and their bone forming capacity when implanted subcutaneously in immunodeficient mice (B). hMSC: human marrow stromal cells, hMSC-TERT: hMSC transfected with human telomerase catalytic subunit (hTERT) gene. BV/TV: bone volume per total volume of the implants [from ref. 72].*

Aging is also associated with a multitude of changes in production and serum levels of calcitropic hormones (decrease in growth hormone, IGF-I, calcitriol and sex steroids and an age-related increase in parathyroid hormone). Furthermore aging is associated with changes in life style and decreased physical activity which may directly or indirectly affect bone remodeling, osteoblast and osteoclast functions and thus contribute to the age-related bone loss.

Concluding remarks

In spite of the presence of a clear age-related deterioration of the skeleton, the cellular and molecular mechanisms underlying these effects remain to be determined. Human studies demonstrate that there is maintenance of osteogenic cell population and its differentiation potential with age. Also, cells from aged donors seem to be able to respond adequately (though not optimally) to osteogenic stimuli. Studies of gene-modified animal models provide a more complex picture of interplay between intrinsic age-related defects in osteoblastic cell functions and age-related defects in bone microenvironment and endocrine system. Both these groups of factors lead to impaired osteoblast functions and decreased bone formation. It is hoped that determining the relative contribution of these factors to the age-related phenotype of the skeleton will lead to a more effective preventive and therapeutic approaches to abolish the age-related bone loss.

References

1. Frost HM (2001). Why should skeletal scientists, and clinicians learn the Utah paradigm of skeletal physiology? *J Musculoskel Neuron Interact.* 2: 121–30.
2. Parfitt AM (1991). Bone forming cells in clinical conditions. In: Hall BK, ed. *Bone, The Osteoblast and Osteocyte.* London: The Telford Press, pp. 351–426.
3. Hauge E, Qvesek D, Eriksen EF, Mosekilde L, Melsen F (2001). Cancellous bone remodelling occurs in specialized compartments lined by cellsa expressing osteoblastic markers. *J Bone Miner Res.* 16: 1575–82.
4. Riggs BL, Melton LJ, III (1986). Involutional osteoporosis. *N Engl J Med.* 314: 1676–86.
5. Eriksen EF, Langdahl B, Vesterby A, Rungby J, Kassem M (1999). Hormone replacement therapy prevents osteoclastic hyperactivity: A histomorphometric study in early post-menopausal women. *J Bone Miner Res.* 14: 1217–21.
6. Parfitt AM, Villanueva AR, Foldes J, Rao DS (1995). Relations between histologic indices of bone formation: implications for the pathogenesis of spinal osteoporosis. *J Bone Miner Res.* 10: 466–73.
7. Brockstedt H, Kassem M, Eriksen EF, Mosekilde L, Melsen F (1993). Age- and sex-related changes in iliac cortical bone mass and remodeling. *Bone* 14: 681–91.
8. Friedenstein AJ, Ivanov-Smolenski AA, Chajlakjan RK, *et al.* (1978). Origin of bone marrow stromal mechanocytes in radiochimeras and heterotopic transplants. *Exp Hematol.* 6: 440–4.
9. Friedenstein AJ (1991). Osteogenic stem cells in the bone marrow. *Bone Miner.* 7: 243-272.
10. Friedenstein AJ, Piatetzky-Shapiro II, Petrakova KV (1966). Osteogenesis in transplants of bone marrow cells. *J Embryol Exp Morphol.* 16: 381–90.

11. Owen M (1988). Marrow stromal stem cells. *J Cell Sci.* 10: 63–76.
12. Bruder SP, Jaiswal N, Haynesworth SE (1997). Growth kinetics, self-renewal, and the osteogenic potential of purified human mesenchymal stem cells during extensive subcultivation and following cryopreservation. *J Cell Biochem.* 64: 278–94.
13. Bianco P, Costantini M, Dearden LC, Bonucci E (1988). Alkaline phosphatase positive precursors of adipocytes in the human bone marrow. *Br J Haematol.* 68: 401–3.
14. Bianco P, Riminucci M, Gronthos S, Robey PG (2001). Bone marrow stromal stem cells: nature, biology, and potential applications. *Stem Cells* 19: 180–92.
15. Simmons PJ, Torok-Storb B (1991). Identification of stromal cell precursors in human bone marrow by a novel monoclonal antibody, STRO-1. *Blood* 78: 55–62.
16. Satomura K, Krebsbach P, Bianco P, Gehron RP (2000). Osteogenic imprinting upstream of marrow stromal cell differentiation *J Cell Biochem.* 78: 391–403.
17. Park SR, Oreffo RO, Triffitt JT (1999). Interconversion potential of cloned human marrow adipocytes *in vitro. Bone* 24: 549–54.
18. Lip P, Coupron P, Meunier PJ (1978). Mean wall thickness of trabecular bone packets in human iliac crest: changes with age. *Calcif Tissue Res.* 26: 13–17.
19. Rao MS, Mattson MP (2001). Stem cells and aging: expanding the possibilities. *Mech Ageing Dev.* 122: 713–34.
20. Castro-Malaspina H, Gay RE, Resnick G, *et al.* (1980). Characterization of human bone marrow fibroblast colony-forming cells (CFU-F) and their progeny. *Blood* 56: 289–301.
21. Kuznetsov SA, Krebsbach PH, Satomura K, *et al.* (1997). Single-colony derived strains of human marrow stromal fibroblasts form bone after transplantation *in vivo. J Bone Miner Res.* 12: 1335–47.
22. Haynesworth SE, Goldberg VM, Caplan A (1994). Diminution of the number of mesenchymal stem cells as a cause for skeletal aging. In: Buckwalter JA, Goldberg VM, Woo SLY, eds. *Musculoskeletal Soft-tissue Aging: Impact on Mobility.* Rosemont, IL, pp. 80–6.
23. Egrise D, Martin D, Vienne A, Neve P, Schoutens A (1992). The number of fibroblastic colonies formed from bone marrow is decreased and the *in vitro* proliferation rate of trabecular bone cells increased in aged rats. *Bone* 13: 355–61.
24. Quarto R, Thomas D, Liang CT (1995). Bone progenitor cell deficits and the age-associated decline in bone repair capacity. *Calcif Tissue Int.* 56: 123–9.
25. Kahn A, Gibbons R, Perkins S, Gazit D (1995). Age-related bone loss. A hypothesis and initial assessment in mice. *Clin Orthop.* 69–75.
26. Xu CX, Hendry JH, Testa NG, Allen TD (1983). Stromal colonies from mouse marrow: characterization of cell types, optimization of plating efficiency and its effect on radiosensitivity. *J Cell Sci.* 61: 453–66.
27. Brockbank KG, Ploemacher RE, van Peer CM (1983). An *in vitro* analysis of murine hemopoietic fibroblastoid progenitors and fibroblastoid cell function during aging. *Mech Ageing Dev.* 22: 11–21.
28. Bellows CG, Aubin JE (1989). Determination of numbers of osteoprogenitors present in isolated fetal rat calvaria cells *in vitro. Dev Biol.* 133: 8–13.
29. Tsuji T, Hughes FJ, McCulloch CA, Melcher AH (1990). Effects of donor age on osteogenic cells of rat bone marrow *in vitro. Mech Ageing Dev.* 51: 121–32.
30. Nishida S, Endo N, Yamagiwa H, Tanizawa T, Takahashi HE (1999). Number of osteoprogenitor cells in human bone marrow markedly decreases after skeletal maturation 5. *J Bone Miner Metab.* 17: 171–7.

31. Muschler GF, Nitto H, Boehm CA, Easley KA (2001). Age- and gender-related changes in the cellularity of human bone marrow and the prevalence of osteoblastic progenitors. *J Orthop Res.* 19: 117–25.
32. D'ippolito G, Schiller PC, Ricordi C, Roos BA, Howard GA (1999). Age-related osteogenic potential of mesenchymal stromal stem cells from human vertebral bone marrow. *J Bone Miner Res.* 14: 1115–22.
33. Mets T, Verdonk G (1981). Variations in the stromal cell population of human bone marrow during aging. *Mech Ageing Dev.* 15: 41–9.
34. Oreffo RO, Bord S, Triffitt JT (1998). Skeletal progenitor cells and ageing human populations. *Clin Sci (Lond).* 94: 549–55.
35. Stenderup K, Justesen J, Eriksen EF, Rattan SI, Kassem M (2001). Number and proliferative capacity of osteogenic stem cells are maintained during aging and in patients with osteoporosis. *J Bone Miner Res.* 16: 1120–9.
36. Justesen J, Stenderup K, Ebbesen EN, Mosekilde L, Steiniche T, Kassem M (2001). Adipocyte tissue volume in bone marrow is increased with aging and in patients with osteoporosis. *Biogerontology* 2: 165–71.
37. Meunier P, Aaron J, Edouard C, Vignon G (1971). Osteoporosis and the replacement of cell populations of the marrow by adipose tissue. A quantitative study of 84 iliac bone biopsies. *Clin Orthop.* 80: 147–54.
38. Burkhardt R, Kettner G, Bohm W, *et al.* (1987). Changes in trabecular bone, hematopoiesis and bone marrow vessels in aplastic anemia, primary osteoporosis, and old age: a comparative histomorphometric study. *Bone* 8: 157–64.
39. Beresford JN, Bennett JH, Devlin C, Leboy PS, Owen ME (1992). Evidence for an inverse relationship between the differentiation of adipocytic and osteogenic cells in rat marrow stromal cell cultures. *J Cell Sci.* 102 : 341–51.
40. Bennett JH, Joyner CJ, Triffitt JT, Owen ME (1991). Adipocytic cells cultured from marrow have osteogenic potential. *J Cell Sci.* 99: 131–9.
41. Kajkenova O, Lecka-Czernik B, Gubrij I, *et al.* (1997). Increased adipogenesis and myelopoiesis in the bone marrow of SAMP6, a murine model of defective osteoblastogenesis and low turnover osteopenia. *J Bone Miner Res.* 12: 1772–9.
42. Nuttall ME, Gimble JM (2000). Is there a therapeutic opportunity to either prevent or treat osteopenic disorders by inhibiting marrow adipogenesis? *Bone* 27: 177–84.
43. Justesen J, Stenderup K, Eriksen EF, Kassem M (2002). Maintenance of osteoblastic and adipocytic differentiation potential with age and osteoporosis in human marrow stromal cell cultures. *Calcif Tissue Int.* 71: 36–44.
44. Tornvig L, Mosekilde LI, Justesen J, Falk E, Kassem M (2001). Troglitazone treatment increases bone marrow adipose tissue volume but does not affect trabecular bone volume in mice. *Calcif Tissue Int.* 69: 46–50.
45. Campisi J (2001). From cells to organisms: can we learn about aging from cells in culture? *Exp Gerontol.* 36: 607–18.
46. Cristofalo VJ, Allen RG, Pignolo RJ, Martin BG, Beck JC (1998). Relationship between donor age and the replicative lifespan of human cells in culture: a reevaluation. *Proc Natl Acad Sci USA* 95: 10614–19.
47. Kassem M, Ankersen L, Eriksen EF, Clark BF, Rattan SI (1997). Demonstration of cellular aging and senescence in serially passaged long-term cultures of human trabecular osteoblasts. *Osteoporos Int.* 7: 514–24.
48. Kveiborg M, Rattan SI, Clark BF, Eriksen EF, Kassem M (2001). Treatment with 1,25-dihydroxyvitamin D3 reduces impairment of human osteoblast functions during cellular aging in culture. *J Cell Physiol.* 186: 298–306.

49. Dimri GP, Lee X, Basile G, *et al.* (1995). A biomarker that identifies senescent human cells in culture and in aging skin *in vivo*. *Proc Natl Acad Sci USA* 92: 9363–7.
50. Allsopp RC, Chang E, Kashefi-Aazam M, *et al.* (1995). Telomere shortening is associated with cell division *in vitro* and *in vivo*. *Exp Cell Res*. 220: 194–200.
51. Roholl PJ, Blauw E, Zurcher C, Dormans JA, Theuns HM (1994). Evidence for a diminished maturation of preosteoblasts into osteoblasts during aging in rats: an ultrastructural analysis. *J Bone Miner Res*. 9: 355–66.
52. Long MW, Ashcraft EK, Normalle D, Mann KG (1999). Age-related phenotypic alterations in populations of purified human bone precursor cells. *J Gerontol A Biol Sci Med Sci*. 54: B54–62.
53. Katzburg S, Lieberherr M, Ornoy A, Klein BY, Hendel D, Somjen D (1999). Isolation and hormonal responsiveness of primary cultures of human bone- derived cells: gender and age differences. *Bone* 25: 667–73.
54. Fedarko NS, Vetter UK, Weinstein S, Robey PG (1992). Age-related changes in hyaluronan, proteoglycan, collagen, and osteonectin synthesis by human bone cells. *J Cell Physiol*. 151: 215–27.
55. Koshihara Y, Honda Y (1994). Age-related increase in collagen production in cultured human osteoblast-like periosteal cells. *Mech Ageing Dev*. 74: 89–101.
56. Inoue K, Ohgushi H, Yoshikawa T, *et al.* (1997). The effect of aging on bone formation in porous hydroxyapatite: biochemical and histological analysis. *J Bone Miner Res*. 12: 989–94.
57. Huibregtse BA, Johnstone B, Goldberg VM, Caplan AI (2000). Effect of age and sampling site on the chondro-osteogenic potential of rabbit marrow-derived mesenchymal progenitor cells. *J Orthop Res*. 18: 18–24.
58. Pfeilschifter J, Diel I, Pilz U, Brunotte K, Naumann A, Ziegler R (1993). Mitogenic responsiveness of human bone cells *in vitro* to hormones and growth factors decreases with age. *J Bone Miner Res*. 8: 707–17.
59. Ankrom MA, Patterson JA, d'Avis PY, *et al.* (1998). Age-related changes in human oestrogen receptor alpha function and levels in osteoblasts. *Biochem J*. 333: 787–94.
60. d'Avis PY, Frazier CR, Shapiro JR, Fedarko NS (1997). Age-related changes in effects of insulin-like growth factor I on human osteoblast-like cells. *Biochem J*. 324: 753–60.
61. Mohan S, Baylink DJ (1991). Bone growth factors. *Clin Orthop*. 30–48.
62. Seck T, Bretz A, Krempien R, Krempien B, Ziegler R, Pfeilschifter J (1999). Age-related changes in insulin-like growth factor I and II in human femoral cortical bone: lack of correlation with bone mass. *Bone*. 24: 387–93.
63. Bismar H, Kloppinger T, Schuster EM, *et al.* (1999). Transforming growth factor beta (TGF-beta) levels in the conditioned media of human bone cells: relationship to donor age, bone volume, and concentration of TGF-beta in human bone matrix *in vivo*. *Bone* 24: 565–9.
64. Irving JT, LeBolt SA, Schneider EL (1981). Ectopic bone formation and aging. *Clin Orthop*. 249–53
65. Nishimoto SK, Chang CH, Gendler E, Stryker WF, Nimni ME (1985). The effect of aging on bone formation in rats: biochemical and histological evidence for decreased bone formation capacity. *Calcif Tissue Int*. 37: 617–24.
66. Hosny M, Sharawy M (1985). Osteoinduction in young and old rats using demineralized bone powder allografts. *J Oral Maxillofac Surg*.43: 925–31.
67. Syftestad GT, Urist MR (1982). Bone aging. *Clin Orthop*. 288–97
68. Nyssen-Behets C, Delaere O, Duchesne PY, Dhem A (1996). Aging effect on inductive capacity of human demineralized bone matrix. *Arch Orthop Trauma Surg*. 115: 303–6.

69. de Boer J, Andressoo JO, de Wit J, *et al.* (2002). Premature aging in mice deficient in DNA repair and transcription. *Science* 296: 1276–9.
70. Tyner SD, Venkatachalam S, Choi J, *et al.* (2002). p53 mutant mice that display early ageing-associated phenotypes. *Nature* 415: 45–53.
71. Kveiborg M, Kassem M, Langdahl B, Eriksen EF, Clark BF, Rattan SI (1999). Telomere shortening during aging of human osteoblasts *in vitro* and leukocytes *in vivo*: lack of excessive telomere loss in osteoporotic patients. *Mech Ageing Dev.* 106: 261–71.
72. Simonsen JL, Rosada C, Serakinci N, *et al.* (2002). Telomerase expression extends the proliferative life-span and maintains the osteogenic potential of human bone marrow stromal cells. *Nat Biotechnol.* 20: 592–6.
73. Shi S, Gronthos S, Chen S, *et al.* (2002). Bone formation by human postnatal bone marrow stromal stem cells is enhanced by telomerase expression. *Nat Biotechnol.* 20: 587–91.

T Cell Immunosenescence

Graham Pawelec
Center for Medical Research, University of Tubingen Medical School, Waldhornlestr. 22, D-72072 Tubingen, Germany

Introduction

Provided that one is aware of the limitations, studying cell aging outside the body using culture techniques can yield important information which, however, always has to be validated by reference to the situation *in vivo* and the particular cell and organ system being examined. As I shall limit my remarks in this chapter mostly to the human system we also need to distinguish between the use of the term "*in vivo*" to mean what happens inside the organism and "*ex vivo*" to mean the behavior of freshly isolated cells subjected at most to only short-term culture during the assay. In humans, it is obviously next to impossible to perform controlled experiments *in vivo*, so that most validation assays have to be performed *ex vivo*. In the case of T lymphocyte aging, certain assays can in fact be carried out *in vivo*, especially measurement of delayed-type hypersensitivity (DTH) reactions and some other tests in the reasonably-accessible skin. Thus, for example, one of the first pieces of evidence for age-associated decreased immune responsiveness in humans was the observation of reduced delayed-type hypersensitivity reactions to recall antigens in the elderly [1]. Other than this type of assay, however, *in vivo* experiments are difficult in humans. One could also argue that decreased efficacy of vaccination in the elderly is an *in vivo* test of immune function, but this is of course a clinical outcome rather than a controlled test of function [2]. In general, it remains quite difficult to bridge the gap between aspects of immunosenescence measured *in vitro* or *ex vivo* and the clinical relevance thereof. Nonetheless, cell culture systems clearly offer a model in which it is possible to examine age-associated changes and interventions to influence these which would be very difficult or impossible *in vivo*. Alterations identified in long-term cultures can then be sought *ex vivo*. In this way, "biomarkers" of aging discovered *in vitro* can be validated *ex vivo*. This chapter will consider available information in this area, and discuss the relevance of T cell culture models of aging to

Sunil C. Kaul and Renu Wadhwa (eds.), Aging of Cells In and Outside the Body, 85–100.

organismal T cell immunosenescence. Because of space constraints, the direct clinical relevance of immunosenescence will not be discussed in detail here; the reader is referred to our recent reviews on this subject [2–4].

Characteristics of immunosenescence *in vivo*

First, what do we know about the state of the immune system in the elderly? This is currently a rapidly advancing area in which many disparate data are being generated and published. However, at the moment, it is probable that most investigators would agree to the following brief overall state-of-the-art consensus: We know that hematopoiesis is not severely affected, and that innate immunity is not markedly affected [for review, see ref. 5]; also, alterations to B cell function are thought to be mostly secondary to alterations to T cells [3]. We know that output of naive T cells by the thymus decreases drastically with age but does not cease entirely or at least not until extreme old age [6]. However, we also know that in an outbred species like humans, there is a great deal of inter-individual variation, the exact implications of which are not yet clear. We know that probably as a result of this decreased thymic output, the proportion of naive cells present in the periphery decreases with age and reciprocally, memory cells increase (i.e., the number of cells available to recognize new antigenic challenges decreases, while the number previously exposed to foreign antigens over the lifetime of the individual increases) [3]. We know that many parameters of T cell function change with age but agreement on exactly what these changes are and what they mean is less widespread [7]. We know that these parameters are affected by the state of health of the individual, which makes it difficult to distinguish between changes that are caused by age rather than illness [2]. Finally, it seems that many measurable immune parameters in the very elderly and in the oldest old (i.e., >90 years or centenarians and older) are more similar to those of the average middle-aged person than to the average 75 year-old person [8]. Apart from the above general statements, it would probably be difficult to achieve consensus on the impact of aging on the immune system in humans, and the clinical relevance thereof. There are data strongly suggesting that immune function impacts upon survival, primarily due to decreased capacity to ward off infectious disease (in mice, possibly cancer rather than infection [9]) and, further, that clusters of immune parameters can predict mortality [10], probably for this reason.

Because T cell immunity relies upon the extensive clonal expansion of a limited number of T cells carrying specific antigen receptors, and because even naive cells are continuously dividing *in vivo* (see section on "Longevity, immunosenescence and memory"), there is a school of thought hypothesizing that replicative senescence of T cells might impact upon the ability of the individual to respond to antigen challenge or especially rechallenge [11]. As described in several of the other chapters in this book, replicative senescence as defined *in vitro* is a state of permanent growth arrest of cells after they have completed a certain number of cell divisions (the Hayflick limit, often observed to be ca. 50 population doublings, PD), which is thought to represent a tumor suppressor mechanism. It is not clear whether replicative senescence in this sense occurs in T cells, although there is evidence both from *in vitro* cultures of CD8+ T cells and some *ex vivo* studies of CD8+ cells that this may be the case [12]. However, it is clear that T cells cannot continue to proliferate *in vitro* beyond a certain (from clone to clone rather variable) number of population doublings (PD) as discussed in detail in the following section; but usually cultures apoptose and are lost, especially cultures of CD4+ cells, rather than persisting as post-mitotic cells (as commonly observed using fibroblasts). There may therefore be some differences between CD4 and CD8 subsets of antigen-specific T cells.

Culture models for T cell aging

Peripheral blood T cells from young and old donors can be successfully cultured *in vitro* provided that they are not only supplied with growth factors but also intermittently stimulated via their surface receptors. Experiments wherein T cells are stimulated only once at the beginning of the cultures and not thereafter do not measure the potential lifespan of the cells in culture. This is because on initial stimulation, T cells upregulate growth factor receptors which are thereafter down-regulated again regardless of the presence of growth factors [13]. Restimulation via the antigen receptor or, under certain circumstances, other surface receptors, is required thereafter for re-upregulation of growth factor receptors and continued growth. Restimulation too soon after antigen exposure, on the other hand, may lead to activation-induced cell death by apoptosis [14]. Hence, T cells must be maintained in culture by balancing provision of sufficient antigen to avoid growth cessation and excess antigen, which increases apoptosis. Under appropriate culture conditions, all major subsets of T cells can be cultured (i.e., CD4+ TCR2, CD8+ TCR2, CD4,8-negative TCR1, and NKT cells). However, the latter two types tend to be more difficult to maintain. Early assessments of longevity of human T cell cells in culture, even if restimulation conditions were appropriate, commonly measured culture duration of mixed populations of cells, i.e., they measured only the longevity of the longest-lived clone [15]. Assessments of clonal longevity remain rare because of the laborious nature of the experiments, but some data are available on the behavior of human T cell clones from different sources. Early data from McCarron *et al.* suggested that clonal longevity was influenced by age of the donor: clones derived from neonates averaged 52 PD, those derived from young adults (20–30 years) managed 40 PD, but those from the elderly (70–90 years) only 32 PD, as calculated from the data displayed in that publication [16]. Our own more extensive results are somewhat different. It is obvious from our summarized data in Table 1 that although individual T cell clones do have very varied lifespans, the overall patterns for T cells of quite different origins and donors of different ages are remarkably similar.

T cell clones from young healthy donors

The data on cloning of peripheral T cells from young healthy donors, the usual source of material for this type of experiment, show that these cells can be cloned with a high efficiency (cloning efficiency, CE, nearly 50%). Most of the T cell clones (TCC) obtained are CD4+ not CD8+, suggesting that these particular cloning conditions are favourable for T cell outgrowth, but only of CD4+ cells. A CE of

50% for a mixture of CD4 and CD8 cells therefore implies that almost all CD4+ are capable of clonal outgrowth in culture (a clone being defined as the accumulation of ca. 1000 cells from one original cell). We can therefore conclude that the CD4 T cell repertoire is present more or less in its entirety in the starting clonal populations under these conditions. However, after 20 PD, when the clone size has increased from 10^3 to 10^6, about half of the clones have already been lost (mean of 53% for cloning of young donor CD3 cells, Table 1). At 30 PD, another half has been lost so that only one quarter of the originally clonable cells remains. At 40 PD (which theoretically represents a really large clone size of 10^{12} cells), only 15% of the original starting clonal population remains. The longest-lived clone in all of our experiments thus far achieved ca. 170 PD, exceeding the number of PD sometimes used to indicate immortalization [17, 18], although this was not the case for this clone. These results document a constant attrition of the T cell population at the clonal level in culture, but with retention of perhaps up to 5% of the original CD4 repertoire up to 40 PD and with retention of very rare clones for considerably longer. This implies that the T cell repertoire may become restricted but not negated even after a very large degree of clonal expansion under pressure of chronic antigenic stimulation.

***Table 1**. Longevities of human T cell clones under standard culture conditions*

	Percentage of clones reaching population size					
Origin	CE%	*n* clones/expts	20 PD (10^6)	30 PD (10^9)	40 PD (10^{12})	Longest -lived
CD34+ (peripheral)	55	533/6	31	17	6%	60 PD
CD34+ (cord)	43	94/2	29	15	nyt	57
CD3+ (young)	47	1355/15	47	24	15	170
CD3+ (old SENIEUR)	52	116/2	55	22	16	72
CD3+ (old OCTO)	33	32/3	48	28	19	61
CD3+ (CML)	49	35/1	60	35	14	51

CE, cloning efficiency (calculated from percentage of wells positive in cloning plates). Longevity is expressed as a percentage of established clones (i.e., those counted as positive in calculating the CE) which survive to 20, 30 or 40 PD. Origins: CD34+, positively-selected hematopoietic stem cells from peripheral or cord blood; CD3+, normal peripheral T cells; young, apparently healthy donors under 30 years; old SENIEUR, donors over 90 years, in excellent health according to the criteria of the SENIEUR protocol; old OCTO, donors over 80 years participating in the longitudinal OCTO study, selected only by survival; CML, a middle-aged donor with chronic myelogenous leukemia in chronic phase treated with interferon-γ.

T cell clones from other sources

The questions arises as to how these longevity data compare with those on T cell clones derived from the old or reciprocally, for stem cells or progenitor cells. Our starting hypothesis was that T cells generated in situ from CD34+ hematopoietic progenitors would have greater longevities due to their lack of a previous proliferative history (as T cells at least) [19]. Reciprocally, we anticipated that CE of cells from the elderly would be reduced as would their clonal expansion capacity, and that this would be more prevalent in less healthy donors than in those rigorously selected for good health. Finally, we predicted that T cells from situations of chronic *in vivo* antigenic stress, such as in cancer, would also have lower CE and longevities for the same reasons. As can be seen from Table 1, none of these expectations was fulfilled. CE were not lower in the elderly than in the young, and neither was the state of health of the donor critical (although there may be a tendency towards slightly lower CE in the donors from the Swedish study, who were not specifically selected for good health, compared to the SENIEUR donors (see section on "Longitudinal studies and the concept of the 'immune risk phenotype,' IRP" and Table 1). The CE was equally high when the cells were from a CML patient, a donor subjected to a putative high level of antigenic stress. CE were also similar in the extra-thymic T cell differentiation cultures employed [20]. Bearing in mind that the system selected for CD4 cells, and that the CE are so high that we cannot be dealing with rare contaminants, it has to be concluded that CE of the T cells from these markedly different sources are nonetheless not markedly different. So are their longevities different ? Old cells with a proliferative history *in vivo* might manage the ca. 10 PD needed to be counted as a clone, but not the 20, 30 or more as an established TCC. However, it is clear from the collected data in Table 1 that this is not the case either. Very similar proportions of TCC achieve 20, 30 and 40 PD regardless of their origin. In addition, the maximum longevity of the longest-lived clone is also similar and very close to the Hayflick Limit of 50–60 PD. Exceptional clones, such as that from the young donor which achieved 170 PD, have not been seen yet, but this may merely be due to the smaller number of clones studied. Overall, it seems that CE and longevities of T cells in culture are essentially identical regardless of whether the cells to be cloned are derived from progenitors, young or old donors, healthy or sick.

Alterations to the cells during culture

Surface phenotype and function

This *in vitro* clonal culture model may also be informative for longitudinal studies of age-related changes to CD4 cells under chronic antigenic stimulation. Because of the constraints of the cloning procedure, "young" cells have already undergone at least ca. 22, 23 PD, but at this stage, we still have a good representation of the starting repertoire (see discussion above). However, those clones that can be studied over an extended time do of course represent only the small fraction of maybe 10% which are capable of that degree of longevity. One of the simplest analytical techniques to study these cells over time is to use flow cytometry with monoclonal antibodies to molecules expressed on the cell surface. We have examined a large range of different surface markers and found relatively few which change with age (i.e., with increasing PD). One of the most reproducible patterns of age-associated alterations involves a reduction of the level of expression of the costimulatory receptor CD28, as well as the putative costimulatory receptors CD134 and CD154, whereas despite the increased susceptibility to apoptosis [21], surface CD95 expression remains constant. The cells have a memory effector phenotype (CD45RA-negative, CCR7-negative, $CD45RO^{+}$) as would be expected from chronically-stimulated memory effector cells. The level of the TCR also remains stable, suggesting that these cells may retain the ability to recognise and respond to antigen. This typical pattern is occasionally not observed; a few exceptions show that CD134 and/or CD154 expression can also be retained by old cells. However, the most consistent pattern is CD28 reduction which is essentially universal in the aging TCC. Nonetheless, in certain clones, although they show this age-associated decrease at first, CD28 re-expression can occur later. We have correlated this CD28 re-expression with a decreased ability of the TCC to secrete TNF-α, at least in certain clones. This is consistent with the observation that TNF-α downregulates CD28 expresssion [22] and with out earlier observations that TNF-α can directly inhibit some TCC [23].

Major functional changes may follow from the decreased level of expression of costimulatory receptors, CD28 as defined here, and in all likelihood, others yet to be investigated. The most dramatic changes are in the patterns of cytokines secreted, commonly resulting in decreased levels of IL-2 and increased levels of IL-10 [24]. Because the expression of the TCR is maintained and antigen-specific signalling still occurs, it is likely that these differences are caused by differences in the delivery of costimuli to the T cells, ever-increasing numbers of which are now being defined [25]. The balance of these, and their function, will determine the consequences of TCR ligation, apparently more so more than this antigen-specific "signal one" itself [24], although this may also be changed with aging [25]. Not very much is known about the regulatory mechanisms for CD28 expression, let alone for the expression of these other positive and negative costimulatory receptors. This is likely to be an area of intensive investigation by immunogerontologists over the next few years.

Telomere length and genome integrity

Although more data are still required, measurements of telomere lengths for the CD4 clones described above may show differences between clones derived from the young and elderly. This is intriguing and would be the first hint of differences in the long-term cultured cells from young and old. Telomere lengths as assessed by flow-FISH in early-passage clones from the young are in the region of 10–12 kb, generally decreasing to below 5 kb at the end of their lifespan. However, although telomere lengths of clones derived from two very old (>90 years) extremely healthy (SENIEUR-compliant) donors were shorter in early passage cells (around 7.5 kb) they remained at this length over the lifespan of the clones (T. Brummendorf, Tübingen, unpublished results). Whether telomere lengths per se, and/or variations in the ability of the cells to upregulate telomerase [26], control the longevity of the clones is still open. Efforts to resolve this question using some of the same T cell clones are ongoing in several different laboratories as part of the work of the EU-supported network "Immunology and Aging in Europe," ImAginE [27]. Initial results suggest that a major change with age is the increasing susceptibility of the clones to oxidative DNA damage [28] which can be partially prevented using anti-oxidants such as PBN that also extend the longevity of the cells in culture (P Hyland, unpublished results). The impact of enforced expression of telomerase [29] in such cells is being assessed, along with DNA repair mechanisms.

In vivo parallels?

Longevity, immunosenescence and memory

Accepting that thymic output does generally decrease markedly with age, and that T cells cannot be derived extra-thymically, the aging individual will become increasingly dependent on the T cells formed earlier on in life. Under natural conditions, humans are exposed to and must successfully respond to large numbers of pathogens early in life, building and maintaining immunological memory for later life. Assuming that no new pathogens are introduced into the environment, and that memory can be maintained over the entire lifespan, there may be no reason to maintain a constant supply of new T cells (this explanation has in fact been put forward as the reason for the occurrence of thymic involution, see ref. 30). However, both of these assumptions may be unlikely. We know that immunological memory can be maintained for very long periods, and that the T cell population within which those responsible for maintaining memory reside continues to divide slowly. There has been controversy as to whether persistence of antigen is required for this process, but whether or not the division is antigen-driven, the fact remains that such cells do divide. This has been elegantly demonstrated recently by means of some important experiments in humans which are very difficult to perform and will be discussed here in some detail. The technique employed relies on pulsing volunteers with deuterated glucose, which is incorporated into the DNA of cells that happen to be dividing at the time. At later time points, the amount of ^{2}H-glucose found in different T cell subpopulations can be assessed by gas chromatography and the number of cell divisions undergone calculated from the dilution factor. In this manner, Prof. Peter Beverley's group at the Jenner Institute in the UK have examined lymphocyte turnover in several healthy young and old donors [see ref. 4]. They examined CD4 and CD8 cells separately; moreover, they subdivided the cells according to their expression of CD45R0 (majority of cells in this subset being antigen-experienced) or CD45RA (mostly naive cells). In young donors, CD4 CD45RA cells were estimated to divide on average about once every 140 days, whereas CD4 CD45RO cells divided once every 40 days. For CD8 cells, the picture was similar, but division was somewhat slower, at 250 and 55 days, respectively. These results clearly suggest that both naive and memory cells in both major T cell subsets are not quiescent in normal healthy donors, but are constantly dividing. The situation was similar in the elderly, except that in general, cell division was slower, being estimated at 250 days for CD4 CD45RA cells (compared to 140 in the young) and 50 days for CD4 CD45RO cells (40 in the young). However, in CD8 cells, turnover of naive cells was faster in the elderly than in the young (140 days-versus 250 days), although this was not the case for memory cells (140 days in the elderly, 55 days in the young). These preliminary results raise some very interesting points. They suggest that if the *in vitro* lifespan (Hayflick limit) applies to T cells (as we have shown above) and if it occurs *in vivo*, then CD4 memory cells would already reach a 50 PD limit after only around 7 years, even when calculated on the basis of the average (slower) turnover in the elderly (and being only about 5 years in the young). For CD8 memory cells, these figures would be ca. 19 years and 7.5 years in the old and young, respectively. As expected, naive cells would persist for longer, between a minimum of 19 years and maximum of 35 years. However, it must be borne in mind that all values mentioned above refer to the average cell division dynamics of the entire T cell subpopulation. It is possible that each antigen-specific clone actually responsible for maintaining memory divides more slowly. To correlate these findings with immunological memory, it would be necessary to analyse antigen-specific cells separately. Recently, technology has become available in the shape of soluble MHC/peptide tetrameric complexes which would enable this to be accomplished.

Although memory persists for long periods, it may not persist indefinitely and may also differ from antigen to antigen and according to genetic and other variation in different individuals. The maintenance of immunological memory may begin to be problematic in the very old or in younger individuals subjected to chronic antigenic stress (e.g. tumor-bearers, parasite-infested people, persistent viral infection). There is little direct evidence for T cell clonal exhaustion in these conditions, but quite a lot of circumstantial evidence. Because CD8 cells may become more resistant to apoptosis at the end of their proliferative lifespan, whereas CD4 cells may become more susceptible, one might expect decreases of CD4 cell numbers and increased numbers of apoptosis-resistant CD8 cells under these circumstances. This is indeed what is commonly seen in cancer patients, in the aged and of course in AIDS and other chronic infections.

Biomarkers of aging

At present, it is impossible to assess the age of T cells in the periphery; there are no appropriate "biomarkers." The ability to assess the number of times a particular cell population has divided *in vitro* by measuring the dilution factor of "T cell receptor excision circles" (TRECS) [6] is thought to have a sensitivity of only around 10 divisions, and is therefore not useful for studying longer-term events. A second marker intensively investigated in various cell types recently, namely telomere length, can also not be used to accurately assess the number of cell divisions previously undergone by T cells *in vivo*, because on activation T cells upregulate telomerase and maintain telomere lengths, at least for a certain (probably variable) period of time [31]. The use of certain immunological biomarkers of aging, such as decreased expression of CD28, is fraught with difficulty and has never been employed as a way of measuring the actual number of cell divisions undergone by a particular population. Nonetheless, for what it is worth, it is possible to compare parameters measured on cultured T cells with those found in the elderly versus the young. Because of the great variations in the published data on the latter, however, no very accurate comparison can be made and it may be best to view Table 2 as a personal opinion. Perhaps for this very reason, it may be appropriate to concentrate on those parameters which do agree between the *in vitro* model system and what is observed in the elderly *ex vivo.*

Longitudinal studies and the concept of the "immune risk phenotype" (IRP)
The data summarized in Table 2 and indeed the majority of studies in humans are conducted on separate groups of different ages. These groups will differ in health, nutrition, environmental influences and genetic background, as well as age, possibly obscuring age-associated differences. Gerontological donor selection procedures, such as the "Senior European" protocol [32], may help obviate the differences between groups in terms of health status but cannot approach the other variables. A closer approximation to an ideal situation is to perform longitudinal studies, following the same population over time [33]. Although clearly logistically and financially problematic, such studies may yield important data otherwise unavailable. By studying the same donors from the same geographical area, some environmental variables may also be reduced, although it is obviously difficult to control for nutritional and lifestyle differences. Unsurprisingly there are very few longitudinal studies from which data are currently available and even fewer including immunological data. One such study series is the Swedish OCTO (over 80 years old) and current NONA (over 90) project being carried out in Jönköping in the very elderly [34–37]. Obviously, these people are already selected as longer-than-average survivors, but with this proviso data collected every two years over the last decade can be very valuable, perhaps not only for the already-successfully aged. The OCTO and NONA studies have established that significantly lower numbers of CD4+ and CD8+CD28+ cells, and larger numbers of CD8+CD28-negative apoptosis-resistant cells, as well as altered CD4/8 ratios, are characteristic of the elderly compared to the healthy middle-aged. Clustering these with certain other parameters has led to the emergence of the concept of the IRP as a tool with power to predict incipient mortality in the very elderly (summarized in Table 3). One of the current challenges is to refine the constitution of the IRP and extend the concept to younger individuals to assess whether IRP parameters established in the aged are informative for younger people [4]. Currently, we are exploring the apparent obsession of the immune system in the elderly with the recognition of CMV. Using tetramer technology, it can be demonstrated that large fractions of the CD8 repertoire are taken up by cells bearing receptors for CMV. Such cells appear to be compromized in their function, at least as far as secretion of the anti-viral cytokine interferon-γ is concerned: They also seem to be unable to undergo further clonal expansion, as reflected in their high level expression of a negative regulatory receptor, KLRG1, associated with inability to proliferate [37a]. These cells, which are likely to be resistant to apoptosis, may represent senescent CD8 clones which can neither divide nor delete – and hence may result in dysregulated responses. Deleting such cells, by analogy with a transgenic mouse model, may "rejuvenate" the immune system [38].

Table 2. *Examples of changes in human T cells in the elderly ex vivo and in cultures in vitro*

Decreased				
CD28 expression	*Ex vivo*:	CD8		[39–42]
		CD4		[43, 44]
	In vitro:	CD8		[40, 45]
		CD4		[24, 46]
CD28 promoter binding activity	*Ex vivo*		Binding activities of 2 factors reduced	[43]
	In vitro		Binding activity of one of these reduced	[44]
CD154	*Ex vivo*			[47–49]
	In vitro			[24]
CD134	*Ex vivo*		Not known	
	In vitro			[24]
IL 2R (high affinity) expression	*Ex vivo*			[50, 51]
	In vitro			[24]
IL 2 production	*Ex vivo*			[52]
	In vitro			[24]
bcl-2 expression	*Ex vivo*			[53, 54]
	In vitro		May apply only to CD4 cells:	[55]
			CD8 cells may have increased bcl-2	[56]
Telomere lengths	*Ex vivo*			[57–59]
	In vitro			[60, 61]
Telomerase induction	*In vitro*			[26]
	Ex vivo			[62]

Table 2. *(continued)*

DNA repair	*Ex vivo*		[63, 64]
	In vitro	Not tested in T cells	
Stress resistance and HSP expression	*Ex vivo*		[65, 66]
	In vitro		[66, 67]
Increased			
p27 (i.e., delayed downregulation)	*Ex vivo*		[68, 69]
	In vitro		[28, 70]
CD152 expression	*Ex vivo*		[71]
	In vitro		[24]
IL 10 production	*Ex vivo*		[72–74]
	In vitro		[24]
DNA damage	*Ex vivo*		[75, 76]
	In vitro		[77]
Apoptosis (AICD) of CD4 cells	*Ex vivo*		[21]
CD95 and CD95 L expression	*Ex vivo*		[53]
	In vitro		[80, 81]
CD45RO+ cells	*Ex vivo*		[82, 83]
	In vitro		[55]

Table 3*. The "immunological risk phenotype"*

defined as:

* CD4:8 ratio of <1,
* poor T cell proliferative responses
* increased CD8+, CD28-negative, CD57+ cells
* low B cells
* CMV-seropositivity, and, possibly,
* clonal expansions of CD8 cells carrying receptors for CMV or EBV antigens
* a high proportion of dysfunctional cells amongst the CMV-specific CD8 cells which are CD28-negative but positive for an NK negative receptor

Note that the IRP consists of a cluster of these parameters, not each individually; which are the most important and which additional factors are involved remains to be determined and that, importantly, this may be independent of current clinical health status of the donor [4].

Acknowledgments

Experimental work from the Tübingen lab is currently supported by DFG grant Pa 361/7-1 and by a grant from the VERUM Foundation; the ImAginE network is supported by the European Commission through contract QLK6-CT-1999-02031 (www.medizin.uni-tuebingen.de/imagine/). Particular thanks are due to Qin Ouyang, Karin Hähnel and Arnika Rehbein, currently in the lab, for their contributions to the experimental work and to Prof. A. Wikby, Jönköping, for critical review of this manuscript.

References

1. Dworsky R, Paganini-Hill A, Arthur M, Parker J (1983). Immune responses of healthy humans 83–104 years of age. *J Nat Cancer Inst.* 71: 265–8.
2. Remarque E, Pawelec G (1998). T cell immunosenescence and its clinical relevance in man. *Rev Clin Gerontol.* 8: 5–25.
3. Pawelec G, Barnett Y, Effros R, *et al.* (2002). T cells and aging. *Front Biosci.* 7: D1056–183.
4. Pawelec G, Ouyang Q, Colonna-Romano G, Candore G, Lio D, Caruso C (2002). Is human immunosenescence clinically relevant? Looking for "immunological risk phenotypes." *Trends Immunol.* 23: 330–2.
5. Pawelec G, Solana R, Remarque E, Mariani E (1998). Impact of aging on innate immunity. *J Leukocyte Biol.* 64: 703–12.
6. Jamieson BD, Douek DC, Killian S, *et al.* (1999). Generation of functional thymocytes in the human adult. *Immunity* 10: 569–75.
7. Pawelec G, Solana R (2001). Immunoaging: the cause or effect of morbidity? *Trends Immunol.* 22: 348–50.

8. Franceschi C, Monti D, Sansoni P, Cossatizza A (1995). The immunology of exceptional individuals: the lesson of centenarians. *Immunol Today* 16: 12–16.
9. Miller RA, Chrisp C (2002). T cell subset patterns that predict resistance to spontaneous lymphoma, mammary adenocarcinoma, and fibrosarcoma in mice. *J Immunol.* 169: 1619–25.
10. Ferguson FG, Wikby A, Maxson P, Olsson J, Johansson B (1995). Immune parameters in a longitudinal study of a very old population of Swedish people: a comparison between survivors and nonsurvivors. *J Gerontol A Biol Sci Med Sci.* 50: B378–82.
11. Effros RB, Pawelec G (1997). Replicative senescence of T lymphocytes: does the Hayflick Limit lead to immune exhaustion? *Immunol Today* 18: 450–4.
12. Globerson A, Effros RB (2000). Aging of lymphocytes and lymphocytes in the aged. *Immunol Today* 21: 515–21.
13. Gullberg M, Smith KA (1986). Regulation of T cell autocrine growth. T4+ cells become refractory to interleukin 2. *J Exp Med.* 163: 270–84.
14. Wesselborg S, Janssen O, Kabelitz D (1993). Induction of activation-driven death (apoptosis) in activated but not resting peripheral blood T-cells. *J Immunol.* 150: 4338–45.
15. Perillo NL, Walford RL, Newman MA, Effros RB (1989). Human T lymphocytes possess a limited *in vitro* life span. *Exp Gerontol.* 24: 177–87.
16. McCarron M, Osborne Y, Story C, Dempsey JL, Turner R, Morley A (1987). Effect of age on lymphocyte proliferation. *Mech Ageing Dev.* 41: 211–18.
17. Tang DG, Tokumoto YM, Apperly JA, Lloyd AC, Raff MC (2001). Lack of replicative senescence in cultured rat oligodendrocyte precursor cells. *Science* 291: 868–71.
18. Mathon NF, Malcolm DS, Harrisingh MC, Cheng LL, Lloyd AC (2001). Lack of replicative senescence in normal rodent glia. *Science* 291: 872–5.
19. Pawelec G, Muller R, Rehbein A, Hahnel K and Ziegler BL (1999). Finite lifespans of T cell clones derived from CD34(+) human haematopoietic stem cells *in vitro*. *Exp Gerontol.* 34: 69–77.
20. Pawelec G, Muller R, Rehbein A, Hahnel K, Ziegler BL (1998). Extrathymic T cell differentiation *in vitro* from human CD34(+) stem cells. *J Leukocyte Biol.* 64: 733–9.
21. Pawelec G, Sansom D, Rehbein A, Adibzadeh M, Beckman I (1996). Decreased proliferative capacity and increased susceptibility to activation-induced cell death in late-passage human CD4(+) TCR2(+) cultured T cell clones. *Exp Gerontol.* 31: 655–68.
22. Bryl E, Vallejo AN, Weyand CM, Goronzy JJ (2001). Down-regulation of CD28 expression by TNF-alpha. *J Immunol.* 167: 3231–8.
23. Pawelec GP, Rehbein A, Schaudt K, Busch FW (1989). IL-4-responsive human helper T cell clones are resistant to growth inhibition by tumor necrosis factor-α. *J Immunol.* 143: 902–6.
24. Pawelec G, Rehbein A, Haehnel K, Merl A, Adibzadeh M (1997). Human T cell clones as a model for immunosenescence. *Immunol Rev.* 160: 31–43.
25. Pawelec G, Hirokawa K, Fulop T (2001). Altered T cell signalling in aging. *Mech Ageing Dev.* 122: 1613–37.
26. Pawelec G, Adibzadeh M, Rehbein A, Hahnel K, Wagner W, Engel A (2000). *In vitro* senescence models for human T lymphocytes. *Vaccine* 18: 1666–74.
27. Pawelec G (2000). Working together for robust immune responses in the elderly. *Nat Immunol.* 1: 91.
28. Hyland P, Barnett C, Pawelec G, Barnett Y (2001). Age-related accumulation of oxidative DNA damage and alterations in levels of p16(INK4a/CDKN2a), p21(WAF1/CIP1/SDI1) and p27(KIP1) in human CD4+T cell clones *in vitro*. *Mech Ageing Dev.* 122: 1151–67.

29. Hooijberg E, Ruizendaal JJ, Snijders PJF, Kueter EWM, Walboomers JMM, Spits H (2000). Immortalization of human CD8(+) T cell clones by ectopic expression of telomerase reverse transcriptase. *J Immunol.* 165: 4239–45.
30. George AJT, Ritter MA (1996). Thymic involution with aging: obsolescence or good housekeeping? *Immunol Today* 17: 267–72.
31. Weng NP (2001). Interplay between telomere length and telomerase in human leukocyte differentiation and aging. *J Leukocyte Biol.* 70: 861–7.
32. Ligthart GJ, Corberand JX, Fournier C, *et al.* (1984). Admission criteria for immunogerontological studies in man: the SENIEUR protocol. *Mech Ageing Dev.* 28: 47–55.
33. Pawelec G, Ferguson FG, Wikby A (2001). The SENIEUR protocol after 16 years. *Mech Ageing Dev.* 122: 132–4.
34. Wikby A, Maxson P, Olsson J, Johansson B, Ferguson FG (1998). Changes in CD8 and CD4 lymphocyte subsets, T cell proliferation responses and non-survival in the very old: the Swedish longitudinal OCTO-immune study. *Mech Ageing Dev.* 102: 187–98.
35. Olsson J, Wikby A, Johansson B, Lofgren S, Nilsson BO, Ferguson FG (2000). Age-related change in peripheral blood T-lymphocyte subpopulations and cytomegalovirus infection in the very old: the Swedish longitudinal OCTO immune study. *Mech Ageing Dev.* 121: 187–201.
36. Wikby A, Johansson B, Olsson J, Löfgren S, Nilsson B-O, Ferguson FG (2002). Expansion of peripheral blood CD8 T lymphocyte subpopulations and an association with cytomegalovirus positivity in the elderly: the Swedish NONA-immune study. *Exp Gerontol.* 37: 445–53.
37. Wikby A, Johansson B, Ferguson F, Olsson J (1994). Age-related changes in immune parameters in a very old population of Swedish people: a longitudinal study. *Exp Gerontol.* 29: 531–41.
37a. Voehringer D, Koschella M, Pircher H (2002). Lack of proliferative capacity of human effector and memory T cells expressing killer cell lectin-like receptor G1 (KLRG1). Blood 100: 3698–702.
38. Zhou T, Edwards CK, Mountz JD (1995). Prevention of age-related T cell apoptosis defect in CD2-fas-transgenic mice. *J Exp Med.* 182: 129–37.
39. Hoshino T, Yamada A, Honda J, *et al.* (1993). Tissue-specific distribution and age-dependent increase of human CD11b+ T-Cells. *J Immunol.* 151: 2237–46.
40. Effros RB, Boucher N, Porter V, *et al.* (1994). Decline in CD28(+) T cells in centenarians and in long-term T cell cultures: a possible cause for both *in vivo* and *in vitro*/ immunosenescence. *Exp Gerontol.* 29: 601–9.
41. Fagnoni FF, Vescovini R, Mazzola M, *et al.* (1996). Expansion of cytotoxic CD8(+) CD28(-) T cells in healthy aging people, including centenarians. *Immunology* 88: 501–7.
42. Boucher N, DufeuDuchesne T, Vicaut E, Farge D, Effros RB, Schachter F (1998). CD28 expression in T cell aging and human longevity. *Exp Gerontol.* 33: 267–82.
43. Vallejo AN, Nestel AR, Schirmer M, Weyand CM, Goronzy JJ (1998). Aging-related deficiency of CD28 expression in CD4+ T cells is associated with the loss of gene-specific nuclear factor binding activity. *J Biol Chem.* 273: 8119–29.
44. Vallejo AN, Brandes JC, Weyand CM, Goronzy JJ (1999). Modulation of CD28 expression: Distinct regulatory pathways during activation and replicative senescence. *J Immunol.* 162: 657–9.
45. Fiorentini S, Malacarne F, Ricotta D, *et al.* (1999). Generation of CD28(-) cells from long-term-stimulated CD8(+)CD28(+) T cells: a possible mechanism accounting for the increased number of CD8(+)CD28(–) T cells in HIV-1-infected patients. *J Leukocyte Biol.* 65: 641–8.

46. Adibzadeh M, Pohla H, Rehbein A, Pawelec G (1995). Long-term culture of monoclonal human T lymphocytes: models for immunosenescence? *Mech Ageing Dev.* 83: 171–83.
47. Lio D, D'Anna C, Gervasi F, *et al.* (1998). Interleukin 12-release by mitogen-stimulated mononuclear cells in the elderly. *Mech Ageing Dev.* 102: 211–19.
48. Weyand CM, Brandes JC, Schmidt D, Fulbright JW, Goronzy JJ (1998). Functional properties of CD4+CD28- T cells in the aging immune system. *Mech Ageing Dev.* 102: 131–47.
49. FernandezGutierrez B, Jover JA, DeMiguel S, *et al.* (1999). Early lymphocyte activation in elderly humans: impaired T and T-dependent B cell responses. *Exp Gerontol.* 34: 217–29.
50. Hara H, Tanaka T, Negoro S, *et al.* (1988). Age-related changes of expression of IL-2 receptor subunits and kinetics of IL-2 internalization in T cells after mitogenic stimulation. *Mech Ageing Dev.* 45: 167–75.
51. Froelich CJ, Burkett JS, Guiffaut S, Kingsland R, Brauner D (1988). Phytohemagglutinin-induced proliferation by aged lymphocytes: reduced expression of high affinity interleukin 2 receptors and interleukin 2 secretion. *Life Sci.* 43: 1583–90.
52. Bernstein ED, Murasko DM (1998). Effect of age on cytokine production in humans. *Age* 21: 137–51.
53. Aggarwal S, Gupta S (1998). Increased apoptosis of T cell subsets in aging humans: Altered expression of Fas (CD95), Fas ligand, Bcl-2, and Bax. *J Immunol.* 160: 1627–37.
54. Gupta S (2000). Molecular and biochemical pathways of apoptosis in lymphocytes from aged humans. *Vaccine* 18: 1596–601.
55. Salmon M, Pilling D, Borthwick NJ, *et al.* (1994). The progressive differentiation of primed T cells is associated with an increasing susceptibility to apoptosis. *Eur J Immunol.* 24: 892–9.
56. Spaulding C, Guo W, Effros RB (1999). Resistance to apoptosis in human CD8+T cells that reach replicative senescence after multiple rounds of antigen-specific proliferation. *Exp Gerontol.* 34: 633–44.
57. Slagboom PE, Droog S, Boomsma DI (1994). Genetic determination of telomere size in humans: a twin study of three age groups. *Am J Hum Genet.* 55: 876–82.
58. Weng NP, Levine BL, June CH, Hodes RJ (1995). Human naive and memory T lymphocytes differ in telomeric length and replicative potential. *Proc Natl Acad Sci USA*, 92: 11091–4.
59. Frenck RW, Blackburn EH, Shannon KM (1998). The rate of telomere sequence loss in human leukocytes varies with age. *Proc Natl Acad Sci USA*, 95: 5607–10.
60. Bodnar AG, Kim NW, Effros RB, Chiu CP (1996). Mechanism of telomerase induction during T cell activation. *Exp Cell Res.* 228: 58–64.
61. Rufer N, Dragowska W, Thornbury G, Roosnek E, Lansdorp PM (1998). Telomere length dynamics in human lymphocyte subpopulations measured by flow cytometry. *Nat Biotechnol.* 16: 743–7.
62. Hiyama K, Hirai Y, Kyoizumi S, *et al.* (1995). Activation of telomerase in human lymphocytes and hematopoietic progenitor cells. *J Immunol.* 155: 3711–15.
63. Barnett YA, King CM (1995). An investigation of antioxidant status, DNA repair capacity and mutation as a function of age in humans. *Mutat Res.* 338: 115–28.
64. Frasca D, Barattini P, Tirindelli D, *et al.* (1999). Effect of age on DNA binding of the ku protein in irradiated human peripheral blood mononuclear cells (PBMC). *Exp Gerontol.* 34: 645–58.

65. Faasen AE, O'Leary JJ, Rodysill KJ, Bergh N, Hallgren HM (1989). Diminished heat shock protein synthesis following mitogen stimulation of lymphocytes from aged donors. *Exp Cell Res.* 183: 326–34.
66. Jurivich DA, Qiu L, Welk JF (1997). Attenuated stress responses in young and old human lymphocytes. *Mech Ageing Dev.* 94: 233–49.
67. Effros RB, Zhu XM, Walford RL (1994). Stress response of senescent T lymphocytes – reduced Hsp70 is independent of the proliferative block. *J Gerontol.* 49: B65–70.
68. Arbogast A, Boutet S, Phelouzat MA, Plastre O, Quadri R, Proust JJ (1999). Failure of T lymphocytes from elderly humans to enter the cell cycle is associated with low Cdk6 activity and impaired phosphorylation of Rb protein. *Cell Immunol.* 197: 46–54.
69. Tamir A, Miller RA (1999). Aging impairs induction of cyclin-dependent kinases and down-regulation of p27 in mouse CD4+cells. *Cell Immunol.* 198: 11–20.
70. Pawelec G, Wagner W, Adibzadeh M, Engel A (1999). T cell immunosenescence *in vitro* and *in vivo*. *Exp Gerontol.* 34: 419–29.
71. Wakikawa A, Utsuyama M, Hirokawa K (1997). Altered expression of various receptors on T cells in young and old mice after mitogenic stimulation: a flow cytometric analysis. *Mech Ageing Dev.* 94: 113–22.
72. Cakman I, Rohwer J, Schutz RM, Kirchner H, Rink L (1996). Dysregulation between TH1 and TH2 T cell subpopulations in the elderly. *Mech Ageing Dev.* 87: 197–209.
73. Castle S, Uyemura K, Wong W, Modlin R, Effros R (1997). Evidence of enhanced type 2 immune response and impaired upregulation of a type 1 response in frail elderly nursing home residents. *Mech Ageing Dev.* 94: 7–16.
74. Castle SC, Uyemura K, Crawford W, Wong W, Klaustermeyer WB, Makinodan T (1999). Age-related impaired proliferation of peripheral blood mononuclear cells is associated with an increase in both IL-10 and IL-12. *Exp Gerontol.* 34: 243–52.
75. Turner DR, Morley AA, Seshadri RS, Sorrell JR (1981). Age-related variations in human lymphocyte DNA. *Mech Ageing Dev.* 17: 305–9.
76. Hartwig M, Korner IJ (1987). Age-related changes in DNA unwinding and repair in human peripheral lymphocytes. *Mech Ageing Dev.* 39: 73–8.
77. Hyland P, Duggan O, Hipkiss A, Barnett C, Barnett Y (2000). The effects of carnosine on oxidative DNA damage levels and *in vitro* lifespan in human peripheral blood derived CD4+T cell clones. *Mech Ageing Dev.* 121: 203–15.
78. Lechner H, Amort M, Steger MM, Maczek C, Grubeckloebenstein B (1996). Regulation of CD95(APO-1). expression and the induction of apoptosis in human T cells: changes in old age. *Int Arch Allergy Immunol.* 110: 238–43.
79. Aggarwal S, Gollapudi S, Gupta S (1999). Increased TNF-alpha-induced apoptosis in lymphocytes from aged humans: changes in TNF-alpha receptor expression and activation of caspases. *J Immunol.* 162: 2154–61.
80. Soares MVD, Maini MK, Beverley PCL, Salmon M, Akbar AN (2000). Regulation of apoptosis and replicative senescence in CD8(+) T cells from patients with viral infections. *Biochem Soc Trans.* 28: 255–8.
81. McLeod JD (2000). Apoptotic capability in aging T cells. *Mech Ageing Dev.* 121: 151–9.
82. Hannet I, Erkeller-Yuksel F, Lydyard P, Deneys V, De Bruyere M (1992). Developmental and maturational changes in human blood lymphocyte subpopulations. *Immunol Today* 13: 215–18.
83. Utsuyama M, Hirokawa K, Kurashima C, *et al.* (1992). Differential age-change in the numbers of CD4+CD45RA+ and CD4+CD29+ T-cell subsets in human peripheral blood. *Mech Ageing Dev.* 63: 57–68.

Aging of Premature-Aging Syndrome Cells

Jan O. Nehlin

Novo Nordisk A/S, Novo Allé, 2880 Bagsværd, Denmark and Laboratory of Molecular Gerontology and Dermatology, Copenhagen University Hospital, Department 6311, 2100 Copenhagen, Denmark

Natural aging is regarded as the gradual deterioration of body functions throughout the lifetime of an individual. The aging process results from a contribution of genetic factors as well as environmental agents acting on the organism since the time of conception. Lifespan can be envisioned as an out come of multiple metabolic processes, each one of them acting simultaneously to facilitate an optimal body, tissue and cell function. In order to understand in more detail what metabolic processes are involved during the aging process, several models are being explored. One such model is the study of premature aging phenotypes. Insights into the molecular mechanisms leading to accelerated aging in progeria-like disorders have led to the identification and functional characterization of several genes thought to be associated with symptoms of accelerated aging. These findings are helping to understand the possible cellular processes at fault during natural aging. The best-characterized human premature aging disorder, to date, is the Werner syndrome, with many symptoms reminiscent of natural aging. Tremendous efforts have been undertaken, aiming to elucidate the biochemical role of the proteins involved in accelerated aging. This chapter will present a current vision of this approach to understand the aging process.

Introduction

Several models of aging have been used during the last decades to facilitate the description of the aging process at the organism, tissue and cellular levels. It is becoming increasingly evident that natural aging is the result of multiple processes acting at different levels and at different rates in the body. Lifespan is influenced by inherited factors i.e., polymorphisms as well as by environmental impact [1, 2]

Sunil C. Kaul and Renu Wadhwa (eds.), Aging of Cells In and Outside the Body, 101–119.

(Figure 1). In order to understand the individual contribution of each one of such processes it is rather important to dissect the role of those various molecular pathways or mechanisms. Thus, research work has led to the identification of genetic traits linked to particular loci whose deletion or overexpression results in diverse aging phenotypes. Most premature aging syndromes are very rare and recessive, and result from genetic defects in at least one gene. It is imperative to make distinctions between different types of premature aging. Besides classic progerias, many inherited diseases in humans, resemble in one or more conditions to that of senescence. Such disorders should be considered segmental progeroid syndromes, because they are early onset and because no one condition encompasses all of the common signs of aging that normal human people would exhibit during the natural aging process [1, 2]. One needs to differentiate between aging occurring at an accelerated speed and showing most of the signs of old age and, signs of old age restricted to only a few tissues or organs. As presented in the following section, a model of classic premature human aging that most closely resembles natural aging is the Werner syndrome.

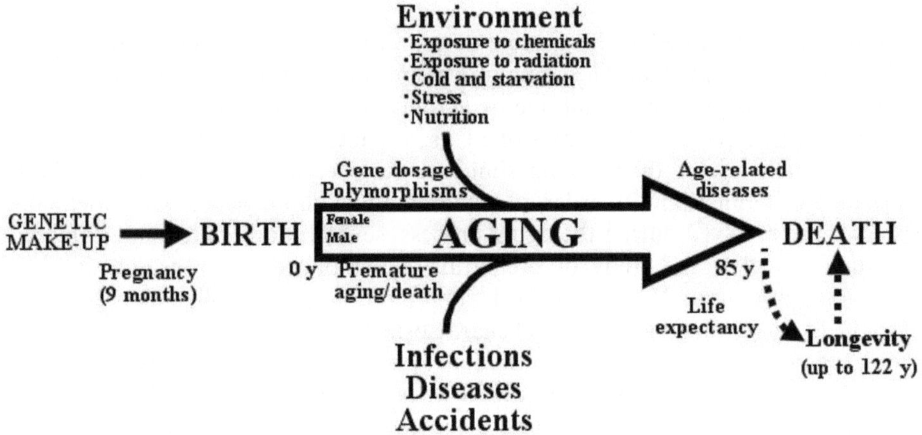

Figure 1. *Human lifespan. Successful longevity is the net effect of many different factors acting simultaneously at different times. The genetic make-up, environmental factors and stochastic events influence the length of life. Gene dysfunction and detrimental environmental factors lead to accelerated senescence, and hence a shorter lifespan.*

Normal replicating cells that age in culture undergo a finite number of population doublings after which they enter senescence, a terminally growth-arrested state, showing an altered phenotype, and expressing an age-associated β-galactosidase, which is considered a biomarker of aging [3.] Senescent cells remain in G_1 and are unable to enter S phase in response to mitogens. They are able to undergo Fas-mediated apoptosis showing chromatin condensation and dependence upon caspase activation, but are refractory to apoptosis in response to serum withdrawal [4.] As discussed below premature senescent cells share many of the molecular features found in senescent cells grown in culture (*in vitro* replicative senescence) [5]. Aging has evolved as a tumor-suppresive mechanism [3]. There is a striking correlation between physiological age and the probability of cancer. Gene variants (alleles) regulating fundamental biological processes ensure that the organism can reach at least a post-reproductive age. In premature aging syndromes, however, metabolic defects arise even before reproduction [2]. The aim of this chapter is to present a current view of the field of premature aging research and the molecular mechanisms thought to be responsible for accelerating aging.

Models of premature aging

Several model organisms have been used to study the genetic regulation of lifespan. During genetic screens, both short-lived and long-lived mutants have been identified. Most work has been focused on the characterization of long-lived gene mutants that would offer hints to understand what metabolic processes are necessary for lifespan extension. Important lessons have been obtained from the study of aging in model organisms such as yeast, roundworms, fruit flies and mice. The present revision will focus on the analysis of senescent mouse models and human premature aging syndromes, always bearing in mind that there are good examples of mechanisms regulating lifespan in lower eukaryotes that are analogous to those described in mammals.

Aging in yeast, roundworms and fruitflies

The budding yeast *Saccharomyces cerevisiae*, the roundworm *Caenorhabditis elegans* and the fruitfly *Drosophila melanogaster* have been used successfully to study the aging process. Yeast aging occurs when the mother cell exhausts gradually its potential to divide asymmetrically by giving rise to daughter cells. The end of yeast lifespan occurs when the mother cell cannot longer divide. In a way, yeast aging resembles, in principle, aging of replicating mammalian cells [6]. Several yeast mutants have been identified whose lifespan is dramatically shortened or lengthened [7, 8]. Such tools have enabled researchers to understand what pathways and mechanisms play a role in the yeast aging process, which could then serve as learning examples to study the aging process in higher organisms. Examples of mechanisms [7–10] involved in yeast lifespan include:

a) silencing of chromatin, where chromosomal regions become transcriptionally inactive

b) accumulation of extrachromosomal rDNA circles

c) link between carbon-source utilization and energy levels

d) retrograde regulation signaling mitochondrial dysfunction

e) mutations in DNA replication genes.

Studies on multicellular organisms, *C. elegans* and *Drosophila sp.* have allowed researchers to isolate a number of mutants that have been used to identify genes relevant to the aging process in these organisms. However, a major difference between these organisms and mammals, their post-mitotic nature, would mean that the aging process observed occurs post-mitotically and there is no involvement of exhaustion of replicative potential of the cells. These models are therefore good models to study the post-mitotic aging process and could be correlated with aging of human post-mitotic cells. Examples of mechanisms involved in the roundworm lifespan include:

a) somatic signaling through an insulin/insulin-like growth factor 1 endocrine system

b) sensory-transduction pathways in the nervous system that relay environmental signals

c) reproductive signaling where the germ-line coordinates the schedule of reproduction with the rate of aging

d) oxidative damage and the activity of antioxidant enzymes

e) a metabolic sensor such as *clk-1* whose role is to report to the nucleus on the metabolic state of the mitochondria.

f) several other single-gene mutations have been found to influence lifespan in *C. elegans* [6, 7, 11].

Similarly, studies on lifespan of *Drosophila* mutants have led to the identification of several single-gene mutations that confer longevity. These include the mutants with ability to resist different forms of stress such as starvation, high temperature and free radicals and thus have an increased likelihood to live longer. Examples include InR, where insulin-like hormones can trigger aging [12], Methuselah, Indy, Chico, and overexpression of antioxidant enzymes [11].

Mouse premature-aging models

Among mammals, the study of aging in mice has offered fundamental insights into the regulation of lifespan. Their relatively short lifespan ($\sim$4.5 y max.) and their well-established genetics, make them an ideal choice to study aging. Some mouse models of aging have been found serendipitously while others have been created as best-guess candidates. Several loss of function gene mutations have been characterized in mice that result in lifespan extension and several mice strains have been identified whose lifespan is much shorter than average. The latter are presented below:

Telomerase RNA-deficient mice

Progressive shortening of telomere length during replicative senescence is thought to be one key mechanism that leads to permanent cell cycle arrest. Telomeres at the chromosome ends are essential for chromosomal stability and cell viability. Telomerase is the enzyme responsible for synthesizing telomeric repeats. It is composed of a catalytic protein subunit TERT and a RNA component, TR. Third generation $mTR^{-/-}$ mice already show evidence of shortened telomeres but appear to be phenotypically normal. Mouse TR knock-outs in the 6th generation, show several symptoms that are reminiscent of aging including infertility, graying of the hair, alopecia (hair loss), skin lesions, impaired stress response, neoplasms, delayed wound healing and impaired hematopoesis. Their average lifespan was 18 months in the 6th generation [13]. In a different genetic background, generation 4 $mTR^{-/-}$ mice die at only 5 months of age and coincides with very short telomeres, sterility, splenic atrophy, reduced proliferative capacity of B and T cells, abnormal hematology and atrophy of the small intestine [14]. Thus, the loss of telomeres has been shown to correlate with loss of viability and whole-body aging.

p53 transgenic mice

p53 transgenic mice express the carboxy-terminal region of the tumor suppressor p53, and exhibit enhanced resistance to spontaneous tumors and display a premature onset of phenotypes associated with aging, including reduced lifespan, osteoporosis, organ atrophy and reduced tolerance to stress [15]. These findings provide a link between p53 activation, cellular senescence and organismal aging. Aging has evolved as a tumor suppressive mechanism and p53 is a key regulator in this process [16].

Ku-mutant mice

The Ku autoantigen is an abundant nuclear protein that consists of two subunits, Ku86 and Ku70. Ku binds to the double-stranded DNA ends, nicks/breaks and hairpin loops, and recruits DNA-dependent-Serine/Threonine protein kinase (DNA-PK). Ku and DNA-PK play essential roles in V(D)J recombination of immunoglobulin genes and double-strand break repair by non-homologous end joining. Activation of DNA-PK triggers a signal transduction cascade that activates DNA repair genes and arrests cell-cycle progression. Genetic ablation of Ku80 (also called Ku86) in mice, results in increased sensitivity to ionizing radiation, defective lymphocyte development, early onset of an age-related phenotype, and premature replicative senescence [17]. In addition, if p53 levels are reduced in $ku80^{-/-}$ cells, they become hypersensitive to gamma radiation and reactive oxygen species, and rescue the cells from premature senescence and allowing spontaneous immortalization [18]. These results suggest that $ku80^{-/-}$ cells accumulate double-strand breaks, in response to spontaneous DNA damage, which are recognized by a p53-mediated pathway that triggers senescence. Interestingly, expression of Ku and DNA-PK is downregulated during replicative senescence of normal human fibroblasts [19].

Klotho mutant mice

Klotho mice exhibit phenotypes that resemble human aging, including a short lifespan, growth retardation, infertility, arteriosclerosis, skin atrophy, ectopic calcification, premature thymic involution, osteoporosis and emphysema [20, 21]. The genetic defect leading to such phenotypes was due to an insertional mutation in the *klotho* gene (kl). Since kl/kl mice develop normally up to at least 3 weeks of age, the phenotypes observed cannot simply be a result of incomplete development. Thereafter, they almost stop growing and gradually die, with an average lifespan of 9 weeks [21]. The klotho gene is predominantly expressed in the kidneys and encodes secreted and cell membrane-bound proteins containing β-glucosidase-like domains. Moreover, *klotho* mutant mice almost lack white adipose tissue indicating less energy storage, they have increased glucose tolerance and insulin sensitivity and have less energy expenditure. These findings may explain the growth retardation observed in these mice [22]. Klotho may play a role in maintenance of energy homeostasis.

Growth-hormone transgenic, Snell, Ames and Laron dwarf mice

Overexpression of growth hormone in transgenic mice results in accelerated aging, as well as shortened lifespan, scoliosis, weight loss, decline of reproductive capacity, insulin resistance, hyperinsulinemia and histopathological changes. Fibroblasts derived from transgenic mice have a reduced replicative potential. Catalase activity is reduced in GH transgenics, suggesting that hormonal status modulates antioxidative mechanisms. Catalase is important to eliminate peroxides and its activity might influence lifespan [23]. Long-lived mice strains are worth mentioning here, since they reflect good models of what hormonal processes are involved in aging. Ames and Snell dwarf mice lack anterior pituitary cells producing growth hormone, prolactin and thyroid-stimulating hormone. They exhibit enhanced resistance to stress and lower temperature, suggesting that they could have a lower metabolic rate. Long-lived Laron mouse dwarfs, contain knockouts of growth factor receptor (GHBP) [24, 25].

Senescence-accelerated mice (SAMP)

Early-onset of age-related changes in outbred mice strains led to the characterization of at least 14 senescent-accelerated- mice-prone (SAMP) strains [26]. Aging criteria included shortened lifespan, loss of activity, hair loss, lordokyphosis, skin coarseness at 8 months of age, and incidence and degree of age-related disorders. The various aging phenotypes are observed throughout the different strains, being good examples of tissue-specific premature aging [2, 21, 26, 27]. Pathological phenotypes range from deficits in learning and memory to hearing impairment [26]. Fibroblasts derived from SAMP6 strains exhibit accelerated cellular senescence [28].

Other mouse models

Several mice outbred populations have been reported to show early signs of aging such as atherosclerosis susceptibility strains (Ath-1), mouse strains showing early onset autoimmune and kidney diseases (NZN/NZW), obese mice, etc. [2.] One example describes premature immunosenescence and early decline of proper functioning of the nervous system. Mice with higher emotional response to stress have a decreased immune function [29]. It is possible that immunological parameters are affected negatively by inappropriate hormonal signals in such mice.

Mouse models of human premature aging such as trichothiodystrophy (TTD), Cockayne's syndrome (CS) will be discussed in the human models section.

A splice variant from the mouse proto-oncogene SHC, involved in growth factor- and stress-signalling, was found to induce stress resistance and to prolong life. $p66^{shc}$

mutant mice have an extended lifespan due to resistance to apoptosis following oxidative stress [30]. It is possible that in wild type $p66^{shc}$ mice, the loss of cells due to stress-induced apoptosis could accelerate aging.

Human models

Progeroid disorders mimic aspects of the natural aging process but they also comprise features that normally would not occur during normal aging such as lack of gonadal activity and unusually short stature. Other genetic diseases that lead to developmental defects also could have phenotypes of aging, but they are more restricted to particular organs or tissues. Thus, progeroid disorders are exact models of accelerated aging [31]. Several premature aging syndromes involve genes whose function is involved with DNA repair. Deficiencies in DNA repair lead to excessive DNA damage, genetic instability, senescence or apoptosis [32].

RecQ protein family and premature aging syndromes

The RecQ gene family encodes highly conserved DNA helicases that fulfill important functions in DNA metabolism. RecQ members in humans include WRN, defective in Werner syndrome, BLM, defective in Bloom syndrome, RecQ4, defective in Rothmund-Thompson, RecQL and RecQ5b [33, 34].

Werner syndrome (WS)

WS is an autosomal recessive disease, wherein cells and tissues display features of accelerated aging. Symptoms include short stature, soft tissue calcification, laryngial atrophy, graying, loss of hair, skin ulcers of legs, neoplasms of mesenchymal origin, hyaluronic aciduria, atherosclerosis, hypogonadism, osteoporosis, cataracts, diabetes mellitus type II, hypertension and central nervous degeneration. It is considered an adult form of progeria. The average age at death is 47 years [35, 36]. This syndrome represents so far, one of the best models of premature aging since it mimics a wide range of signs that would eventually appear during the natural aging process.

At the cellular level, WS cells exhibit a range of abnormalities as described in Table 1.

Mutations or deletions in a gene encoding a DNA helicase were found in WS patients. The gene, termed WRN, encodes a 3′-5′ DNA helicase with a 3′-5′ exonuclease activity at its N-terminus. Mutations and deletions in the gene lead to an inability to target WRN to the nucleus, where it normally functions. WRN is able to interact with a number of different proteins. Such interactions take place during different cellular events and link the function of WRN protein to DNA metabolic processes such as DNA replication, DNA repair, DNA recombination, DNA repair, apoptosis, and telomere maintenance [36, 37]. Each new interaction identified has opened new clues towards defining the functional roles of the WRN protein.

The human Ku heterodimer associates with WRN and stimulates its exonuclease activity. The DNA-dependent protein kinase (DNA-PKcs) is able to bind to and phosphorylate WRN in a Ku-dependent manner, and thereby inhibit WRN exonuclease activity [38, 39]. Also, a telomere binding protein, TRF2, can stimulate WRN helicase activity [40]. Several mechanisms are required to repair double strand

Table 1 *. Molecular defects in Werner syndrome cells: a classical premature aging syndrome [35, 36]*

DNA replication
- Delayed
- Extended S-phase
- Non-cycling cells accumulate in G_1
- Premature replicative senescence

DNA recombination and genomic instability
- Elevated number of chromosomal rearrangements
- Elevated somatic mutation rate, and particularly deletions
- Accelerated telomere shortening
- Compromised DNA-ligation fidelity
- Hyperrecombination

DNA repair
- Elevated number of double strand breaks after X-ray irradiation
- Subtle transcription-coupled DNA repair defect
- Subtle telomeric repair defect
- Hypersensitivity to 4-nitroquinoline-N-oxide
- Hypersensitivity to DNA-crosslinking drugs
- Sensitivity to topoisomerase inhibitors
- Reduced repair after arrest/DNA damage

DNA transcription
- Deficient RNA pol II-dependent transcription
- Global and gene-specific transcription defects

Apoptosis
- Attenuated p53-mediated apoptosis

breaks including homologous recombination (HR), non-homologous-end-joining (NHEJ) and a RAD52-dependent pathway for maintaining telomere length [33]. WS cells display a homologous recombination resolution defect: WRN appears to resolve recombination products generated during DNA recombination, DNA repair, or replication restart, etc. in a RAD51-dependent HR pathway [41, 42] (see below).

The apparent lack of all symptoms of natural aging in WS patients is explained by compensatory mechanisms in some selected cell types [43].

Bloom syndrome (BS)

BS is a rare, autosomal recessive disorder that predisposes affected individuals to a wide range of cancers and genomic instability. BS patients also exhibit stunted growth, facial sun sensitivity, immunodeficiency and reduced fertility. Germline mutations in the BLM gene are responsible for BS. BLM, encodes a member of the RecQ family of DNA helicases [34]. It has recently been proposed that aberrant function of the non-homologous end-junction (NHEJ) pathway is responsible for the chromosomal instability characterized by elevated sister chromatid exchanges (SCEs), as well as chromosomal breaks, deletions and rearrangements observed in Bloom syndrome. BS cells display aberrant end-joining of double-strand breaks [44]. BLM protein associates with multiple proteins [34]. Even BLM and WRN proteins co-localize with each other at nuclear foci. BLM protein was found to inhibit the exonuclease activity of WRN [45]. BLM helicase is involved in recombinational repair via interaction with RAD51. Also, BLM is associated with ATM (gene defective in ataxia, see below). The ATM phosphorylation sites in BLM are important for radiosensitivity status in BS cells but not for SCE frequency [46]. BLM forms part of the BASC complex that seems to play a role as damage sensor. It has also been found that BLM can associate with MLH1, a protein required for mismatch repair. The function of BLM in this process appears to be in some aspect of genetic recombination [47]. p53 binds to BLM and WRN DNA helicases and attenuates their ability to unwind synthetic Holliday junctions *in vitro*. BLM and WRN function to suppress inappropriate homologous recombination during DNA replication. Following blockage of DNA replication, phosphorylated p53, BLM, and RAD51 colocalize in nuclear foci at sites likely to contain DNA replication intermediates in cells [42].

Rothmund-Thompson syndrome (RTS)

RTS is a rare, autosomal recessive, disorder with various features of accelerated aging. RTS is characterized by early onset of progressive poikiloderma. RTS patients show growth deficiency, photosensitivity with poikilodermatous skin changes, early graying and hair loss, juvenile cataracts, hypogonadism, bone defects and predisposition to cancer, especially osteogenic sarcomas. An association between RTS and mutations in the RECQL4 gene has been reported [48, 49]. RECQL4 encodes another DNA helicase member of the RecQ helicase family. Some RTS patients show diverse karyotypic abnormalities and some have reduced DNA repair [50].

Hutchinson-Gilford Progeria syndrome (HGPS)
HGPS is an extremely rare (1 in 4–8 million births), developmental disorder that exhibits a very rapid aging rate. Progeria patients show symptoms of aging, as those observed in older humans, at a very early age. They die at an average age of 13 years due mainly to cardiovascular disease. Patients show an increase in the production of urinary hyaluronic acid, early scleroderma-like cutaneous changes, several skin and hair abnormalities, baldness, loss of subcutaneous adipose tissue, osteporosis, wrinkled skin and short stature. The gene(s) deficient in HGPS remain still unknown. It is assumed that the syndrome is inherited in an autosomal recessive fashion in a subset of families. No karyotypic abnormalities have been detected [51] but there is some evidence suggesting accelerated telomere shortening in HGPS cells. Microarray analysis of progeria cells led to the identification of genes whose expression is associated with age-related phenotypes. A comparison of mRNA levels between progeria and normal young cells revealed a change in the expression of 76 genes. Genes involved in cell division and DNA or RNA synthesis are commonly down-regulated in progeria and normal old age, and whose deregulation could lead to increasing errors in the mitotic machinery of dividing cells [52].

Ataxia telangiectasia (AT)
Ataxia is a rare autosomal recessive syndrome characterized by progressive cerebellar ataxia, cancer predisposition, radiosensitivity and genome instability. The average lifespan is 40 years [2]. Cells derived from AT patients exhibit telomere shortening, premature senescence, and defects in the activation of cell cycle checkpoints in response to double strand breaks induced by ionizing radiation. Mutations in the ATM gene are responsible for such abnormalities. ATM encodes a phosphatidyl inositol kinase-like kinase with many targets. One substrate is the telomere binding protein TRF1/Pin2, which mediates ATM effects on telomere length and the DNA damage response [53]. Targeted disruption of the ATM gene results in growth retardation, neurological dysfunction, infertility, malignant thymic lymphoma, hypersensitivity to X-ray irradiation [54].

Cockayne syndrome (CS)
DNA damage caused by UV, bulky chemical adducts and oxidative damage, is repaired by the nucleotide excision repair (NER) mechanism. Xeroderma pigmentosum XPA-G genes are involved in NER, and among them, XPB and XPD, encode helicases whose deficiency can lead to CS and to a photosensitive form of the brittle hair disorder trichothiodistrophy (TTD) [55] (see below). Also, mutations in the human XPG gene could cause early onset CS [56]. CS and TTD cases exhibit postnatal growth failure, progressive neurological dysfunction, impaired sexual development, skeletal abnormalities, and a strongly reduced life expectancy [55]. CS also is characterized by sensitivity to sunlight and accelerated aging. CS afflicted individuals have an average lifespan of 50 years. CS has two complementation groups, CS-A and CS-B. Cells derived from CS patients have a defect in transcription-coupled DNA repair (TCR), which is a DNA repair mechanism that rapidly corrects certain DNA lesions in the transcribed strand of active genes. CS cells are

unable to remove 8-oxo-guanine in a transcribed sequence. Thus, CS results from defects in TCR of oxidative lesions [57]. The best-studied CS gene, CSB, encodes a Swi/Snf-like DNA-dependent ATPase. CSB protein is involved in cellular repair of 8-hydroxyguanine and 8-hydroxyadenine, lesions found in oxidatively damaged DNA [58]. CSB-deficient mice exhibit all of the CS repair characteristics: ultraviolet (UV) sensitivity, inactivation of TCR, unaffected global genome repair, and inability to resume RNA synthesis after UV exposure [59].

Trichothiodystrophy (TTD)

TTD patients show photosensitivity and sulphur-deficient brittle hair and nails, ichtyosis, they fail to thrive, leading to cachexia and susceptibility to infections, which is the most frequent cause of death. The repair-deficient form of trichothiodystrophy (TTD) most often results from mutations in the genes XPB or XPD, encoding helicases of the transcription/repair complex TFIIH, involved in NER (see CS section). The genetic defect in a third group, TTD-A, is unknown, but is also caused by dysfunctioning TFIIH. TTD-A appears to result from sublimiting amounts of TFIIH affecting its repair function [60].

Mice with mutations in XPD (TTD mice) develop premature aging features caused by DNA damage. Such mice have brittle hair that is normally pigmented, female mice lose fertility over time, they exhibit osteoclerosis, as well as osteoporosis and prominent kyphosis (curvature of the spin column) both of which are hallmarks of aging in humans. Thus, unrepaired DNA damage seems to compromise transcription, causing aging [55].

Down's syndrome (DS)

Individuals affected with Down's syndrome show a wide range of anatomical and biochemical abnormalities resulting of trisomy at chromosome 21. Forty percent of them are born with congenital heart defects, most have small brains and many are at increased risk for developing cataracts or other vision impairments. They have a lifespan of 40–70 years [2.] It is a rather common progeroid syndrome and shows pathologies typically found in older people. Features include glucose intolerance, diabetes mellitus, vascular disease, higher incidence of cancer, hair loss, degenerative cone disease, premature death, CNS impairment, retardation and accelerated appearance of neuritic plaques and neurofibrillary tangles found typically in Alzheimer's disease. Aberrant expression of Cu-Zn superoxide dismutase leads to deficient repair of oxidative damage in the mitochondrial DNA of DS patients [61]. The exact gene(s) involved remain at large.

Seip-Berardinelli syndrome (lipodistrophy)

Lipodistrophies are characterized by the loss of subcutaneous tissue, including insulin resistance, hyperlipidemia and diabetes, resulting in a very short lifespan [62]. Congenital generalized lipodystrophy, or Berardinelli-Seip syndrome (BSCL), is a rare autosomal recessive disease characterized by a near-absence of adipose tissue from birth or early infancy and severe insulin resistance. Other clinical and biological features include increased basal metabolic rate, acanthosis nigricans, hyperandrogenism, muscular hypertrophy, hepatomegaly, altered glucose tolerance or diabetes mellitus, and hypertriglyceridemia. A BSCL1 locus was mapped to 9q34 with evidence of heterogeneity. Another locus, BSCL2, at 11q13, contain mutations in the gene encoding a protein "seipin," homologous to the murine guanine nucleotide-binding protein (G protein), gamma3-linked gene (Gng3lg). BSCL2 might be involved in metabolic disturbances and diabetes [63].

Wiedemann-Rautenstrauch syndrome (NPS)
Wiedemann-Rautenstrauch (neonatal progeroid) syndrome, NPS, is a rare, autosomal recessive, condition with characteristic appearance of premature aging even at birth (old-looking face, natal teeth, and wrinkled skin). Other features of the syndrome are premature scleroderma, baldness, intrauterine growth retardation, pseudo-hydrocephalus, a tiny face and mouth, large hands and feet with long fingers and toes, generalized lipoatrophy with specific fat accumulation in the lateral suprabuttock region, macrocephaly (pseudo-hydrocephalus), hypotrichosis of the scalp hair, eyebrows, and eyelashes, prominent scalp veins, hypothyroidism, and mild to moderate mental retardation. Some cases also show abnormalities in their endocrine and lipid metabolism [64, 65, 66]. NPS is usually lethal by 7 months; however, on rare occasions, patients have survived into the teens. The genetic causes of this syndrome remain unclear. The telomere length in cultured fibroblasts obtained from a 16-year-old NPS patient did not show shortened telomeres as compared to that of normal fibroblasts [66].

Dyskeratosis congenital (DKC)
DKC patients live between 16 and 50 years, and exhibit pulmonary disease, graying and loss of hair, osteoporosis, increase incidence of cancers, etc. DKC cells have a defect in cellular proliferative capacity. The gene affected in DKC is known as dyskerin, a nucleolar protein required for normal processing of RNA. In DKC cells, telomere length is abnormally short and telomerase activity low, suggesting that a telomere component is involved in aging [67].

Progeroid Ehlers-Danlos syndrome (E-D)
Patients with E-D syndrome show an aged appearance, developmental delay, dwarfism, defective wound healing, hypotonic muscles, loose but elastic skin and other abnormalities. Mutations in the gene encoding galactosyl transferase I (XGalT-1), involved in the synthesis of common linkage regions of proteoglycans, cause E-D. Proteoglycans such as glycosaminoglycans are essential in the regulation of cell proliferation, tissue development, organogenesis, and infections [68].

Huntington's chorea (HC)
It is an autosomal dominant disease leading to the premature deterioration of neurons on the caudate and putamen, leading to memory loss and dementia. The disease appears first at mid-life and causes death within 15 years. Peripheral signs of aging are also recognized including diabetes mellitus, brittle bones, weight loss, and pale, wrinkled skin [2]. HC is caused by mutations in the gene Huntingtin [69].

Premature ovarian failure (POF)
Female humans have a finite number of oocytes within primordial follicles. As they age, an atretic degeneration process takes place, whereby the oocyte stock is exhausted, leading to menopause, normally by the fifth decade. However, a number of individuals have been diagnosed with POF, which represents a unique example of rapid organ senescence [70].

Glucose-6-phosphate dehydrogenase (G6PD) deficiency
G6PD is a key regulatory enzyme in the hexose monophosphate shunt and generates NAPDH necessary for reductive biosynthesis and maintenance of the cellular redox status. Defects in G6PD affects more than 200 million people worldwide with various degrees of hemolysis. Human foreskin fibroblasts that were deficient in G6PD showed retarded growth and accelerated cellular senescence. The similarity in the rates of telomere shortening in G6PD-deficient and normal cells suggested that signals other than telomere loss trigger the senescence program in G6PD-deficient cells. It is possible that G6PD deficiency leads to an increase production of free radicals. The resulting oxidative stress and reduced antioxidant capacity may cause damage to be accumulated at an increased rate, leading to the onset of premature senescence. A deficiency in G6PD activity results in a lower NADPH/NADP+ ratio and a diminished glutathione level [71].

Other clinically relevant premature aging disorders
The human homologue of mouse *Klotho* also has been described. Mice with defective *Klotho* exhibit a phenotype of accelerated aging (see above). Alleles of human *KLOTHO* have been characterized, and it was found that a variant of *KLOTHO* in homozygous elderly individuals, contributes to the onset and severity of human age-related phenotypes. Thus, human aging could be influenced by changes in the trafficking and catalytic activity of functional variants of Klotho [72].

Polycystic kidney disease (PKD) may be a useful model for the age-degeneration of kidney function [2]. A patient described as showing symptoms of premature aging, including cataracts, diabetes mellitus, osteoporosis, growth deficiency, was described. None of the other described premature syndromes seems to overlap precisely with the symptoms observed in this case, suggesting that a novel premature aging syndrome might have been found [73].

Cushing's syndrome, caused by an excess production of the stress hormone cortisol, exhibits many signs and symptoms reminiscent of aging, including weakening muscles and neuropsychiatric disorders [2].

An aged-skin appearance is found in many genetic disorders (see above) and also in diseases such as Grotton's acrogeria, De Barsy syndrome, etc.

Gene/protein-induced premature senescence in human cells
Premature senescence in human cells growing in culture can be induced by a variety of ways. Gene deregulation can often result in changes of cellular homeostasis, cell cycle arrest, apoptosis and cancer. Activation of certain oncogenes such as Ras, Raf, or Mek in human fibroblasts triggers a state of permanent cell cycle arrest that

resembles replicative senescence. It is thought that senescence is induced to prevent tumorigenesis, and relies on the CDK inhibitors $p21^{Cip1/WAF1/Sdi1}$, $p16^{INK4a}$, and the activity of p53 and Rb tumor suppressor proteins. Ras-induced senescence occurs in the absence of telomere shortening and shows hallmarks of replicative senescence including senescence-associated β-galactosidase activity, flat and enlarged morphology and the up-regulation of extracellular matrix proteins [74, 75]. The PML isoform IV nuclear protein [76], caveolin-1 [77], etc. are other examples of genes whose expression can induce premature senescence. Interestingly, the human SIR2 protein, homologous to yeast Sir2, a modulator of yeast lifespan, can antagonize PML/p53-induced cellular senescence [78]. Human endothelial cells, grown on glycated collagen, express hallmarks of premature senescence, but without attrition of telomeres or decrease in telomerase activity. Early aging of the vascular endothelium could lead to diabetic vasculopathy [79].

Stress-induced premature senescence (SIPS)
Human cells display a senescent-like phenotype (SIPS) if they become exposed to non-cytotoxic oxidative stress such as hyperoxia, UV light, radioactivity, hydrogen peroxide, etc. Many biomarkers of senescence are found in such cells including altered cell morphology, irreversible growth arrest, overexpression of $p21^{Cip1/WAF1/Sdi1}$, $p16^{INK4a}$, decrease in DNA synthesis and increased senescent-β-galactosidase activity. Molecular analyses of cells exposed to subcytotoxic stress indicate that premature senescence can be triggered by a pRB-mediated release of transforming growth factor β1 (TGF-β1) [5]. SIPS could be induced by pathways alternative to telomere shortening [80].

Concluding remarks
Much information has been gained in the past few years regarding the involvement of particular genes in metabolic pathways whose deregulation could eventually lead to symptoms of premature aging. We now have a better understanding about how natural aging possibly can arise from the study of the causes that lead to symptoms of accelerated aging in several genetic diseases. Accelerated aging arises from defects in DNA repair and DNA recombination, DNA transcription, apoptosis, events leading to excessive telomere shortening, deregulation of tumor suppressor activity i.e., p53-family of proteins [15, 81], overexpression of proteins with anti-proliferative activities $p16^{INK4a}$, $p27^{Kip1}$ and $p2^{Cip1/WAF1/Sdi1}$, oxidative damage, etc. Germinal homozygous mutations in a gerontogene can lead to phenotypes of rapid aging in the whole body, especially if expression of the normal gene is ubiquitous to all tissues. If the normal gene is only expressed in a given tissue, then it is expected that defects in its expression would lead to phenotypic changes in only that tissue. Homozygous mutations affecting both copies of a single gene is more detrimental than heterozygous mutations.

One could well argue that certain genes whose mutation or deletion in mice or humans, in a genetic model of premature aging, could have an important role in the normal aging process, but one needs to recognize whether such role is restricted to the tissue level per se or to the whole organism. It has been reported that the genes associated so far with premature aging are not expressed in all tissues. Therefore, one cannot conclude necessarily that a gene whose mutation leads to some phenotypes of aging is involved in any way with aging of tissues or organs, where its expression is absent. This notion supports the idea of different aging processes acting in parallel, in different tissues. No single gene deficiency in an animal model can recapitulate all of the phenotypes normally observed in natural aging, since aging is a collection of different symptoms caused by different effectors, at various tissue levels. Thus, segmental progeroid disorders would be of help, in particular with respect to the study of tissue-specific age-related symptoms. In retrospective, the closest example of a gene whose loss of function could lead to many age-related symptoms is the WRN gene, causative of Werner syndrome.

Many of the cellular features and aging-like body symptoms observed in premature aging syndromes are due to germinal gene defects. It is likely, though, that in some genetic conditions, rapid aging is exacerbated by metabolic changes that arise because of genomic instability leading to other gene defects, which must therefore be considered indirect causes. Since such changes might occur at random during a long period of time, any single patient afflicted by a premature-aging disorder may not show an identical pathology to another given patient. Whether some of the gene defects and their resulting metabolic changes described for premature aging syndromes might occur somatically during normal aging is a matter of intense research, and some evidence towards their role in natural aging is beginning to emerge [72].

The major challenge ahead of us, in the field of premature aging, is to be able to extrapolate the above findings to search for analogies within the process of natural aging, that is, to be able to correlate the rate of aging of a normal person to given changes in the function of genes whose deficiency in the germinal cell lineage can lead to premature aging. The possibility that some changes also could happen in single somatic cells, that would lead them to become senescent cells, awaits the improvement of methodologies to isolate single senescent cells from living tissues, to characterize their genotype and to figure out what pathway of aging led to their senescent state. Perhaps, one day, it would be possible to determine which one of the pathways of aging is going to affect more a given individual, and thus, design intervention therapies aiming at de-accelerating the effects of gerontogenes, whether they act at the whole body level or at the tissue level.

Acknowledgments

I thank the Danish Medical Research Council, Prof. Vilhelm Bohr and Prof. Gunhild Lange Skovgaard for support.

References

1. Martin GM (1997). Genetics and pathobiology of ageing. *Phil Trans R Soc Lond B* 352: 1773–80.

2. Finch CE (1994). *Longevity, Senescence, and the Genome.* Chicago: The University of Chicago Press.
3. Campisi J (2000). Cancer, aging and cellular senescence. *In Vivo* 14: 183–8.
4. Tepper CG, Seldon MF, Mudryj M (2000). Fas-mediated apoptosis of proliferating, transiently growth-arrested, and senescent normal human fibroblasts. *Exp Cell Res.* 260: 9–19.
5. Frippiat C, Chen QM, Zdanov S, Magalhaes J-P, Remacle J, Toussaint O (2001). Subcytotoxic H_2O_2 stress triggers a release of transforming growth factor-β1, which induces biomarkers of cellular senescence of human diploid fibroblasts. *J Biol Chem.* 276: 2531–7.
6. Tissenbaum HA, Guarente L (2002). Model organisms as a guide to mammalian aging. *Dev Cell* 1: 9–19.
7. Guarente L, Kenyon C (2000). Genetic pathways that regulate ageing in model organisms. *Nature* 408: 255–62.
8. McVey M, Kaeberlein M, Tissenbaum HA, Guarente L (2001). The short lifespan of *Saccharomyces cerevisiae sgs1* and *srs2* mutants is a composite of normal aging processes and mitotic arrest due to defective recombination. *Genetics* 157: 1531–42.
9. Defossez P-A, Lin S-J, McNabb DS (2001). Sound silencing: the Sir2 protein and cellular senescence. *BioEssays* 23: 327–32.
10. Mays-Hoopes LL, Budd M, Choe W, Weitao T, Campbell JL (2002). Mutations in DNA replication genes reduce yeast lifespan. *Mol Cell Biol.* 22: 4136–46.
11. Arking R (1998). *Biology of Aging*, 2nd edn. Sunderland, Massachusetts: Sinauer Associates.
12. Tatar M, Kopelman A, Epstein D, Tu M-P, Yin C-M, Garofalo RS (2001). A mutant *Drosophila* insulin receptor homolog that extends life-span and impairs neuroendocrine function. *Science* 292: 107–10.
13. Rudolph KL, Chang S, Lee H-W, *et al.* (1999). Longevity, stress response, and cancer in aging telomerase-deficient mice. *Cell* 96: 701–12.
14. Herrera E, Samper E, Marin-Caballero J, Flores JM, Lee H-W, Blasco MA (1999). Disease states associated with telomerase deficiency appear earlier in mice with short telomeres. *EMBO J.* 18: 2950–60.
15. Tyner SD, Venkatachalam S, Choi J, *et al.* (2002). p53 mutant mice that display early ageing-associated phenotypes. *Nature* 415: 45–53.
16. Campisi J (2002). Between Scylla andCharybdis: p53 links tumor suppression and aging. *Mech Ageing Dev.* 123: 567–73.
17. Vogel H, Lim DS, Karsenty G, Finegold M, Hasty P (1999). Deletion of Ku86 causes early onset of senescence in mice. *Proc Natl Acad Sci USA* 19: 10770–5.
18. Lim D-S, Vogel H, Willerford DM, Sands AT, Platt KA, Hasty P (2000). Analysis of ku80-mutant mice and cells with deficient levels of p53. *Mol Cell Biol.* 20: 3772–80.
19. Salminen A, Helenius M, Lahtinen T, *et al.* (1997). Down-regulation of Ku autoantigen, DNA-dependent protein kinase, and poly(ADP-ribose) polymerase during cellular senescence. *iochem Biophys Res Comm,.* 238: 712–16.
20. Kuro-o M, Matsumura Y, Aizawa H, *et al.* (1997). Mutation of the mouse *klotho* gene leads to a syndrome resembling ageing. *Nature* 390: 45–51.
21. Kuro-o M (2001). Disease model: human aging. *Trends Mol Med.* 7: 179–81.
22. Mori K, Yahata K, Mukoyama M, *et al.* (2000). Disruption of *klotho* gene causes an abnormal energy homeostasis in mice. *Biochem Biophys Res Comm.* 278: 665–70.
23. Brown-Borg HM, Rakoczy SG (2000). Catalase expression in delayed and premature aging mouse models. *Exp Gerontol.* 35: 199–212.

24. Bartke A, Coshigano K, Kopchick J, *et al.* (2001). Genes that prolong life: relationships of growth hormone and growth to aging and lifespan. *J Gerontol.* 56A: B340–9.
25. Gems D, Partridge L (2001). Insulin/IGF signaling and ageing: seeing the bigger picture. *Curr Opin Genet Dev.* 11: 287–92.
26. Takeda T (1999). Senescence-accelerated mouse (SAM): a biogerontological resource in aging research. *Neurobiol Aging* 20: 105–10.
27. Yabuki A, Suzuki S, Matsumoto M, Nishinakagawa H (2002). Lifespan, renal morphological characterization of the SAMP1//Ka mouse. *Exp Anim.* 51: 75–81.
28. Lecka-Czernik B, Moerman EJ, Reis RJ, Lipschitz DA (1997). Cellular and molecular biomarkers indicate precocious *in vitro* senescence in fibroblasts from SAMP6 mice. Evidence supporting a murine model of premature senescence and osteopenia. *J Gerontol A Biol Sci Med.* 52: B331–6.
29. Guayerbas N, Puerto M, Victor VM, Miguel J, De la Fuente M (2002). Leukocyte function and lifespan in a murine model of premature immunosenescence. *Exp Gerontol.* 37: 249–56.
30. Migliaccio E, Giorgio M, Mele S, Pelicci G, Reboldi P, Pandolfi PP, Lanfrancone L, Pelicci PG (1999). The $p66^{shc}$ adaptor protein controls oxidative stress response and lifespan in mammals. *Nature* 402: 309–13.
31. Martin GM, Oshima J (2000). Lessons from human progeroid syndromes. *Nature* 408: 263–66.
32. Bernstein C, Bernstein H, Payne CM, Garewal H (2002). DNA repair/pro-apoptotic dual-role proteins in five major DNA repair pathways: fail-safe protection against carcinogenesis. *Mutat Res.* 511: 145–78.
33. Wu L, Hickson IA (2001). DNA ends RecQ-uire attention. *Science* 292: 229–30.
34. Mohaghegh P, Hickson I (2002). Premature aging in RecQ helicase-deficient human syndromes. *Int J Biochem Cell Biol.* 34: 1496–501.
35. Nehlin JO, Lange Skovgaard G, Bohr VA (2000). The Werner syndrome: a model to study human aging. *Ann NY Acad Sci.* 908: 167–79.
36. Brosh RM Jr., Bohr VA (2002). Roles of the Werner syndrome protein in pathways required for maintenance of genome stability. *Exp Gerontol.* 37: 491–506.
37. Shen JC, Loeb L (2001). Unwinding the molecular basis of the Werner syndrome. *Mech Ageing Dev.* 122: 921–44.
38. Karmakar P, Piotrowski J, Brosh RM Jr, *et al.* (2002). Werner protein is a target of DNA-dependent protein kinase *in vivo* and *in vitro* and its catalytic activities are regulated by phosphorylation. *J Biol Chem.* 277: 18291–302.
39. Li B, Comai L (2002). Displacement of DNA-PKcs from DNA ends by the Werner syndrome protein. *Nucleic Acids Res.* 30: 3653–61.
40. Opresko PL, Von Kobbe C, Laine JP, Harrigan J, Hickson ID, Bohr VA (2002). Telomere binding protein TRF2 binds to and stimulates the Werner and Bloom syndrome helicases. *J Biol Chem.* 277: 41110–9.
41. Saintigny Y, Makienko K, Swanson C, Emond MJ, Monnat, RJ, Jr. (2002). Homologous recombination resolution defect in Werner syndrome. *Mol Cell Biol.* 22: 6971–8.
42. Yang Q, Zhang R, Wang XW, *et al.* (2002). The processing of Holliday junctions by BLM and WRN helicases is regulated by p53. *J Biol Chem.* 277: 31980–7.
43. Ostler EL, Wallis CV, Sheerin AN, Faragher RGA (2002). A model for the phenotypic presentation of Werner's syndrome. *Exp Gerontol.* 37: 285–92.
44. Gaymes TJ, North PS, Brady N, Hickson ID, Mufti GJ, Rassool FV (2002). Increased error-prone non homologous DNA end-joining – a proposed mechanism of chromosomal instability in Bloom's syndrome. *Oncogene* 21: 2525–33.

45. von Kobbe C, Karmakar P, Dawut L, *et al.* (2002). Colocalization, physical, and functional interaction between Werner and Bloom syndrome proteins. *J Biol Chem.* 277: 22035–44.
46. Beamish H, Kedar P, Kaneko H, *et al.* (2002). Functional link between BLM defective in Bloom's syndrome and the ataxia-telangiectasia-mutated protein, ATM. *J Biol Chem.* 277: 30515–23.
47. Pedrazzi G, Perrera C, Blaser H, *et al.* (2002). Direct association of Bloom's syndrome gene product with the human mismatch repair protein MLH1. *Nucleic Acids Res.* 29: 4378–86.
48. Kitao S, Shimamoto A, Goto M, *et al.* (1999). Mutations in RECQL4 cause a subset of cases of Rothmind-Thomson syndrome. *Nat Genet.* 22: 82–4.
49. Lindor NM, Furuichi Y, Kitao S, Shimamoto A, Arndt C, Jalal S (2000). Rothmund-Thomson syndrome due to RECQ4 helicase mutations: report and clinical and molecular comparisons with Bloom syndrome and Werner syndrome. *Am J Med Gen.* 90: 223–8.
50. Durand F, Castorina P, Morant C, Delobel B, Barouk E, Modiano P (2002). Rothmund-Thomson syndrome, trisomy 8 mosaicism and RECQ4 gene mutation. *Ann Dermatol Venereol.* 129: 892–5.
51. Uitto J (2002). Searching for clues to premature aging. *Trends Mol Med.* 8: 155–7.
52. Ly DH, Lockhart DJ, Lerner RA, Schultz PG (2000). Mitotic misregulation and human aging. *Science* 287: 2486–92.
53. Kishi S and Lu KP (2002). A critical role for Pin2/TRF1 in ATM-dependent regulation. *J Biol Chem.* 277: 7420–9.
54. Barlow C, Hirotsune S, Paylor R, *et al.* (1996). Atm-deficient mice: a paradigm of ataxia telangiectasia. *Cell.* 86: 159–71.
55. de Boer J, Andressoo JO, de Wit J, *et al.* (2002). Premature aging in mice deficient in DNA repair and transcription. *Science* 296: 1276–9.
56. Lee SK, Yu SL, Prakash L, Prakash S (2002). Requirement of yeast RAD2, a homolog of human XPG gene, for efficient RNA polymerase II transcription. implications for Cockayne syndrome. *Cell* 109: 823–34.
57. Le Page F, Kwoh EE, Avrutskaya A, *et al.* (2000). Transcription-coupled repair of 8-oxoguanine: requirement for XPG, TFIIH, and CSB and implications for Cockayne syndrome. *Cell* 101: 159–71.
58. Tuo J, Jaruga P, Rodriguez H, Dizdaroglu M, Bohr VA (2002). The Cockayne syndrome group B gene product is involved in cellular repair of 8-hydroxyadenine in DNA. *J Biol Chem.* 277: 30832–7.
59. van der Horst GT, van Steeg H, Berg RJ, *et al.* (1997). Defective transcription-coupled repair in Cockayne syndrome B mice is associated with skin cancer predisposition. *Cell* 89: 425–35.
60. Vermeulen W, Bergmann E, Auriol J, *et al.* (2000). Sublimiting concentration of TFIIH transcription/DNA repair factor causes TTD-A trichothiodystrophy disorder. *Nat Genet.* 26: 307–13.
61. Druzhyna N, Nair RG, LeDoux SP, Wilson GL (1998). Defective repair of oxidative damage in mitochondrial DNA in Down's syndrome. *Mutat Res.* 409: 81–9.
62. Bhayana S, Hegele RA (2002). The molecular basis of genetic lipodystrophies. *Clin Biochem.* 35: 171–7.
63. Magre J, Delepine M, Khallouf E, *et al.* (2001). Identification of the gene altered in Berardinelli-Seip congenital lipodystrophy on chromosome 11q13. *Nature Genet.* 28: 365–70.

64. Pivnick EK, Angle B, Kaufman RA, *et al.* (2002). Neonatal progeroid (Wiedemann-Rautenstrauch) syndrome: report of five new cases and review. *Am J Med Genet.* 90: 131–40.
65. Hoppen T, Hausser I, Theile U, Ferrari R, Muller W, Rister M (2000). Neonatal progeroid syndrome (Wiedemann-Rautenstrauch syndrome): case report and review of the literature. *Klin Padiatr.* 212: 71–6.
66. Korniszewski L, Nowak R, Okninska-Hoffmann E, Skorka A, Gieruszczak-Bialek D, Sawadro-Rochowska M (2001). Wiedemann-Rautenstrauch (neonatal progeroid) syndrome: new case with normal telomere length in skin fibroblasts. *Am J Med Genet.* 103: 144–8.
67. Marciniak RA, Johnson FB, Guarente L (2000). Dyskeratosis congenita, telomeres and human ageing. *TIG* 16: 193–5.
68. Okajima T, Fukumoto S, Furukawa K, Urano T, Furukawa K (1999). Molecular basis for the progeroid variant of Ehlers-Danlos syndrome. *J Biol Chem.* 274: 28841–4.
69. Zuccato C, Ciammola A, Rigamonti D, *et al.* (2001). Loss of Huntingtin-mediated BDNF gene transcription in Huntington's disease. *Science* 293: 493–8.
70. Pal L, Santoro N (2002). Premature ovarian failure (POF): discordance between somatic and reproductive aging. *Ageing Res Rev.* 1: 413–23.
71. Ho H-Y, Cheng M-L, Lu F-J, *et al.* (2000). Enhanced oxidative stress and accelerated cellulkar senescence in glucose-6-phosphate dehydrogenase (G6PD)-deficient human fibroblasts. *Free Rad Biol Med.* 29: 156–69.
72. Arking AE, Krebsova A, Macek M Sr, *et al.* (2002). Association of human aging with a functional variant of klotho. *Proc Natl Acad Sci USA* 99: 856–61.
73. Okamoto N, Satomura K, Hatsukawa Y, *et al.* (1997). Premature aging syndrome with osteosarcoma, cataracts, diabetes mellitus, osteoporosis, eythroid macrocytosis, severe growth and developmental deficiency. *Am J Med Gen.* 69: 169–70.
74. Barradas M, Gonos ES, Zebedee Z, *et al.* (2002). Identification of a candidate tumor-suppressor gene specifically activated during Ras-induced senescence. *Exp Cell Res.* 273: 127–37.
75. Peeper DS, Shvarts A, Brummelkamp T, *et al.* (2002). A functional screen identifies hDRIL1 as an oncogene that rescues RAS-induced senescence. *Nature Cell Biol.* 4: 148–53.
76. Bischof O, Kirsh O, Pearson M, Itahana K, Pelicci PG, Dejean A (2002). Deconstructing PML-induced premature senescence. *EMBO J.* 21: 3358–69.
77. Volonte D, Zhang K, Lisanti MP, Galbiati F (2002). Expression of caveolin-1 induces premature cellular senescence in primary cultures of murine fibroblasts. *Mol Biol Cell* 13: 2502–17.
78. Langley E, Pearson M, Faretta M, *et al.* (2002). Human SIR2 deacetylates p53 and antagonizes PML/p53-induced cellular senescence. *EMBO J.* 21: 2383–96.
79. Chen J, Brodsky SV, Goligorsky DM, *et al.* (2002). Glycated collagen I induces premature senescence-like phenotypic changes in endothelial cells. *Circ Res.* 90: 1290–8.
80. Magalhaes JP, Chainiaux F, Remacle J, Toussaint O (2002). Stress-induced premature senescence in BJ and hTERT-BJ1 human foreskin fibroblasts. *FEBS Lett.* 523: 157–62.
81. Djelloul S, Tarunina M, Barnouin K, Mackay A, Jat PS (2002). Differential protein expression, DNA binding and interaction with SV40 large tumour antigen implicate the p63-family of proteins in replicative senescence. *Oncogene* 21: 981–9.

The p53 Pathway, Cancer and Aging

Antony W. Braithwaite and Sara J. Edwards
Pathology Department, Dunedin School of Medicine, University of Otago, Box 913, Dunedin, New Zealand

Introduction

The p53 tumor suppressor is one of the most intensely studied proteins in biomedical science. The reason for this is that mutations in the p53 gene itself, or defects in its regulation, are probably the most common genetic markers of all human cancers [1]. Despite the fact that tumor etiologies differ widely, this evidence shows that there are common pathways in tumor development. A key step in tumor development is the conversion of a cell from one with a limited lifespan to one that is immortalized and inactivation of p53 function appears to be an important part of this immortalization process. This immediately suggests that p53 may be important in the control of cellular senescence. Indeed, the evidence suggests that this is so, although there is clearly at least one other senescence pathway not involving p53 [2, 3].

In this chapter we first review the area of p53 biology, addressing the issues of how p53 functions, how it is controlled and how it regulates its responsiveness to various input signals. Following this, we discuss the role p53 plays in cellular senescence and then we address the issue of a role for p53 in aging at the level of the organism. The relationship between senescence and cancer is discussed throughout. This review is not intended to be a comprehensive analysis of each area but attempts to give an overview of p53 function and how this is linked with cancer and cellular aging. Excellent recent reviews on p53 function [4–6] and cellular senescence [2, 7, 8] are available.

The function of the p53 tumor suppressor

The p53 protein is a nuclear transcription factor, although there is evidence that it carries out activities not involving transcription. It is generally modular in structure having DNA binding (DBD) and acidic transactivator (TA) domains (Figure 1). To

Sunil C. Kaul and Renu Wadhwa (eds.), Aging of Cells In and Outside the Body, 121–149.

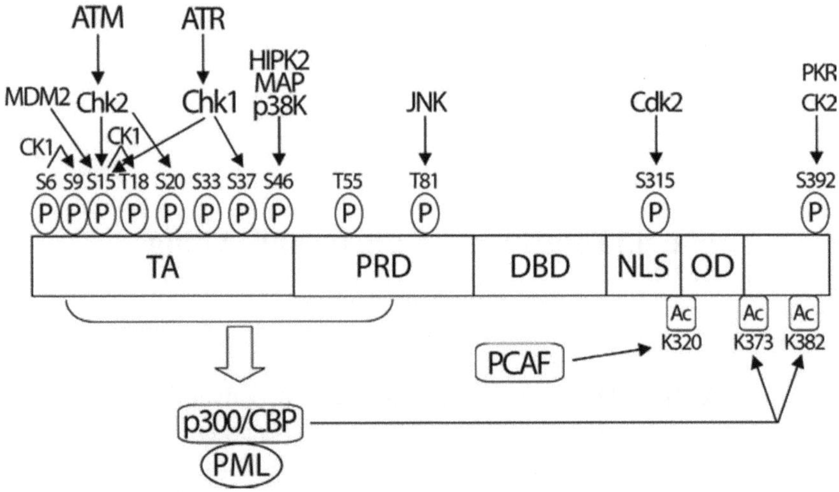

Figure 1. Structure of the p53 protein. A schematic diagram of the p53 protein showing the functional domains. TA = transactivation domain, PRD = proline rich domain, DBD = DNA binding domain, NLS = nuclear localization signal, OD = oligomerization domain. Sites of post-translational modification through phosphorylation and acetylation are indicated and described in the text. Not to scale. CK1 = casein kinase 1; CK2 = casein kinase 2; Chk1,2 = checkpoint kinase 1,2; ATM = ataxia telangiectasia mutant kinase; ATR = AT-related kinase; HIPK2 = home-odomain interacting protein kinase 2; MAP = mitogen activated kinase; JNK = Jun N-terminal kinase; Cdk2 = cyclin dependent kinase; PKR = RNA activated kinase; p300/CBP = CREB binding protein; PML = Promyelocytic leukaemia; PCAF = p300/CBP associated factor.

transcriptionally activate genes, p53 forms a homo-tetramer through its oligomerization domain (OD) [9] in the C-terminus (Figure 1), which then allows it to efficiently bind specific sequences in the promoters of a considerable number of genes [10, 11]. A number of these genes are listed in Table 1, but more than 150 candidate genes have been reported to be transactivated by p53 [12].

No doubt more will be added to the list as array technology becomes more common-place. The precise details of how p53 stimulates transcription are not clear, but p53 has been shown to interact with several common transcription factors associated with the pre-initiation complex, such as the TATA-box binding protein (TBP) and Sp1 [13, 14], so it probably stabilizes transcription complexes on promoters. In addition, p53 forms complexes with other transcriptional regulators, such as acetyl transferases including the CREB binding protein (p300/CBP) [15]. This binding is important to allow the acetylation of the histone proteins surrounding

Table 1. *Genes regulated by p53*

	Reference
Activated by p53	
p21 WAF1/CIP1	[22]
MDM2	[186]
GADD45	[187]
cyclin G	[188, 189]
BAX	[26]
IGF-BP3	[190]
PIG-3	[191]
14-3-3σ	[24]
NOXA	[32]
APAF1	[27–29]
KILLER/DR5	[192]
PUMA	[30, 31]
TSP (thrombospondin 1)	[193]
FAS (APO1)	[194]
Cathepsin D	[195]
caveolin	[160]
FDXR	[196]
p53AIP1	[33]
p53DINP	[197]
PERP	[198]
PIDD	[199]
PTEN	[200]
WIP1	[201]
BTG2	[202]
p53R2	[203, 204]
GD-AIF	[205]
BAI1	[206]
MMP2	[207]
MASPIN	[208]
KAI1	[209]
TP73	[210]
PCNA	[211]
Repressed by p53	
MAP4	[212]
PRESENILIN	[213]
VEGF	[214]
Bcl-2	[215]

p53 binding sites which may permit p53 access to the basal transcription machinery necessary for gene activation [16]. As well as activating transcription, p53 also represses transcription of a small, although growing, number of cellular gene promoters (Table 1) and several viral promoters [17–19]. Transcriptional repression may occur by p53 directly binding transcription factors off the promoter, thereby preventing them from binding to their cognate response element [13, 14]. More recent experiments suggest that transcriptional repression may also involve an inhibition of histone deacetylase activity needed for chromatin accessibility [20]. Repression mainly occurs with promoters that do not have a p53-binding site although again more recent data suggest that repression may be mediated by novel p53 binding sites [21]. The biological significance of transcriptional repression of most viral promoters is not known.

Although p53 transcriptionally regulates a diverse range of genes, for the most part these genes are associated with the control of cell division and cell survival (Table 1), although p53 can also regulate expression of some genes involved in DNA repair. In general p53 activates (or represses) genes to cause an arrest of cell cycle progression or apoptosis. The most ubiquitous of the growth arrest genes is $p21^{WAF1/CIP1}$, the protein product of which inhibits the phosphorylation of the pRb105 protein thereby inducing arrest of cell cycle progression at the G_1/S phase border [22, 23]. p53 may also inhibit cell cycle progression at the G_2/M border by increasing transcription of the *14-3-3σ* gene [24], the product of which binds to the cyclin B/cdc2 complex, preventing its translocation to the nucleus [25]. Although, as indicated, p53 can activate many other genes to cause growth arrest, many experimental observations may be explained by activation of the $p21^{WAF1/CIP1}$ and/or 14-3-3σ proteins.

p53 also induces a variety of genes that promote cell death or apoptosis, such as *BAX*, *APAF1* (apoptosis protease-activating factor 1), *PUMA* (p53 upregulated modulator of apoptosis), *p53AIP1* (p53-regulated apoptosis inducing protein 1) and *NOXA* [26–33]. With the exception of the p53AIP1 gene product, these proteins all contain the Bcl-2 homology domain BH3 which has been shown to be required for cell killing [34, 35]. During apoptosis, these proteins are found at the mitochondria and are involved in triggering the caspase cascade [reviewed in ref. 36]. Of these, BAX was the first pro-apoptotic target of p53 discovered [26, 37], and was isolated by its ability to bind Bcl-2 [38]. In normal tissues, the BAX protein is localized in the cytosol but is specifically translocated to the mitochondria in response to apoptotic stress [39]. At the mitochondria BAX, together with other pro-apoptotic proteins, triggers cytochrome C release, although the precise mechanism by which this occurs is unclear. Another BH3 domain protein is NOXA which has been shown to be activated in a p53-dependent manner in response to X-ray irradiation [32]. NOXA has also been shown to be localized in the mitochondria during the cell's apoptotic response. Although the role of all of these proteins in apoptosis remains to be determined, they are clearly involved at some level in triggering the caspase cascade [36, 40].

How p53 "decides" to induce cell cycle arrest or apoptosis is still far from clear. One possibility is that the promoters of pro-apoptotic genes have lower affinity p53 binding sites than the promoters of genes that regulate the cell cycle or DNA repair.

Thus, they would only be activated at higher p53 concentrations and usually only after growth arrest genes. There is in fact some experimental support for this model [41–43]. However, this cannot explain all observations as, for example, the promoter of the apoptotic gene *PUMA* has a similar affinity for p53 as the cell cycle arrest genes $p21^{WAF1/CIP1}$ and *MDM2* [44]. Moreover, the human p53 mutant p53(AD1-) [45] induces apoptosis nearly as well as wild type p53 but is unable to induce cell cycle arrest or activate $p21^{WAF1/CIP1}$. We have obtained similar results with a mouse p53 mutant deleted for residues 14–43 [45a]. Both mutants are no more stable than wild type p53. This would not be possible if cell cycle arrest genes had to be induced first.

Recent and compelling work suggests that transactivation of apoptosis genes may require co-factor proteins. The ASPP family proteins bind directly with p53 [46] and ASPP1 and 2 specifically enhance the interaction of p53 with the promoters of the apoptotic genes [47]. Conversely, the inhibition of *ASPP* expression selectively blocks p53 dependent apoptosis. How the association of p53 and ASPP proteins is itself regulated is not known, but it may involve post-translational modification of key amino acid residues in the proline rich domain (PRD) of p53 [48] (Figure 1). It has also been shown that p53 is unable to induce apoptosis in cells that are genetically deficient in either of the p53 family members, p63 or p73, and this failure of cell death is associated with a failure of p53 to bind the promoters of apoptotic genes [49]. As indicated above, p53 also binds acetyltransferases such as p300/CBP and PCAF (p300/CBP associated factor) which may be important in promoter discrimination [50–52; reviewed in ref. 6]. Although these interactions would generally enhance p53 dependent transactivation, it appears some may be able to confer promoter specificity. JMY (Junction-Mediating regulatory protein), for example, associates with p300/CBP to enhance the ability of p53 to activate the *BAX* promoter whilst having no effect on the $p21^{WAF1/CIP1}$ promoter [15]. Other important p53 co-factors are members of the PML family. The PML gene is fused with the retinoic acid receptor α as a result of the t(15:17) translocation in patients with acute promyelocytic leukemia [53]. All PML isoforms bind p53 and at least PML IV confers promoter specificity on p53 [54–57]. The interactions with PML appear to be very important in senescence induction [55, 58]. Thus, the binding of protein co-factors may be important in determining whether p53 induces cell cycle arrest or apoptosis.

In addition to transactivating apoptosis genes, p53 can in some circumstances, for example in response to hypoxic stress, induce apoptosis by non-transcriptional means [59, 60]. This appears to involve a translocation of p53 to the mitochondrial membrane [61, 62] and facilitation of cytochrome C release. This results in the activation of caspases 3 and 9. The mechanism of cytochrome C release is not clear, but over-expression of the anti-apoptotic protein Bcl-2 can inhibit this process, suggesting that p53 can interfere with Bcl-2 function. The non-transcriptional induction of apoptosis by p53 has been recently reviewed [63].

In summary, although a great deal is known about how p53 carries out its various biological activities, precisely how it works is still not clear. Nonetheless, it is clearly a most important protein crucially involved in the regulation of life and death of a cell. Because of this, p53 presents a potentially serious problem for a cell. p53 protein levels must be therefore very tightly regulated.

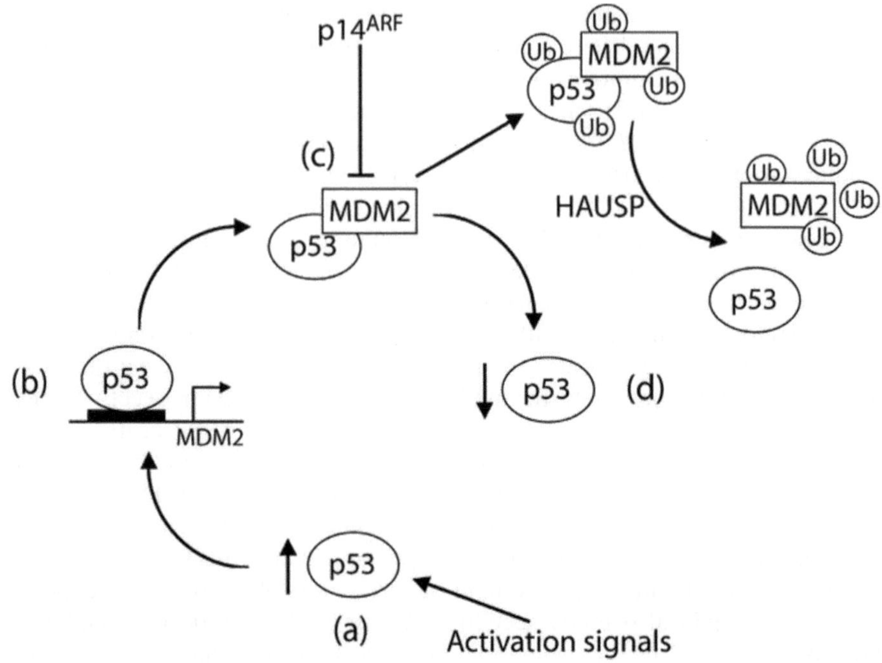

Figure 2. The p53/MDM2 autoregulatory loop. (a) As p53 levels rise above a threshold p53 protein molecules bind to their cognate site in the MDM2 promoter (b). p53 then acts as a transcription factor leading to MDM2 protein production. (c) MDM2 protein then binds p53 and catalyses its ubiquitination and therefore degradation, leading to a lowering of p53 protein level (d). In response to stress-induced modifications, the actions of p14^{ARF} and HAUSP, the loop is disrupted.

p53 protein levels are stringently regulated

Although p53 is constitutively expressed in all cells, its level is kept very low by another protein known as MDM2 [64]. MDM2 is an E3 ubiquitin ligase [65–68] that binds p53 and by catalyzing the addition of ubiquitin molecules to p53, targets it for degradation via the ubiquitin pathway. The *MDM2* gene however is also expressed at low levels, but can be transactivated by p53 as it contains a p53-responsive element in its promoter [64, 69]. Thus, as p53 levels rise, the *MDM2* gene is transactivated. MDM2 protein concomitantly rises which then degrades p53. This in turn results in a reduction in *MDM2* transcription. Thus, MDM2 and p53 exist in an auto-regulatory feedback loop (Figure 2), maintaining p53 at a level that enables cells to survive and to progress through the cell cycle.

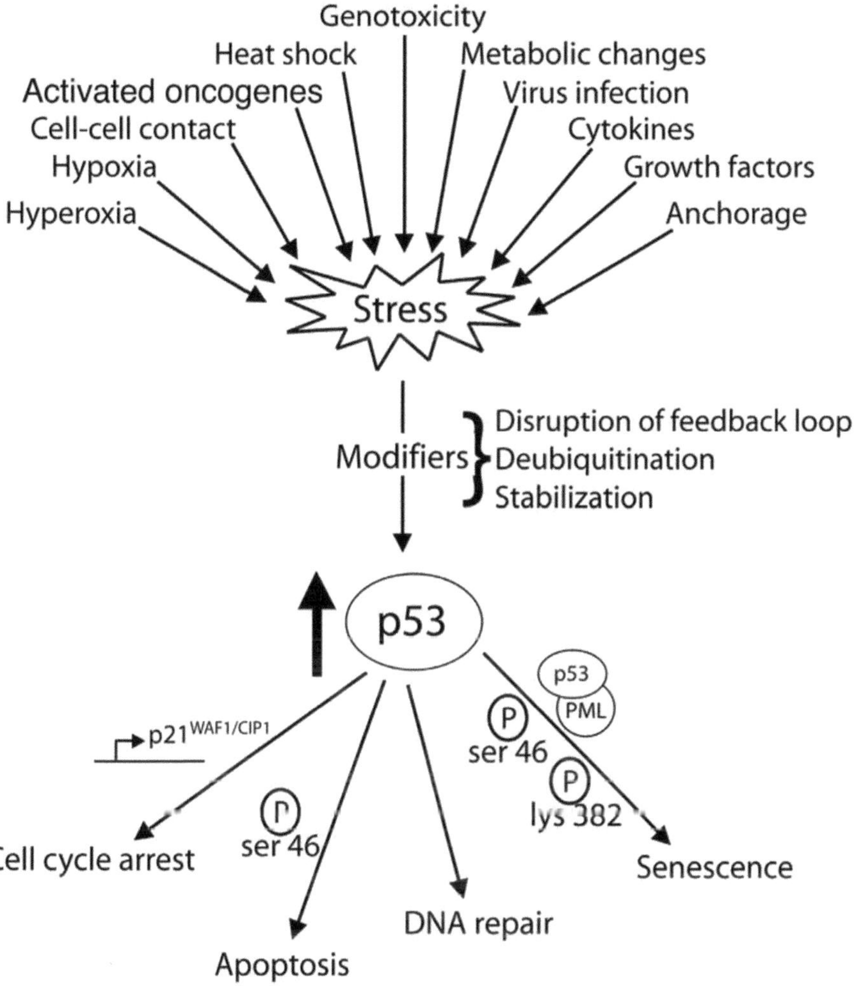

Figure 3. The p53 response to stress. In response to the indicated stresses, p53 protein levels rise due to stabilization, as described in the text. This allows p53 to efficiently function as a transactivator of its various downstream target genes.

p53 protein is stabilized in response to many cellular stress signals

Initially it was thought that p53 only responded to the stress of DNA damage [70], but now it is known to respond to a wide variety of cellular stresses (Figure 3) [reviewed in ref. 4]. Many of these stresses disrupt the auto-regulatory loop, which leads to a rise in the level of p53 protein. This rise then enables p53 to efficiently transactivate its various down-stream target genes.

There appear to be two distinct pathways that disrupt the p53/MDM2 regulatory loop. One involves covalent modification of one or both proteins and the other involves the action of another tumor suppressor protein known as $p14^{ARF}$. In addition, p53 function is blocked or altered by direct interaction with other proteins.

Covalent modification

The most well-studied covalent modification of p53 is phosphorylation although recent evidence suggests that acetylation at specific residues is also important. Phosphorylation has been reported to occur at 11 different sites in human p53 whilst other modifications occur at another 7 sites [reviewed in ref. 6]. These sites are predominantly located in the N- and C- terminals associated with important regulatory domains of p53 (Figure 1). A number of protein kinases have been shown to phosphorylate p53 (Figure 1) [6].

An attractive and elegant model is that phosphorylation (and other modifications) of specific residues represents a "signature" of the upstream activating signals for p53 which, at least in part, determines p53's biological response. However, strong evidence for this model is still lacking. Indeed, most stresses show overlapping patterns of phosphorylation [reviewed in ref. 6]. One of the clearer examples of signal specific phosphorylation comes from studies of DNA damage induced by ionizing radiation and UV. Ionizing radiation induces phosphorylation of p53 at serine 15 that requires Checkpoint kinase 2 (Chk2) and ataxia telangiectasia mutant (ATM) kinase [71, 72]. UV induces phosphorylation at serines 15 and 37, which are dependent on Chk1 and AT related (ATR) kinase [73, 74] and Chk1 also specifically phosphorylates p53 at serine 20 [73, 75]. However, UV also induces phosphorylation at 5 other N-terminal serines and 2 threonine residues and in the C-terminus at serine 392 [reviewed in ref. 6]. Phosphorylation at serine 15 and threonine 18 interfere with the interaction of MDM2 with p53 as they are within the MDM2 binding domain [76–78]. Thus phosphorylation at these sites causes p53 levels to rise.

An interesting example of what may be promoter specific activation by phosphorylation has been reported. The apoptosis inducing gene *p53AIP1* has been reported to be transactivated by p53 in a serine 46 dependent manner [33], a site phosphorylated in response to UV by mitogen activated (MAP) kinase [79] and homeodomain interacting protein (HIP) kinase 2 [80, 81]. Mutation of serine 46 abolished activation of *p53AIP1* and inhibited p53 dependent apoptosis in response to DNA damaging agents, but did not abolish the ability of p53 to induce transcription of the $p21^{WAF1/CIP1}$, *MDM2* or other p53 target genes [33]. Phosphorylation at this site by MAP kinase has also been shown to cause p53 stabilization [79] although the mechanism of stabilization is not clear. Similarly, Jun N-terminal kinase (JNK) phosphorylates p53 at threonine 81 in response to DNA damage and other stresses [82] and this too leads to p53 protein stabilization and activation of a transcriptional response. Once again, however, the mechanism of stabilization is not clear.

At the C terminus cyclin B dependent kinase (p34 cdc2) phosphorylates p53 at serine 315 [83] and RNA activated kinase (PKR) phosphorylates p53 at serine 392 [84]. These modifications also lead to enhanced p53 transactivation ability and an increase in protein level. Of particular interest, particularly in the context of cellular senescence is acetylation. p300/CBP and PCAF are responsible for acetylation at lysines 320, 373, and 382 [51, 85]. Ionizing radiation and UV stimulate acetylation at lysine 382 [85, 86] and this correlates with transcriptional activation of p53 [87]. p300/CBP binds at the N-terminus of p53 and probably requires phosphorylation of serine 15 [88–90]. It may therefore be competitive with the MDM2/p53 interaction thereby contributing to p53 stabilization.

Another recently described mechanism by which p53 levels can rise is through the action of the newly discovered protein HAUSP (herpes-virus associated ubiquitin-specific protease) [91]. This protein has been shown to strongly stabilize p53 even in the presence of excess MDM2, through the specific deubiquitination of p53 (Figure 2). This suggests a mechanism by which p53 levels can rapidly rise in a cell simply by deubiquitinating those p53 molecules already tagged for degradation.

In addition to direct post-translational modification of p53, the regulators of p53 may also be modified, affecting their interactions with p53. This may be illustrated by MDM2 which is phosphorylated by ATM kinase in response to DNA damage [92]. This also affects p53 stability, as does the sumoylation of MDM2 [93]. A detailed discussion of the effects of modifications to p53 regulatory proteins is, however, outside the scope of this review, but such observations serve to illustrate that the regulation of the p53 response is very complex.

The p14^{ARF} pathway

p14^{ARF} is one of two protein products encoded by the INK4A locus, the other being p16^{INK4A} [94], a negative regulator of pRb105. p14^{ARF} levels rise when pRb105 becomes hyperphosphorylated, or inactivated by the binding of other proteins (e.g., adenovirus E1a [95]), which results in the release of transcription factors such as E2F. E2F transcriptionally activates *p14^{ARF}* [96]. p14^{ARF} binds to MDM2 and inhibits its E3 ubiquitin ligase activity (Figure 2) [97] thereby preventing the MDM2 mediated degradation of p53 [97–101]. p53 levels can then rise. Thus, p14^{ARF} probably links many signals that impact on pRb105 function to p53 and is therefore a crucial regulator of p53 responsiveness. It should be noted, however, that p53 can be stabilized in cells not expressing p14^{ARF}, so there has to be at least one other mechanism [102–104]. There is also evidence that p53 may down-regulate p14^{ARF} [100], analogous to the p53/MDM2 autoregulatory loop, although as the p53 target gene *p21$^{WAF1/CIP1}$* activates pRb105, which would result in a reduction in p14^{ARF}, this would appear to be redundant. Furthermore, recent results from our laboratory have shown that p53 does not always inhibit p14^{ARF} expression [102].

Non-covalent protein-protein interactions

A hallmark of p53 is its ability to interact with other viral and cellular proteins, of which there are many [reviewed in ref. 4]. Some of these have been discussed above, which result in covalent modifications to p53, but some appear to alter p53 function by virtue of complex formation alone. These include replication protein A [105]; the tumor suppressor proteins WT1 [106] and Brca1 [107]; the DNA recombination/repair proteins Rad51 [108] and TFIIH [109]; the redox repair protein Ref-1 [110];

the cold shock protein YB1 [111]; and others. Once again, a detailed discussion of the physiological consequences of these interactions is beyond the scope of this review.

Co-ordination of the p53 response

It is clear from the above that p53 can respond to many different intracellular and extracellular signals and must coordinate an appropriate response. Although there is still much uncertainty about how p53 does this, the nature of the p53 response must depend on what p53 is "told" to do by the various post-translational modifications, signaling through the pRb105 pathway, the availability of co-factors and interactions with other proteins. Thus, at any point in time, some p53 molecules must be responding to some upstream signal. That is, its level of expression and activity must be in constant fluctuation, and only when one particular signal dominates, or there are enough smaller but multiple signals, are there sufficient p53 molecules "activated" (recruited to promoters), that a profound biological response ensues (arrest of the cell cycle or apoptosis). This model is outlined in Figure 4. Thus, the p53 response is likely to be stochastic. This may be rather like nerve cell induction of muscle contraction. Sodium channels are constantly opening and closing in the axonal membrane, giving rise to miniature endplate potentials. Only when a threshold of receptor engagement by neurotransmitter is exceeded, is there a coordinated depolarization of the cell membrane resulting in an action potential. Furthermore, only when this occurs in a sufficient number of neurons does it lead to muscle contraction.

If p53 responds in a stochastic manner to stresses, there is no need to think about it having a latent state waiting to be activated upon stress, because it is constantly monitoring the cell for DNA damage and other stresses and constantly responding to these stresses. In this way, the cell is always protected from accumulating genetic lesions that may contribute to the development of a malignant phenotype. This view of p53 functioning also explains why it has been difficult to define specific stress induced modifications to p53. If p53 does respond in a stochastic manner to multiple signals, a proportion of p53 molecules will be always be activated but probably in different ways.

Cellular senescence

Although what characterizes senescence is discussed in detail throughout this series, it seems appropriate to at least provide a working definition of senescence and to outline some of the signals that induce senescence. Only in this way can a role for p53 be properly understood.

Senescence was first described for cultured fibroblasts [112] and was referring to "replicative senescence" in which there is a decline in the replicative capacity of cells in culture (Figure 5A), with characteristic changes in cell morphology. Most studies have in fact focused on fibroblasts, but there are a number of reports describing replicative senescence in other cell types [reviewed in refs. 2, 7, 8]. The reason for replicative senescence is now widely regarded as the result of progressive shortening of telomeres, due to the problem of replicating the ends of linear DNA. Senescence

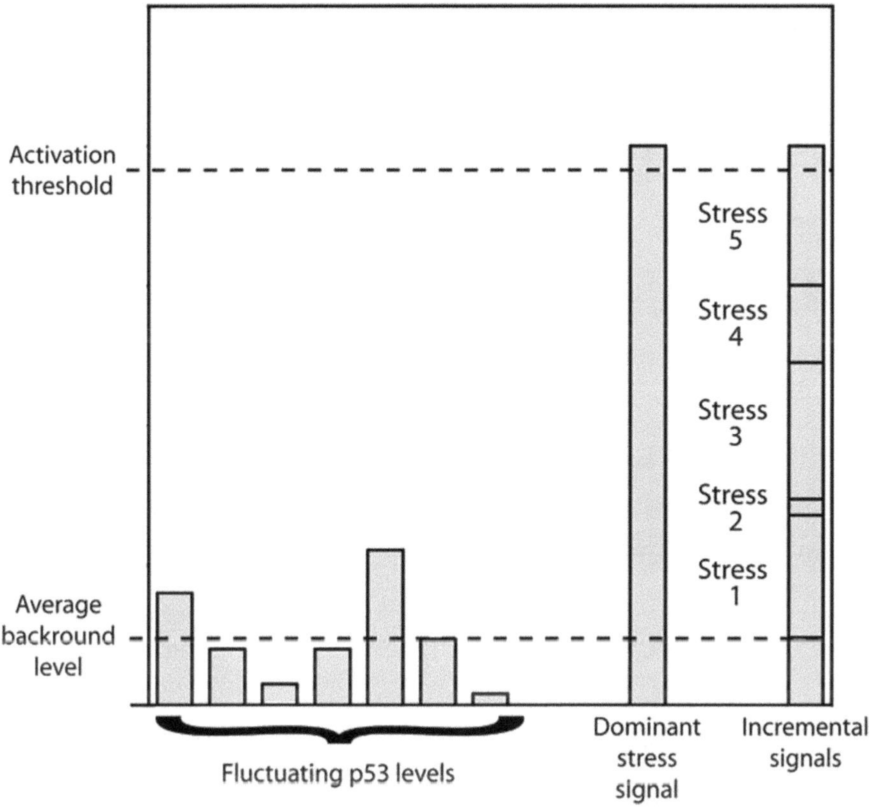

***Figure 4**. A stochastic model for p53 activation. p53 levels constantly fluctuate, controlled by the p53/MDM2 feedback loop. In response to a dominant stress signal, p53 levels are raised above the threshold necessary for activation of p53 function. A similar response may be generated by several small incremental signals, each of which raises the p53 level slightly.*

occurs when critically short telomeres have been attained and the chromosomes are no longer capable of replication, although as is no doubt discussed elsewhere, senescence occurs well before massive telomere loss occurs.

In addition to short telomere induced senescence, a number of other stimuli induce cellular senescence. DNA damage resulting from double strand breaks or oxidation of nucleotide bases [113, 114; reviewed in ref. 7], excessive mitogenic stimulation from for example, MAP kinase [115] and over-expression of activated oncogenes, such as *RAS* and *RAF* [116–118], all induce cellular senescence. Short telomere induced senescence is likely therefore to be just one example of a more general phenomenon of senescence induced by a variety of external signals.

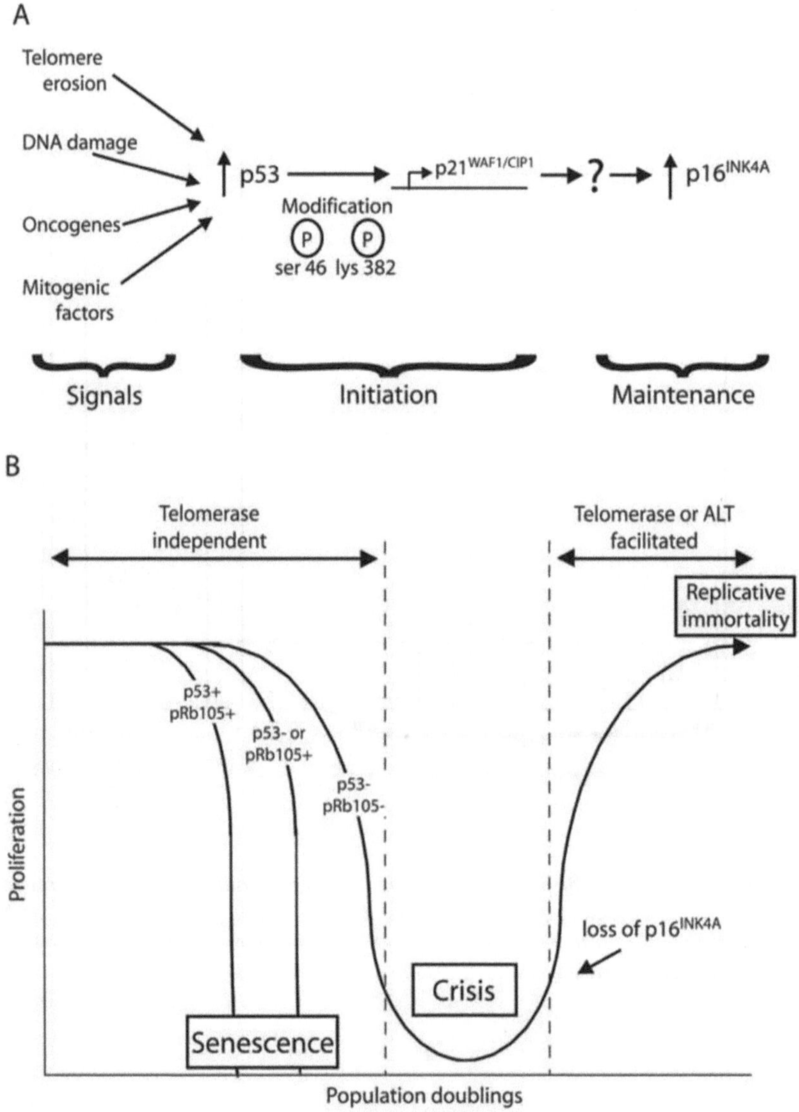

__Figure__ 5. The role of p53 in senescence. A: In response to stress signals, p53 protein is modified, resulting in stabilization and an increase in protein levels. $p21^{WAF1/CIP1}$ is then activated to initiate senescence, which is then maintained by the expression of $p16^{INK4A}$ through an unknown pathway. B: Cells with wild type p53 and pRb105 undergo senescence after a small number of population doublings. This time-span is extended by the loss of either p53 or pRb105, and even further by loss of both proteins. Even then, cells eventually enter "crisis" with a few cells then losing $p16^{INK4A}$, reactivating telomerase (or ALT) and becoming immortalized.

p53 and cellular senescence

Given the above background, one might reasonably guess that p53 does play a role in cellular senescence, as the classes of signals that can lead to senescence are a subset of the signals known to activate p53 [4] (Figure 3). Telomere shortening may simply be seen as a response by p53 to a DNA strand break. Telomeres end in a large duplex loop structure (t-loop) [119]. T-loop formation depends on TRF2, a protein that binds telomeres, and over-expression of a dominant negative TRF2 induces cell death in an ATM/p53 dependent manner [120]. Thus, ATM may well signal telomere abnormalities to p53. In addition to this, several proteins known to be involved in DNA repair, such as the MRE11/RAD50/NBS1 complex and DNA-dependent protein kinase (DNA-PK) are known to bind to telomeres [121, 122]. DNA-PK is made up of three subunits, one of which, Ku86, is bound to TRF1 at the telomere where it protects against end to end chromosome fusion [123]. Although this is perhaps a tenuous link to p53 at this time, it nonetheless places telomeric interactions into the same context as p53, i.e., that of a DNA damage response.

We can also infer that p53 is important in regulating senescence from various lines of evidence examining the early events in oncogenesis, both *in vivo* and *in vitro*. For example, mice engineered for complete loss of the *p53* gene overcome lifespan barriers and as a consequence, develop many different tumors at an early age and mostly die within a few months of birth [124]. And, as was indicated at the beginning of this chapter, over half of all human tumors have functionally inactivating mutations in the *p53* gene [1]. Moreover, transfection of mutant p53 can immortalize cells in culture [125] and cells immortalized by radiation treatment contain p53 mutations [126]. Also, transforming viruses, such as Simian Virus (SV) 40 and human papillomavirus (HPV) in general must inactivate one or both of p53 and pRb105 for cell transformation to occur [127–131]. To do this they have evolved proteins that neutralize p53 activity. Thus it appears that a necessary step in immortalization is the loss of p53 and pRb105 functions [132]. However, loss of p53 alone is not sufficient to immortalize cells [reviewed in refs. 2, 7, 8]. Loss of p53 extends the lifespan of human cells in culture and additional loss of pRb105 extends the lifespan of cells even further (Figure 5B). However, cells without p53 eventually enter a senescent-like state that is distinguishable from true senescence and cells without either p53 or pRb105 enter an unstable phase (crisis) in which most cells eventually die. Subsequently, rare survivors expand into colonies that (theoretically) have an infinite lifespan (Fig 5B). Thus immortalization requires other genetic changes in addition to loss of p53 and pRb105 function. These invariably appear to be loss of the cyclin dependent kinase inhibitor $p16^{INK4A}$ [132–134] and reactivation of telomerase [135; reviewed in ref. 7] or in rare cases, induction of ALT (alternative lengthening of telomeres) [136]. Nonetheless, the fact that p53 loss is required for lifespan extension, implies that p53 plays an important role in the regulation of senescence.

In addition to studies with viruses, a range of other studies in the last few years has shown that p53 plays a direct role in controlling senescence in cultured cells. Such studies include the observations that mutant p53 and inhibition of p53 function with antibodies specific to p53 can both cause loss of senescence [137, 138]. Furthermore, p21$^{WAF1/CIP1}$ expression increases as human diploid cells approach senescence [2] and its highest levels are achieved in senescent cells [139]. Consistent with these observations, senescence appears to be associated with a "switch-on" of the transactivator function of p53 [140, 141]. More recently, controlled expression of p53 using cells containing inducible p53 expression constructs has shown that p53 can directly induce a senescent phenotype [142]. These (and other) data provide compelling evidence that p53 directly causes senescence in some cell types.

Of interest, and of perhaps great importance, although p53 transactivation appears to be required for senescence, it is not [140, 143, 144], or not always [145] accompanied by an increase in p53 protein stabilization, or indeed any increase in protein concentration, normally associated with functional activation of p53. This suggests that senescence induction may require a specialized activation signal. Indeed this may well be true. Induction of senescence by p53 was shown to require acetylation of lysine 382 in human p53 and this acetylation is dependent on one or more of the isoforms of the PML gene [58]. More recently, senescence induction was shown to require the PML-IV isoform and this is associated with both acetylation at lysine 382 and phosphorylation at serine 46 [55]. Finally, co-expression of the deacetylase SIR2 prevents PML dependent, p53 induced senescence [146]. Thus the activating signal for p53 dependent senescence may be quite specific. Whilst this may be true, it is also accompanied by an increase in stabilization of p53 protein which, as indicated, is not usually associated with senescence induction. This might suggest that the stabilization is consequential or that p53 is also carrying out other activities not measured in these experiments.

Induction of cellular senescence by p53

The evidence suggests that p53 must be transcriptionally active in order to induce senescence. That is, it must switch on one or more down-stream genes which in turn induce the senescent phenotype. As already implied, the favorite candidate is the p21$^{WAF1/CIP1}$ cyclin dependent kinase inhibitor, probably the most ubiquitously activated target of p53. Some of the evidence implicating p21$^{WAF1/CIP1}$ as inducing senescence has already been outlined, but there is considerably more. Cells defective in p53 but over-expressing p21$^{WAF1/CIP1}$ undergo senescence [147, 148] and human cells that are homozygous null for *p21$^{WAF1/CIP1}$* do not [149]. Moreover, microinjection of antibodies to p53 rescued from senescence and was accompanied by a decrease in p21$^{WAF1/CIP1}$ expression [137]. Senescence was also rescued [150] by microinjection of antibodies to p21$^{WAF1/CIP}$. These and other data [reviewed in ref. 8] all point to p21$^{WAF1/CIP1}$ being the key inducer of replicative senescence initiated by p53. However, there are reports that are inconsistent with this interpretation.

Studies with Li-Fraumeni fibroblasts that are genetically deficient in p53 and do not therefore show detectable p21$^{WAF1/CIP1}$ expression, are still able to undergo replicative senescence [151] and mouse fibroblasts homozygous null for *p21$^{WAF1/CIP1}$* undergo senescence at the same rate as their wild type counterparts [152]. In this study, clones that had escaped from senescence had mutations in *p53* or *p16^{INK4A}*. These studies suggest that there may be p53 dependent genes other than *p21$^{WAF1/CIP1}$*

that induce senescence, although one has to be aware in interpreting these studies, that strong selection pressure may "force" genes not normally associated with senescence to substitute for *p21*$^{WAF1/CIP1}$. Thus it is much too early to discard p21$^{WAF1/CIP1}$ as the key effector of p53 induced senescence. Furthermore, another study with homozygous mouse fibroblasts came to the opposite conclusion [153], viz, that p21$^{WAF1/CIP}$ is essential for induction of senescence. In addition, array studies have failed to reveal many more candidates, again underscoring the likely importance of p21$^{WAF1/CIP}$ in senescence induction.

However, the recently cloned candidate tumor suppressor gene, *p33*^{ING1} [154], is also a possible senescence inducer. The product of this gene causes a growth arrest [155] and treatment with antisense *p33*^{ING1} increases cell lifespan a similar amount to that induced by p53. Moreover, the cell cycle arrest induced by p33^{ING1} is p53 dependent [155]. *p33*^{ING1} is also induced by DNA damage and associates with deacetylating complexes important in cell cycle control [156].

Another novel protein has been identified that also may play a role in senescence. Caveolin-1, a member of the family of structural proteins involved in the formation of vesicular invaginations of the plasma membrane (caveolae), has been implicated in the induction of senescence [157]. Caveolin-1 is downregulated in human tumors and in cell lines transformed by oncogenes [158, 159]. Overexpression of caveolin-1 is sufficient to block primary mouse cells in G_0/G_1 [160] and this occurs through the p53 pathway. Further, treatment of NIH3T3 cells with low levels of hydrogen peroxide induces senescence and up-regulates caveolin-1 [157]. As, during senescence, changes are seen in the cell membrane as the cell becomes large and misshapen, a role for caveolin proteins in this phenotype may also be inferred.

Observations such as these serve to illustrate that other proteins may be involved in senescence induction, but the evidence that p21$^{WAF1/CIP}$ is necessary for senescence appears to be very convincing and most experimental results can be explained by p21$^{WAF1/CIP}$ alone. Other gene products may serve to modulate the process but may turn out to be not that critical.

How does p21$^{WAF1/CIP}$ function to induce senescence?

In general terms, p21$^{WAF1/CIP}$ functions to arrest cell cycle progression, predominantly at the G_1/S phase boundary, but also in some cases at the G_2/M boundary. How the latter occurs is not well understood, but the G_1/S phase arrest is in large part due to its ability to inhibit cyclin dependent kinases that regulate pRb105. In fact, unlike other inhibitors, p21$^{WAF1/CIP}$ is capable of interacting with essentially all cyclin dependent kinase complexes [reviewed in ref. 161]. However, how does p21$^{WAF1/CIP}$ induce senescence as opposed to simply causing an arrest of the cell cycle? One possibility is that chronic exposure to senescence inducing agents (e.g., telomere erosion) might cause constitutive activation of p21$^{WAF1/CIP}$. However, this appears not to be the case. Although required for the onset of senescence, p21$^{WAF1/CIP}$ levels eventually decline to about those of pre-senescent cells [reviewed in ref. 161]. Thus p21$^{WAF1/CIP}$ is only required in the early stages of senescence, and one or more other genes are required for the maintenance of senescence (Figure 5A). The critical gene appears to be *p16^{INK4}*, which as indicated above, also encodes a cyclin dependent kinase inhibitor that acts on cyclin dependent kinases (CDK) 4 and 6 to inhibit phosphorylation of pRb105. p16^{INK4A} becomes permanently upregulated in senescent cells [162, 163; reviewed in ref. 2] although the mechanism underlying this is not known. This permanent upregulation of p16^{INK4A} appears to distinguish a transient arrest of cell cycle progression from senescence and is consistent with the loss of *p16^{INK4A}* gene function that is associated with escape from senescence during the immortalization process. Thus, there appears to be a sequential activation process: p53 is activated by various signals; it transactivates the *p21$^{WAF1/CIP}$* gene to initiate senescence, which is followed finally by an induction of *p16^{INK4A}* to maintain senescence (Figure 5A).

Evidence for an alternative pathway, not involving either p21$^{WAF1/CIP}$ or p16^{INK4A} has recently been reported [3]. Disruption of the pRb pathway by overexpression of CDK4 or CDK6 delayed senescence by approximately 10 population doublings. Moreover, this occurred in most cases without an increase in p21$^{WAF1/CIP}$ and, surprisingly, was independent of p16^{INK4A} function. When p21$^{WAF1/CIP}$ was entirely downregulated by depletion of p53, senescence escape occurred and cells went on to enter crisis after an extended lifespan. Results such as these serve as a reminder that there may well be other (albeit rare) senescence pathways not involving p53, which can occur under unusual selection pressures.

How might p53 be involved in organism aging?

Although the many observations of replicative senescence of cells *in vitro* may seem to be occurring in a very artificial environment, there is evidence that it may be linked to the aging of an organism [164]. Early data show that cells from older donors senesce earlier that those from younger donors [165, 166]. A commonly used model for aging in humans is Werner syndrome (WRN), an autosomal recessive disease in which patients show premature features of aging, including cataracts, grey hair, hair loss, cancer and diabetes. Cells from patients with the disease rapidly senesce in culture, compared to controls [167], but may be rescued by over-expression of telomerase [168].

The *WRN* gene has been cloned [169] and shown to have both helicase [170] and exonuclease [171] activity. The WRN protein has also been shown to co-immunoprecipitate with PCNA, a gene regulated by p53 [172] (Table 1) and the C-terminal region of the protein has been shown to interact with p53 directly [173, 174]. Functionally, p53-induced apoptosis is attenuated in WRN cells [174] and p53 modulates WRN exonuclease activity [175]. Although WRN is not aging as such, taken together this evidence strongly supports a role for p53 in an aging-like phenotype associated with senescence. Of course there remains the question of whether senescence is a feature or a cause of aging. One piece of evidence for senescence being a marker rather than a cause of aging comes from the cloned sheep "Dolly". Dolly was generated by the transfer of the nucleus from a mammary epithelial cell to an enucleated unfertilized egg [176]. Interestingly, her telomeres were as short as would be expected for the age of the donor of the mammary tissue (6 years) but despite this she was of normal fertility and generally healthy [177] except for arthritis, until she was euthanased at the age of 6 years after developing a virally induced lung tumor.

The use of mice as models for the human aging phenotype has produced mixed results. The most likely reason for this is that mice have longer telomeres than humans and even mice deficient in telomerase have a normal lifespan [178, 179], which is only reduced after 6 consecutive crosses of offspring [179]. These later mice have shortened lifespans and characteristics associated with aging. Interestingly, this phenotype can be rescued by crossing with p53 null mice [180], again suggesting a role for p53 in the aging-related phenotype. Similarly, mice deficient in Ku86 are deficient in nonhomologous end joining and exhibit an aging-related phenotype [181] that is rescued by crossing with p53 null mice [182].

In contrast, mice deleted for *p66^{Shc}* have an extended lifespan of about 30% compared with controls [183]. p66^{Shc} has been shown to be involved in the cell response to reactive oxygen species and *p66^{Shc}* knockout mice are resistant to UV or hydrogen peroxide triggered apoptosis. Importantly, the p53 apoptotic response is impaired in these mice [183]. A rather less straightforward example of the involvement of p53 in aging comes from the p53+/m mouse [184]. This mouse has an extensively rearranged p53 gene that was created by accident [164, 184]. Despite this rearrangement, a C-terminal p53 fragment is produced that appears to have enhanced transcriptional activation capability [184]. Importantly, the mice exhibit a 20% reduction in lifespan compared to controls, with features of aging being apparent by 18 months. Although the authors themselves admit that these results should be interpreted with caution [164, 184], the combined evidence from these mouse models appear to confirm a role for p53 in aging.

In this context, it is tempting to propose a link between cellular senescence, aging and cancer. Age is a significant risk factor for cancer and, as discussed above, cells from older donors senesce early in culture [165, 166], suggesting that senescence could be linked to tumorigenesis. Further, several tumor suppressor genes are involved in triggering senescence, suggesting that it may be a tumor suppressive mechanism. Certainly there is evidence that senescence may be a tumor suppression mechanism *in vitro* [reviewed in ref. 8], but the *in vivo* evidence is lacking. Despite this, a model has been suggested [185] by which senescent cells accumulate with age, disrupting the microenvironment and creating a pro-carcinogenic environment, thereby increasing the risk of cancer with age. An alternative model for aging has been proposed [164], based on the observation that the p53+/m mice with a shortened lifespan are highly resistant to tumors, suggesting a balance between tumor suppression and longevity. This model proposes that there is a reserve functional activity of stem cells which contribute to organ homeostasis and prevent the aging phenotype. p53+/m mice may lose this capacity earlier as more stem cells are depleted by the enhanced p53 tumor suppression mechanism, leading to premature aging. Conversely, wild-type p53 mice maintain organ homeostasis longer and so have longer lifespans but the "trade-off" is an increased incidence of cancer with age.

Clearly the evidence for a role for p53 in cellular senescence is profound and there is growing evidence that it is involved in organismal aging. However, the common assumption that senescence and aging are due to the same mechanism remains to be proven.

Summary

This review has summarized a substantial body of evidence that describes how p53 functions and that it is a most important protein in initiating cellular senescence. To induce senescence it is likely that specific modifications are required, including covalent modification at serine 46 and lysine 382. When activated in this way p53 activates $p21^{WAF1/CIP1}$ which is also required for senescence initiation. A subsequent induction of $p16^{INK4A}$ (by an unknown mechanism) is required in order to maintain senescence. Although p53 is not the only reported inducer of senescence it is clearly very important in most cases. Whether or not other genes are also involved remains to be determined, but a great deal of data can be explained without involving other gene products. Array data may be useful but equally, may disguise the real process by identifying candidate genes that are "consequential" to senescence but not essential for its induction or maintainance. The bulk of the evidence makes it clear that p53, $\mathrm{p21^{WAF1/CIP1}}$ and $\mathrm{p16^{INK4A}}$ are able to account for many of the observations concerning the process of senescence induction.

One of the challenges in the future is to determine how (and if) the cellular senescence mechanism derived from cell culture studies relates to events occurring in an organism. The evidence from tumor studies is consistent with *in vitro* studies but whether there are "true" senescent cells *in vivo* needs to be formally proven. And, furthermore, whether p53 plays a significant role in this process as it does *in vitro*, also needs to be proven. There is still much to learn.

Acknowledgments

This work was supported by funding from the Health Research Council of New Zealand and Lottery Health.

References

1. Soussi T, Dehouche K, Beroud C (2000). p53 website and analysis of p53 gene mutations in human cancer: forging a link between epidemiology and carcinogenesis. *Hum Mutat.* 15: 105–13.
2. Wynford-Thomas D (1999). Cellular senescence and cancer. *J Pathol.* 187: 100–11.
3. Morris M, Hepburn P, Wynford-Thomas D (2002). Sequential extension of proliferative lifespan in human fibroblasts induced by over-expression of CDK4 or 6 and loss of p53 function. *Oncogene* 21: 4277–88.
4. Prives C, Hall PA (1999). The p53 pathway. *J Pathol.* 187: 112–26
5. Vousden KH, Lu X (2002). Live or let die: the cell's response to p53. *Nat Rev Cancer* 2: 594–604.
6. Appella E, Anderson CW (2001). Post-translational modifications and activation of p53 by genotoxic stresses. *Eur J Biochem* 268: 2764–72.

7. Itahana K, Dimri G, Campisi J (2001). Regulation of cellular senescence by p53. *Eur J Biochem.* 268: 2784–91.
8. Marcotte R, Wang E (2002). Replicative senescence revisited. *J Gerontol A Biol Sci Med Sci.* 57: B257–69.
9. Sturzbecher H-W, Brain R, Addison C, *et al.* (1992). A C-terminal α-helix plus basic region motif is the major structural determinant of p53 tetramerization. *Oncogene* 7: 1513–23.
10. Foord O, Navot N, Rotter V (1993). Isolation and characterization of DNA sequences that are specifically bound by wild-type p53 protein. *Mol Cell Biol.* 13: 1378–84.
11. El-Deiry WS, Kern SE, Pietenpol JA, Kinzler KW, Vogelstein B (1992). Definition of a consensus binding site for p53. *Nat Genet.* 1: 45–9.
12. Zhao R, Gish K, Murphy M, *et al.* (2000). Analysis of p53-regulated gene expression patterns using oligonucleotide arrays. *Genes Dev.* 14: 981–93.
13. Ragimov N, Krauskopf A, Navot N, Rotter V, Oren M, Aloni Y (1993). Wild-type but not mutant p53 can repress transcription initiation *in vitro* by interfering with the binding of basal transcription factors to the TATA motif. *Oncogene* 8: 1183–93.
14. Perrem K, Rayner J, Voss T, Sturzbecher H, Jackson P, Braithwaite A (1995). p53 represses SV40 transcription by preventing formation of transcription complexes. *Oncogene* 11: 1299–307.
15. Shikama N, Lee CW, France S, *et al.* (1999). A novel cofactor for p300 that regulates the p53 response. *Mol Cell* 4: 365–76.
16. Zhang W, Kadam S, Emerson BM, Bieker JJ (2001). Site-specific acetylation by p300 or CREB binding protein regulates erythroid Kruppel-like factor transcriptional activity via its interaction with the SWI-SNF complex. *Mol Cell Biol.* 21: 2413–22.
17. Deb S, Jackson CT, Subler MA, Martin DW (1992). Modulation of cellular and viral promoters by mutant human p53 proteins found in tumor cells. *J Virol.* 66: 6164–70.
18. Ginsberg D, Mechta F, Yaniv M, Oren M (1991). Wild-type p53 can down-modulate the activity of various promoters. *Proc Natl Acad Sci USA* 88: 9979–83.
19. Jackson P, Bos E, Braithwaite AW (1993). Wild-type mouse p53 down-regulates transcription from different virus enhancer/promoters. *Oncogene* 8: 589–97.
20. Murphy M, Ahn J, Walker KK, *et al.* (1999). Transcriptional repression by wild-type p53 utilizes histone deacetylases, mediated by interaction with mSin3a. *Genes Dev.* 13: 2490–501.
21. Ori A, Zauberman A, Doitsh G, Paran N, Oren M, Shaul Y (1998). p53 binds and represses the HBV enhancer: an adjacent enhancer element can reverse the transcription effect of p53. *EMBO J.* 17: 544–53.
22. El-Deiry W, Tokino T, Velculescu V, *et al.* (1993). WAF1, a potential mediator of p53 tumour suppression. *Cell* 75: 817–25.
23. Harper W, Adami G, Wei N, Keyomarsi K, Elledge S (1993). The p21 CDK-interacting protein Cip1 is a potent inhibitor of G_1 cyclin-dependent kinases. *Cell* 75: 805–16.
24. Hermeking H, Lengauer C, Polyak K, *et al.* (1997). 14-3-3 sigma is a p53-regulated inhibitor of G2/M progression. *Mol Cell* 1: 3–11.
25. Chan TA, Hermeking H, Lengauer C, Kinzler KW, Vogelstein B (1999). 14-3-3sigma is required to prevent mitotic catastrophe after DNA damage. *Nature* 401: 616–20.
26. Miyashita T, Krajewski S, Krajewska M, *et al.* (1994). Tumor suppressor p53 is a regulator of bcl-2 and bax gene expression *in vitro* and *in vivo*. *Oncogene* 9: 1799–805.
27. Moroni MC, Hickman ES, Denchi EL, *et al.* (2001). Apaf-1 is a transcriptional target for E2F and p53. *Nat Cell Biol.* 3: 552–8.

28. Fortin A, Cregan SP, MacLaurin JG, *et al.* (2001). APAF1 is a key transcriptional target for p53 in the regulation of neuronal cell death. *J Cell Biol.* 155: 207–16.
29. Robles AI, Bemmels NA, Foraker AB, Harris CC (2001). APAF-1 is a transcriptional target of p53 in DNA damage-induced apoptosis. *Cancer Res.* 61: 6660–4.
30. Nakano K, Vousden KH (2001). PUMA, a novel proapoptotic gene, is induced by p53. *Mol Cell* 7: 683–94.
31. Yu J, Zhang L, Hwang PM, Kinzler KW, Vogelstein B (2001). PUMA induces the rapid apoptosis of colorectal cancer cells. *Mol Cell* 7: 673–82.
32. Oda E, Ohki R, Murasawa H, *et al.* (2000). Noxa, a BH3-only member of the Bcl-2 family and candidate mediator of p53-induced apoptosis. *Science* 288: 1053–8.
33. Oda K, Arakawa H, Tanaka T, *et al.* (2000). p53AIP1, a potential mediator of p53-dependent apoptosis, and its regulation by Ser-46-phosphorylated p53. *Cell* 102: 849–62.
34. Chittenden T, Flemington C, Houghton AB, *et al.* (1995). A conserved domain in Bak, distinct from BH1 and BH2, mediates cell death and protein binding functions. *EMBO J.* 14: 5589–96.
35. Hunter JJ, Parslow TG (1996). A peptide sequence from Bax that converts Bcl-2 into an activator of apoptosis. *J Biol Chem.* 271: 8521–4.
36. Antonsson B (2001). Bax and other pro-apoptotic Bcl-2 family "killer-proteins" and their victim the mitochondrion. *Cell Tissue Res.* 306: 347–61.
37. Selvakumaran M, Lin HK, Miyashita T, *et al.* (1994). Immediate early up-regulation of bax expression by p53 but not TGF beta 1: a paradigm for distinct apoptotic pathways. *Oncogene* 9: 1791–8.
38. Oltvai ZN, Milliman CL, Korsmeyer SJ (1993). Bcl-2 heterodimerizes *in vivo* with a conserved homolog, Bax, that accelerates programmed cell death. *Cell* 74: 609–19.
39. Zhang H, Heim J, Meyhack B (1998). Redistribution of Bax from cytosol to membranes is induced by apoptotic stimuli and is an early step in the apoptotic pathway. *Biochem Biophys Res Commun.* 251: 454–9.
40. Schuler M, Green DR (2001). Mechanisms of p53-dependent apoptosis. *Biochem Soc Trans.* 29: 684–8.
41. Chen X, Ko LJ, Jayaraman L, Prives C (1996). p53 levels, functional domains, and DNA damage determine the extent of the apoptotic response of tumor cells. *Genes Dev.*10: 2438–51.
42. Friedlander P, Haupt Y, Prives C, Oren M (1996). A mutant p53 that discriminates between p53-responsive genes cannot induce apoptosis. *Mol Cell Biol.* 16: 4961–71.
43. Ludwig RL, Bates S, Vousden KH (1996). Differential activation of target cellular promoters by p53 mutants with impaired apoptotic function. *Mol Cell Biol.* 16: 4952–60.
44. Kaeser MD, Iggo RD (2002). Chromatin immunoprecipitation analysis fails to support the latency model for regulation of p53 DNA binding activity *in vivo. Proc Natl Acad Sci USA* 99: 95–100.
45. Zhu J, Zhang S, Jiang J, Chen X (2000). Definition of the p53 functional domains necessary for inducing apoptosis. *J Biol Chem.* 275: 39927–34.
45a. Edwards SJ, Hananeia L, Eccles MR, Zhang YF, Braithwaite AW (2003). The proline-rich region of mouse p53 influences transactivation and apoptosis but is largely dispensable for these functions. *Oncogene* (in press).
46. Gorina S and Pavletich NP (1996). Structure of the p53 tumor suppressor bound to the ankyrin and SH3 domains of 53BP2. *Science* 274: 1001–5.
47. Samuels-Lev Y, O'Connor DJ, Bergamaschi D, *et al.* (2001). ASPP proteins specifically stimulate the apoptotic function of p53. *Mol Cell* 8: 781–94.

48. Baptiste N, Friedlander P, Chen X, Prives C (2002). The proline-rich domain of p53 is required for cooperation with anti- neoplastic agents to promote apoptosis of tumor cells. *Oncogene* 21: 9–21.
49. Flores ER, Tsai KY, Crowley D, *et al.* (2002). p63 and p73 are required for p53-dependent apoptosis in response to DNA damage. *Nature* 416: 560–4.
50. Hamamori Y, Sartorelli V, Ogryzko V, *et al.* (1999). Regulation of histone acetyltransferases p300 and PCAF by the bHLH protein twist and adenoviral oncoprotein E1A. *Cell* 96: 405–13.
51. Liu L, Scolnick DM, Trievel RC, *et al.* (1999). p53 sites acetylated *in vitro* by PCAF and p300 are acetylated *in vivo* in response to DNA damage. *Mol Cell Biol.* 19: 1202–9.
52. Scolnick DM, Chehab NH, Stavridi ES, *et al.* (1997). CREB-binding protein and p300/CBP-associated factor are transcriptional coactivators of the p53 tumor suppressor protein. *Cancer Res.* 57: 3693–6.
53. de The H, Lavau C, Marchio A, Chomienne C, Degos L, Dejean A (1991). The PML-RAR α fusion mRNA generated by the t(15;17) translocation in acute promyelocytic leukemia encodes a functionally altered RAR. *Cell* 66: 675–84.
54. Pearson M, Pelicci PG (2001). PML interaction with p53 and its role in apoptosis and replicative senescence. *Oncogene* 20: 7250–6.
55. Bischof O, Kirsh O, Pearson M, Itahana K, Pelicci PG, Dejean A (2002). Deconstructing PML-induced premature senescence. *EMBO J.* 21: 3358–69.
56. Guo A, Salomoni P, Luo J, *et al.* (2000). The function of PML in p53-dependent apoptosis. *Nat Cell Biol.* 2: 730–6.
57. Fogal V, Gostissa M, Sandy P, *et al.* (2000). Regulation of p53 activity in nuclear bodies by a specific PML isoform. *EMBO J.* 19: 6185–95.
58. Pearson M, Carbone R, Sebastiani C, *et al.* (2000). PML regulates p53 acetylation and premature senescence induced by oncogenic Ras. *Nature* 406: 207–10.
59. Haupt Y, Rowan S, Shaulian E, Vousden K, Oren M (1995). Induction of apoptosis in Hela cells by trans-activation deficient p53. *Genes Dev.* 9: 2170–83.
60. Hansen RS, Braithwaite AW (1996). The growth-inhibitory function of p53 is separable from transactivation, apoptosis and suppression of transformation by E1a and Ras. *Oncogene* 13: 995–1007.
61. Marchenko ND, Zaika A, Moll UM (2000). Death signal-induced localization of p53 protein to mitochondria. A potential role in apoptotic signaling. *J Biol Chem.* 275: 16202–12.
62. Sansome C, Zaika A, Marchenko ND, Moll UM (2001). Hypoxia death stimulus induces translocation of p53 protein to mitochondria. Detection by immunofluorescence on whole cells. *FEBS Lett.* 488: 110–15.
63. Moll UM, Zaika A (2001). Nuclear and mitochondrial apoptotic pathways of p53. *FEBS Lett.* 493: 65–9.
64. Momand J, Zambetti G, Olson D, George D, Levine A (1992). The Mdm-2 oncogene product forms a complex with the p53 protein and inhibits p53-mediated transactivation. *Cell* 69: 1237–45.
65. Fang S, Jensen JP, Ludwig RL, Vousden KH, Weissman AM (2000). Mdm2 is a RING finger-dependent ubiquitin protein ligase for itself and p53. *J Biol Chem.* 275: 8945–51.
66. Honda R, Yasuda H (2000). Activity of MDM2, a ubiquitin ligase, toward p53 or itself is dependent on the RING finger domain of the ligase. *Oncogene* 19: 1473–6.
67. Joazeiro CAP, Weissman AW (2000). RING finger proteins: mediators of ubiquitin ligase activity. *Cell* 102: 549–52.

68. Honda R, Tanaka H, Yasuda H (1997). Oncoprotein MDM2 is a ubiquitin ligase E3 for tumor suppressor p53. *FEBS Lett.* 420: 25–7.
69. Barak Y, Juven T, Haffner R, Oren M (1993). mdm2 expression is induced by wild type p53 activity. *EMBO J.* 12: 461–8.
70. Clarke AR, Purdie CA, Harrison DJ, *et al.* (1993). Thymocyte apoptosis induced by p53-dependent and independent pathways. *Nature* 362: 849–52.
71. Banin S, Moyal L, Shieh S, *et al.* (1998). Enhanced phosphorylation of p53 by ATM in response to DNA damage. *Science* 281: 1674–7.
72. Canman CE, Lim DS, Cimprich KA, *et al.* (1998). Activation of the ATM kinase by ionizing radiation and phosphorylation of p53. *Science* 281: 1677–9.
73. Chehab NH, Malikzay A, Stavridi ES, Halazonetis TD (1999). Phosphorylation of Ser-20 mediates stabilization of human p53 in response to DNA damage. *Proc Natl Acad Sci USA* 96: 13777–82.
74. Hirao A, Kong YY, Matsuoka S, *et al.* (2000). DNA damage-induced activation of p53 by the checkpoint kinase Chk2. *Science* 287: 1824–7.
75. Shieh SY, Ahn J, Tamai K, Taya Y, Prives C (2000). The human homologs of checkpoint kinases Chk1 and Cds1 (Chk2). phosphorylate p53 at multiple DNA damage-inducible sites. *Genes Dev.* 14: 289–300.
76. Lin J, Chen J, Elenbaas B, Levine AJ (1994). Several hydrophobic amino acids in the p53 amino-terminal domain are required for transcriptional activation, binding to Mdm-2 and the adenovirus 5 E1B 55-kD protein. *Genes Dev.* 8: 1235–46.
77. Chen J, Marechal V, Levine A (1993). Mapping of the p53 and mdm-2 Interaction Domains. *Mol Cell Biol.* 13: 4107–14.
78. Craig AL, Burch L, Vojtesek B, Mikutowska J, (1999). Thompson A, Hupp TR. Novel phosphorylation sites of human tumour suppressor protein p53 at Ser20 and Thr18 that disrupt the binding of mdm2 (mouse double minute 2) protein are modified in human cancers. *Biochem J.* 342: 133–41.
79. Bulavin DV, Saito S, Hollander MC, *et al.* (1999). Phosphorylation of human p53 by p38 kinase coordinates N-terminal phosphorylation and apoptosis in response to UV radiation. *EMBO J.* 18: 6845–54.
80. D'Orazi G, Cecchinelli B, Bruno T, *et al.* (2002). Homeodomain-interacting protein kinase-2 phosphorylates p53 at Ser 46 and mediates apoptosis. *Nat Cell Biol.* 4: 11–19.
81. Hofmann TG, Moller A, Sirma H, *et al.* (2002). Regulation of p53 activity by its interaction with homeodomain- interacting protein kinase-2. *Nat Cell Biol.* 4: 1–10.
82. Buschmann T, Potapova O, Bar-Shira A, *et al.* (2001). Jun NH2-terminal kinase phosphorylation of p53 on Thr-81 is important for p53 stabilization and transcriptional activities in response to stress. *Mol Cell Biol.* 21: 2743–54.
83. Blaydes JP, Luciani MG, Pospisilova S, Ball HM, Vojtesek B, Hupp TR (2001). Stoichiometric phosphorylation of human p53 at Ser315 stimulates p53-dependent transcription. *J Biol Chem.* 276: 4699–708.
84. Cuddihy AR, Wong AH, Tam NW, Li S, Koromilas AE (1999). The double-stranded RNA activated protein kinase PKR physically associates with the tumor suppressor p53 protein and phosphorylates human p53 on serine 392 *in vitro. Oncogene* 18: 2690–702.
85. Sakaguchi K, Herrera JE, Saito S, *et al.* (1998). DNA damage activates p53 through a phosphorylation-acetylation cascade. *Genes Dev.* 12: 2831–41
86. Ito A, Lai CH, Zhao X, *et al.* (2001). p300/CBP-mediated p53 acetylation is commonly induced by p53-activating agents and inhibited by MDM2. *EMBO J.* 20: 1331–40.
87. Lill NL, Grossman SR, Ginsberg D, DeCaprio J, Livingston DM (1997). Binding and modulation of p53 by p300/CBP coactivators. *Nature* 387: 823–7.

88. Chao C, Saito S, Anderson CW, Appella E, Xu Y (2000). Phosphorylation of murine p53 at ser-18 regulates the p53 responses to DNA damage. *Proc Natl Acad Sci USA* 97: 11936–41.
89. Dumaz N, Meek DW. Serine15 phosphorylation stimulates p53 transactivation but does not directly influence interaction with HDM2. *EMBO J* (1999). 18: 7002–10.
90. Lambert PF, Kashanchi F, Radonovich MF, Shiekhattar R, Brady JN (1998). Phosphorylation of p53 serine 15 increases interaction with CBP. *J Biol Chem.* 273: 33048–53.
91. Li M, Chen D, Shiloh A, *et al.* (2002). Deubiquitination of p53 by HAUSP is an important pathway for p53 stabilization. *Nature* 416: 648–53.
92. Khosravi R, Maya R, Gottlieb T, Oren M, Shiloh Y, Shkedy D (1999). Rapid ATM-dependent phosphorylation of MDM2 precedes p53 accumulation in response to DNA damage. *Proc Natl Acad Sci USA* 96: 14973–7.
93. Buschmann T, Fuchs SY, Lee CG, Pan ZQ, Ronai Z (2000). SUMO-1 modification of Mdm2 prevents its self-ubiquitination and increases Mdm2 ability to ubiquitinate p53. *Cell* 101: 753–62.
94. Quelle DE, Zindy F, Ashmun RA, Sherr CJ (1995). Alternative reading frames of the INK4a tumor suppressor gene encode two unrelated proteins capable of inducing cell cycle arrest. *Cell* 83: 993–1000.
95. de Stanchina E, McCurrach ME, Zindy F, *et al.* (1998). E1A signaling to p53 involves the p19(ARF) tumor suppressor. *Genes Dev.* 12: 2434–42.
96. Bates S, Phillips AC, Clark PA, *et al.* (1998). p14^{ARF} links the tumour suppressors Rb and p53. *Nature* 395: 124–5.
97. Honda R, Yasuda H (1999). Association of p19(ARF) with Mdm2 inhibits ubiquitin ligase activity of Mdm2 for tumor suppressor p53. *EMBO J.* 18: 22–7.
98. Pomerantz J, Schreiber-Agus N, Liegeois NJ, *et al.* (1998). The INK4a tumor suppressor gene product, p19^{ARF}, interacts with MDM2 and neutralizes MDM2's inhibition of p53. *Cell* 92: 713–23.
99. Kamijo T, Weber JD, Zambetti G, Zindy F, Roussel MF, Sherr CJ (1998). Functional and physical interactions of the ARF tumor suppressor with p53 and Mdm2. *Proc Natl Acad Sci USA* 95: 8292–7.
100. Stott FJ, Bates S, James MC, *et al.* (1998). The alternative product from the human CDKN2A locus, p14^{ARF}, participates in a regulatory feedback loop with p53 and MDM2. *EMBO J.* 17: 5001–14.
101. Zhang Y, Xiong Y, Yarbrough WG (1998). ARF promotes MDM2 degradation and stabilizes p53: ARF-INK4a locus deletion impairs both the Rb and p53 tumor suppression pathways. *Cell* 92: 725–34.
102. Edwards SJ, Dix BR, Myers CJ, *et al.* (2002). Evidence that replication of the antitumor adenovirus ONYX-015 is not controlled by the p53 and p14ARF tumor suppressor genes. *J Virol.* 76:12483–90.
103. Tolbert D, Lu X, Yin C, Tantama M, Van Dyke T (2002). p19(ARF) is dispensable for oncogenic stress-induced p53-mediated apoptosis and tumor suppression *in vivo. Mol Cell Biol.* 22: 370–7.
104. Dey D, Dahl J, Cho S, Benjamin TL (2002). Induction and bypass of p53 during productive infection by polyomavirus. *J Virol.* 76: 9526–32.
105. Dutta A, Ruppert J, Aster J, Winchester E (1993). Inhibition of DNA replication factor RPA by p53. *Nature* 365: 79–82.
106. Maheswaran S, Englert C, Bennet P, Heinrich G, Haber D (1995). The WT1 gene product stabilises p53 and inhibits p53 mediated apoptosis. *Genes Dev.* 9: 2143–56.

107. Zhang H, Somasundaram K, Peng Y, *et al.* (1998). BRCA1 physically associates with p53 and stimulates its transcriptional activity. *Oncogene* 16: 1713-1721.
108. Buchhop S, Gibson MK, Wang XW, Wagner P, Sturzbecher HW, Harris CC (1997). Interaction of p53 with the human Rad51 protein. *Nucleic Acids Res.* 25: 3868–74.
109. Lu H, Fisher RP, Bailey P, Levine AJ (1997). The CDK7-cycH-p36 complex of transcription factor IIH phosphorylates p53, enhancing its sequence-specific DNA binding activity *in vitro*. *Mol Cell Biol.* 17: 5923–34.
110. Jayaraman L, Murthy KG, Zhu C, Curran T, Xanthoudakis S, Prives C (1997). Identification of redox/repair protein Ref-1 as a potent activator of p53. *Genes Dev.* 11: 558–70.
111. Okamoto T, Izumi H, Imamura T, *et al.* (2000). Direct interaction of p53 with the Y-box binding protein, YB-1: a mechanism for regulation of human gene expression. *Oncogene* 19: 6194–202.
112. Hayflick L, Moorhead PS (1961). The limited *in vitro* lifetime of human diploid cell strains. *Exp Cell Res* 25: 585–621.
113. Vaziri H (1997). Critical telomere shortening regulated by the ataxia-telangiectasia gene acts as a DNA damage signal leading to activation of p53 protein and limited life-span of human diploid fibroblasts. A review. *Biochemistry (Mosc.)*, 62: 1306–10.
114. Chen QM, Bartholomew JC, Campisi J, Acosta M, Reagan JD, Ames BN (1998). Molecular analysis of H_2O_2-induced senescent-like growth arrest in normal human fibroblasts: p53 and Rb control G_1 arrest but not cell replication. *Biochem J.* 332: 43–50.
115. Lin AW, Barradas M, Stone JC, van Aelst L, Serrano M, Lowe SW (1998). Premature senescence involving p53 and p16 is activated in response to constitutive MEK/MAPK mitogenic signaling. *Genes Dev.* 12: 3008–19.
116. Serrano M, Lin AW, McCurrach ME, Beach D, Lowe SW (1997). Oncogenic Ras provokes premature cell senescence associated with accumulation of p53 and $p16^{INK4a}$. *Cell* 88: 593–602.
117. Ferbeyre G, de Stanchina E, Querido E, Baptiste N, Prives C, Lowe SW (2000). PML is induced by oncogenic ras and promotes premature senescence. *Genes Dev* 14: 2015–27.
118. Zhu J, Woods D, McMahon M, Bishop JM (1998). Senescence of human fibroblasts induced by oncogenic Raf. *Genes Dev.* 12: 2997–3007.
119. Griffith JD, Comeau L, Rosenfield S, *et al.* (1999). Mammalian telomeres end in a large duplex loop. *Cell* 97: 503–14.
120. Karlseder J, Broccoli D, Dai Y, Hardy S, de Lange T (1999). p53- and ATM-dependent apoptosis induced by telomeres lacking TRF2. *Science* 283: 1321–5.
121. Zhu XD, Kuster B, Mann M, Petrini JH, Lange T (2000). Cell-cycle-regulated association of RAD50/MRE11/NBS1 with TRF2 and human telomeres. *Nat Genet.* 25: 347–52.
122. Hsu HL, Gilley D, Blackburn EH, Chen DJ (1999). Ku is associated with the telomere in mammals. *Proc Natl Acad Sci USA* 96: 12454–8.
123. Hsu HL, Gilley D, Galande SA, *et al.* (2000). Ku acts in a unique way at the mammalian telomere to prevent end joining. *Genes Dev.* 14: 2807–12.
124. Donehower L, Harvey M, Slagle B, *et al.* (1992). Mice deficient for p53 are developmentally normal but susceptable to spontaneous tumours. *Nature* 356: 215–21.
125. Gao Q, Hauser SH, Liu XL, Wazer DE, Madoc-Jones H, Band V (1996). Mutant p53-induced immortalization of primary human mammary epithelial cells. *Cancer Res.* 56: 3129–33.

126. Wazer DE, Chu Q, Liu XL, Gao Q, Safaii H, Band V (1994). Loss of p53 protein during radiation transformation of primary human mammary epithelial cells. *Mol Cell Biol.* 14: 2468–78.
127. Vousden KH, Vojtesek B, Fisher C, Lane D (1993). HPV-16 E7 or adenovirus E1A can overcome the growth arrest of cells immortalized with a temperature-sensitive p53. *Oncogene* 8: 1697–702.
128. Nevins JR (1994). Cell cycle targets of the DNA tumor viruses. *Curr Opin Genet Dev.* 4: 130–4.
129. Bryan TM, Reddel RR (1994). SV40-induced immortalization of human cells. *Crit Rev Oncog.* 5: 331–57.
130. Saenz-Robles MT, Sullivan CS, Pipas JM (2001). Transforming functions of Simian Virus 40. *Oncogene* 20: 7899–907.
131. Mantovani F, Banks L (2001). The human papillomavirus E6 protein and its contribution to malignant progression. *Oncogene* 20: 7874–87.
132. Whitaker NJ, Bryan TM, Bonnefin P, *et al.* (1995). Involvement of RB-1, p53, $p16^{INK4}$ and telomerase in immortalisation of human cells. *Oncogene* 11: 971–6.
133. Huschtscha LI, Noble JR, Neumann AA, *et al.* (1998). Loss of $p16^{INK4}$ expression by methylation is associated with lifespan extension of human mammary epithelial cells. *Cancer Res.* 58: 3508–12.
134. Rogan EM, Bryan TM, Hukku B, *et al.* (1995). Alterations in p53 and $p16^{INK4}$ expression and telomere length during spontaneous immortalization of Li-Fraumeni syndrome fibroblasts. *Mol Cell Biol.* 15: 4745–53.
135. Bodnar AG, Ouellette M, Frolkis M, *et al.* (1998). Extension of life-span by introduction of telomerase into normal human cells. *Science* 279: 349–52.
136. Reddel RR, Bryan TM, Colgin LM, Perrem KT, Yeager TR (2001). Alternative lengthening of telomeres in human cells. *Radiat Res.* 155: 194–200.
137. Gire V, Wynford-Thomas D (1998). Reinitiation of DNA synthesis and cell division in senescent human fibroblasts by microinjection of anti-p53 antibodies. *Mol Cell Biol.* 18: 1611–21.
138. Bond JA, Wyllie FS, Wynford-Thomas D (1994). Escape from senescence in human diploid fibroblasts induced directly by mutant p53. *Oncogene* 9: 1885–9.
139. Noda A, Ning Y, Venable SF, Pereira-Smith OM, Smith JR (1994). Cloning of senescent cell-derived inhibitors of DNA synthesis using an expression screen. *Exp Cell Res* 211: 90–8.
140. Atadja P, Wong H, Garkavtsev I, Veillette C, Riabowol K (1995). Increased activity of p53 in senescing fibroblasts. *Proc Natl Acad Sci USA* 92: 8348–52.
141. Bond J, Haughton M, Blaydes J, Gire V, Wynford-Thomas D, Wyllie F (1996). Evidence that transcriptional activation by p53 plays a direct role in the induction of cellular senescence. *Oncogene* 13: 2097–104.
142. Sugrue MM, Shin DY, Lee SW, Aaronson SA (1997). Wild-type p53 triggers a rapid senescence program in human tumor cells lacking functional p53. *Proc Natl Acad Sci USA* 94: 9648–53.
143. Afshari CA, Vojta PJ, Annab LA, Futreal PA, Willard TB, Barrett JC (1993). Investigation of the role of G_1/S cell cycle mediators in cellular senescence. *Exp Cell Res* 209: 231–7.
144. Vaziri H, West MD, Allsopp RC, *et al.* (1997). ATM-dependent telomere loss in aging human diploid fibroblasts and DNA damage lead to the post-translational activation of p53 protein involving poly(ADP-ribose) polymerase. *EMBO J.* 16: 6018–33.

145. Kulju KS, Lehman JM (1995). Increased p53 protein associated with aging in human diploid fibroblasts. *Exp Cell Res* 217: 336–45.
146. Langley E, Pearson M, Faretta M, *et al.* (2002). Human SIR2 deacetylates p53 and antagonizes PML/p53-induced cellular senescence. *EMBO J.* 21: 2383–96.
147. Wang Y, Blandino G, Givol D (1999). Induced p21WAF expression in H1299 cell line promotes cell senescence and protects against cytotoxic effect of radiation and doxorubicin. *Oncogene* 18: 2643–9.
148. Fang L, Igarashi M, Leung J, Sugrue MM, Lee SW, Aaronson SA (1999). p21Waf1/Cip1/Sdi1 induces permanent growth arrest with markers of replicative senescence in human tumor cells lacking functional p53. *Oncogene* 18: 2789–97.
149. Brown JP, Wei W, Sedivy JM (1997). Bypass of senescence after disruption of p21CIP1/WAF1 gene in normal diploid human fibroblasts. *Science* 277: 831–4.
150. Ma Y, Prigent SA, Born TL, Monell CR, Feramisco JR, Bertolaet BL (1999). Microinjection of anti-p21 antibodies induces senescent Hs68 human fibroblasts to synthesize DNA but not to divide. *Cancer Res.* 59: 5341–8.
151. Medcalf AS, Klein-Szanto AJ, Cristofalo VJ (1996). Expression of p21 is not required for senescence of human fibroblasts. *Cancer Res.* 56: 4582–5.
152. Pantoja C, Serrano M (1999). Murine fibroblasts lacking p21 undergo senescence and are resistant to transformation by oncogenic Ras. *Oncogene* 18: 4974–82.
153. Xu Y, Yang EM, Brugarolas J, Jacks T, Baltimore D (1998). Involvement of p53 and p21 in cellular defects and tumorigenesis in $Atm^{-/-}$ mice. *Mol Cell Biol.* 18: 4385–90.
154. Garkavtsev I, Riabowol K (1997). Extension of the replicative life span of human diploid fibroblasts by inhibition of the p33ING1 candidate tumor suppressor. *Mol Cell Biol.* 17: 2014–19.
155. Garkavtsev I, Grigorian IA, Ossovskaya VS, Chernov MV, Chumakov PM, Gudkov AV (1998). The candidate tumour suppressor p33ING1 cooperates with p53 in cell growth control. *Nature* 391: 295–8.
156. Cheung KJ, Jr., Mitchell D, Lin P, Li G (2001). The tumor suppressor candidate p33(ING1) mediates repair of UV-damaged DNA. *Cancer Res.* 61: 4974–7.
157. Volonte D, Zhang K, Lisanti MP, Galbiati F (2002). Expression of caveolin-1 induces premature cellular senescence in primary cultures of murine fibroblasts. *Mol Biol Cell* 13: 2502–17.
158. Koleske AJ, Baltimore D, Lisanti MP (1995). Reduction of caveolin and caveolae in oncogenically transformed cells. *Proc Natl Acad Sci USA* 92: 1381–5.
159. Engelman JA, Lee RJ, Karnezis A, *et al.* (1998). Reciprocal regulation of neu tyrosine kinase activity and caveolin-1 protein expression *in vitro* and *in vivo*. Implications for human breast cancer. *J Biol Chem.* 273: 20448–55.
160. Galbiati F, Volonte D, Liu J, *et al.* (2001). Caveolin-1 expression negatively regulates cell cycle progression by inducing G(0)/G(1) arrest via a p53/p21(WAF1/Cip1)-dependent mechanism. *Mol Biol Cell* 12: 2229–44.
161. Roninson IB (2002). Oncogenic functions of tumour suppressor p21(Waf1/Cip1/Sdi1): association with cell senescence and tumour-promoting activities of stromal fibroblasts. *Cancer Lett.* 179: 1–14.
162. Di Leonardo A, Linke SP, Clarkin K, Wahl GM (1994). DNA damage triggers a prolonged p53-dependent G_1 arrest and long-term induction of Cip1 in normal human fibroblasts. *Genes Dev.* 8: 2540–51.
163. Alcorta DA, Xiong Y, Phelps D, Hannon G, Beach D, Barrett JC (1996). Involvement of the cyclin-dependent kinase inhibitor p16 (INK4a) in replicative senescence of normal human fibroblasts. *Proc Natl Acad Sci USA* 93: 13742–7.

164. Donehower LA (2002). Does p53 affect organismal aging? *J Cell Physiol.* 192: 23–33.
165. Schneider EL, Mitsui Y (1976). The relationship between *in vitro* cellular aging and *in vivo* human age. *Proc Natl Acad Sci USA* 73: 3584–8.
166. Martin GM, Sprague CA, Epstein CJ (1970). Replicative life-span of cultivated human cells. Effects of donor's age, tissue, and genotype. *Lab Invest.* 23: 86–92.
167. Norwood TH, Hoehn H, Salk D, Martin GM (1979). Cellular aging in Werner's syndrome: a unique phenotype? *J Invest Dermatol.* 73: 92–6.
168. Choi D, Whittier PS, Oshima J, Funk WD (2001). Telomerase expression prevents replicative senescence but does not fully reset mRNA expression patterns in Werner syndrome cell strains. *FASEB J.* 15: 1014–20.
169. Yu CE, Oshima J, Fu YH, *et al.* (1996). Positional cloning of the Werner's syndrome gene. *Science* 272: 258–62.
170. Gray MD, Shen JC, Kamath-Loeb AS, *et al.* (1997). The Werner syndrome protein is a DNA helicase. *Nat Genet.* 17: 100–3.
171. Huang S, Li B, Gray MD, Oshima J, Mian IS, Campisi J (1998). The premature ageing syndrome protein, WRN, is a 3′→5′ exonuclease. *Nat Genet.* 20: 114–16.
172. Lebel M, Spillare EA, Harris CC, Leder P (1999). The Werner syndrome gene product co-purifies with the DNA replication complex and interacts with PCNA and topoisomerase I. *J Biol Chem.* 274: 37795–9.
173. Blander G, Kipnis J, Leal JF, Yu CE, Schellenberg GD, Oren M (1999). Physical and functional interaction between p53 and the Werner's syndrome protein. *J Biol Chem.* 274: 29463–9.
174. Spillare EA, Robles AI, Wang XW, *et al.* (1999). p53-mediated apoptosis is attenuated in Werner syndrome cells.*Genes Dev.* 13: 1355–60.
175. Brosh RM, Jr., Karmakar P, Sommers JA, *et al.* (2001). p53 Modulates the exonuclease activity of Werner syndrome protein. *J Biol Chem.* 276: 35093–102.
176. Wilmut I, Schnieke AE, McWhir J, Kind AJ, Campbell KH (1997). Viable offspring derived from fetal and adult mammalian cells. *Nature* 385: 810–13.
177. Shiels PG, Kind AJ, Campbell KH, *et al.* (1999). Analysis of telomere lengths in cloned sheep. *Nature* 399: 316–17.
178. Blasco MA, Lee HW, Hande MP, *et al.* (1997). Telomere shortening and tumor formation by mouse cells lacking telomerase RNA. *Cell* 91: 25–34.
179. Rudolph KL, Chang S, Lee HW, *et al.* (1999). Longevity, stress response, and cancer in aging telomerase-deficient mice. *Cell* 96: 701–12.
180. Chin L, Artandi SE, Shen Q, *et al.* (1999). p53 deficiency rescues the adverse effects of telomere loss and cooperates with telomere dysfunction to accelerate carcinogenesis. *Cell* 97: 527–38.
181. Vogel H, Lim DS, Karsenty G, Finegold M, Hasty P (1999). Deletion of Ku86 causes early onset of senescence in mice. *Proc Natl Acad Sci USA* 96: 10770–5.
182. Lim DS, Vogel H, Willerford DM, Sands AT, Platt KA, Hasty P (2000). Analysis of ku80-mutant mice and cells with deficient levels of p53. *Mol Cell Biol.* 20: 3772–80.
183. Migliaccio E, Giorgio M, Mele S, *et al.* (1999). The p66shc adaptor protein controls oxidative stress response and life span in mammals. *Nature* 402: 309–13
184. Tyner SD, Venkatachalam S, Choi J, *et al.* (2002). p53 mutant mice that display early ageing-associated phenotypes. *Nature* 415: 45–53.
185. Campisi J (1997). Aging and cancer: the double-edged sword of replicative senescence. *J Am Geriatr Soc.* 45: 482–8.
186. Wu X, Bayle JH, Olson D, Levine AJ (1993). The p53-MDM-2 autoregulatory feedback loop. *Genes Dev.* 7: 1126–32.

187. Kastan MB, Zhan Q, El-Deiry WS, *et al.* (1992). A mammalian cell cycle checkpoint pathway utilizing p53 and GADD45 is defective in ataxia-telangiectasia. *Cell* 71: 587–97.
188. Zauberman A, Lupo A, Oren M (1995). Identification of p53 target genes through immune selection of genomic DNA: the cyclin G gene contains two distinct p53 binding site. *Oncogene* 10: 2361–6.
189. Okamoto K, Beach D. Cyclin G is a transcriptional target of the p53 tumor suppressor protein. *EMBO J* (1994). 13: 4816–22.
190. Buckbinder L, Talbott R, Velasco-Miguel S, *et al.* (1995). Induction of the growth inhibitor IGF-binding protein 3 by p53. *Nature* 377: 646–9.
191. Polyak K, Xia Y, Zweier JL, Kinzler KW, Vogelstein B. A model for p53-induced apoptosis. *Nature* (1997). 389: 300–5.
192. Wu GS, Burns TF, McDonald ER, *et al.* (1997). KILLER/DR5 is a DNA damage-inducible p53-regulated death receptor gene. *Nat Genet.* 17: 141–3.
193. Dameron KM, Volpert OV, Tainsky MA, Bouck N (1994). Control of angiogenesis in fibroblasts by p53 regulation of thrombospondin-1. *Science* 265: 1582–4.
194. Owen-Schaub LB, Zhang W, Cusack JC, *et al.* (1995). Wild-type human p53 and a temperature-sensitive mutant induce Fas/APO-1 expression. *Mol Cell Biol.* 15: 3032–40.
195. Wu GS, Saftig P, Peters C, El-Deiry WS (1998). Potential role for cathepsin D in p53-dependent tumor suppression and chemosensitivity. *Oncogene* 16: 2177–83.
196. Hwang PM, Bunz F, Yu J, *et al.* (2001). Ferredoxin reductase affects p53-dependent, 5-fluorouracil-induced apoptosis in colorectal cancer cells. *Nat Med.* 7: 1111–17.
197. Okamura S, Arakawa H, Tanaka T, *et al.* (2001). p53DINP1, a p53-inducible gene, regulates p53-dependent apoptosis. *Mol Cell* 8: 85–94.
198. Attardi LD, Reczek EE, Cosmas C, *et al.* (2000). PERP, an apoptosis-associated target of p53, is a novel member of the PMP-22/gas3 family. *Genes Dev.* 14: 704–18.
199. Lin Y, Ma W, Benchimol S (2000). Pidd, a new death-domain-containing protein, is induced by p53 and promotes apoptosis. Nat Genet, 26: 122–7.
200. Stambolic V, MacPherson D, Sas D, *et al.* (2001). Regulation of PTEN transcription by p53. *Mol Cell* 8: 317–25.
201. Fiscella M, Zhang H, Fan S, *et al.* (1997). Wip1, a novel human protein phosphatase that is induced in response to ionizing radiation in a p53-dependent manner. *Proc Natl Acad Sci USA* 94: 6048–53.
202. Rouault JP, Falette N, Guehenneux F, *et al.* (1996). Identification of BTG2, an antiproliferative p53-dependent component of the DNA damage cellular response pathway. *Nat Genet.* 14: 482–6.
203. Nakano K, Balint E, Ashcroft M, Vousden KH. A ribonucleotide reductase gene is a transcriptional target of p53 and p73. *Oncogene* (2000). 19: 4283–9.
204. Tanaka H, Arakawa H, Yamaguchi T, *et al.* A ribonucleotide reductase gene involved in a p53-dependent cell-cycle checkpoint for DNA damage. *Nature* (2000). 404: 42–9.
205. Van Meir EG, Polverini PJ, Chazin VR, Su Huang HJ, de Tribolet N, Cavenee WK (1994). Release of an inhibitor of angiogenesis upon induction of wild type p53 expression in glioblastoma cells. *Nat Genet.* 8: 171–6.
206. Nishimori H, Shiratsuchi T, Urano T, *et al.* (1997). A novel brain-specific p53-target gene, BAI1, containing thrombospondin type 1 repeats inhibits experimental angiogenesis. *Oncogene* 15: 2145–50.
207. Bian J, Sun Y (1997). Transcriptional activation by p53 of the human type IV collagenase (gelatinase A or matrix metalloproteinase 2) promoter. *Mol Cell Biol.* 17: 6330–8.

208. Zou Z, Gao C, Nagaich AK, *et al.* (2000). p53 regulates the expression of the tumor suppressor gene maspin. *J Biol Chem.* 275: 6051–4.
209. Mashimo T, Watabe M, Hirota S, *et al.* (1998). The expression of the KAI1 gene, a tumor metastasis suppressor, is directly activated by p53. *Proc Natl Acad Sci USA* 95: 11307–11.
210. Chen X, Zheng Y, Zhu J, Jiang J, Wang J (2001). p73 is transcriptionally regulated by DNA damage, p53, and p73. *Oncogene* 20: 769–74.
211. Jackson P, Ridgway P, Rayner J, Noble J, Braithwaite A (1994). Transcriptional regulation of the PCNA promoter by p53. *Biochem Biophys Res Commun.* 203: 133–40.
212. Murphy M, Hinman A, Levine AJ (1996). Wild-type p53 negatively regulates the expression of a microtubule-associated protein. *Genes Dev.* 10: 2971–80.
213. Roperch JP, Alvaro V, Prieur S, *et al.* (1998). Inhibition of presenilin 1 expression is promoted by p53 and p21WAF-1 and results in apoptosis and tumor suppression. *Nat Med.* 4: 835–8.
214. Mukhopadhyay D, Tsiokas L, Sukhatme VP (1995). Wild-type p53 and v-Src exert opposing influences on human vascular endothelial growth factor gene expression. *Cancer Res.* 55: 6161–5.
215. Miyashita T, Harigai M, Hanada M, Reed J (1994). Identification of a p53-dependent negative response element in the bcl-2 gene. *Cancer Res.* 54: 3131–5.

Regulation of Cellular Senescence by the Retinoblastoma Pathway

Therese M. Becker and Helen Rizos
Westmead Institute for Cancer Research at the Westmead Millennium Institute, University of Sydney, Westmead, New South Wales 2145, Australia

Introduction

The progressive decline in proliferative capacity is an intrinsic property of most normal somatic cells. Cells will lose division potential with roughly exponential kinetics and eventually reach a state of permanent growth arrest, referred to as replicative senescence [1]. Senescent cells remain viable and metabolically active, but cannot re-initiate DNA replication in response to physiological mitogenic stimuli [1–3]. Expression of senescence-associated (SA)-β-galactosidase in human skin cells has provided evidence that cells also undergo senescence *in vivo* [4]. One of the proposed consequences of the senescence response is organismal aging; skin biopsies from older individuals have a greater proportion of senescent cells *in situ* and a lesser proliferative capacity in culture [5, 6]. Also, cross-species comparisons suggest an inverse relationship between the proliferative lifespan of fibroblasts in culture and organism life span [7]. Certainly, decrements in cell renewal would compromise tissue function and integrity, but changes in the differentiated function of senescence cells would also contribute to aging. Senescent fibroblasts, for example, switch from matrix-producing to matrix-degrading cells, secreting large amounts of interstitial collagenase and stromelysin [reviewed in ref. 8] which may contribute to thinning of the dermis as observed *in vivo*. Thus, senescence may prevent the perpetual proliferation of cells during early adulthood, but the accumulation of dysfunctional, apoptotic-resistant senescent cells can have deleterious effects later in life and may promote the development of cancer [8].

Replicative senescence is driven by the loss of telomeric repeat DNA at each cell division; once telomeric restriction fragments reach an average size of 4–6 kb, human cells undergo irreversible growth arrest and adopt a characteristic senescent phenotype [reviewed in ref. 9]. Many proliferative cell types also senesce prematurely

Sunil C. Kaul and Renu Wadhwa (eds.), Aging of Cells In and Outside the Body, 151–169.

in response to DNA damage, the expression of certain oncogenes (e.g., activated Ras or Raf), chromatin remodelling, cytokines and supraphysiological mitogenic signals (e.g., overexpression of E2F-1) [reviewed in ref. 10]. These stimuli have the potential to induce neoplastic transformation, and senescence acts as a failsafe mechanism that can inhibit the growth of potentially oncogenic cells. Thus, telomere-dependent replicative senescence is part of a much broader process termed cellular senescence. Consistent with its role in suppressing cancer, the senescence response is controlled and maintained by tumor suppressor genes. The most crucial of these encode the p53 and retinoblastoma (pRb) tumor suppressor proteins. Inactivation of either p53 or pRb is sufficient to extend the replicative lifespan of human cells and inactivation of both tumor suppressors extends the lifespan to a greater extent than inactivation of either alone (Figure 1) [11–13].

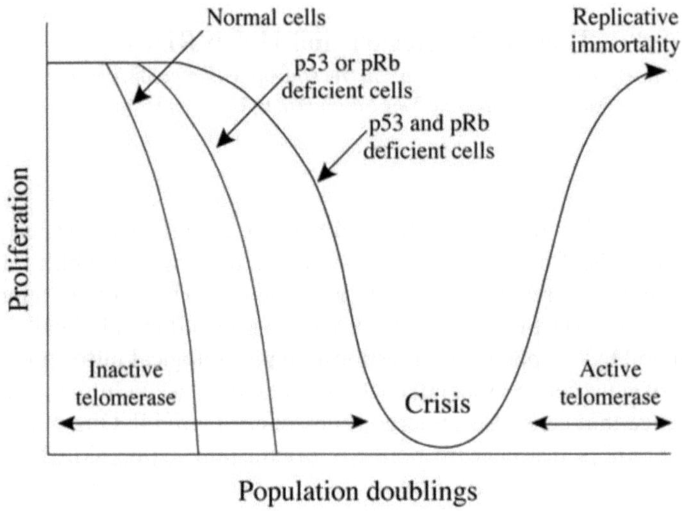

Figure 1. Roles of p53 and pRb in replicative senescence. Inactivation of either p53 or pRb extends the proliferative capacity of normal diploid human cells. Inactivation of both p53 and pRb causes cells to enter crisis, an unstable state in which rare immortal clones may arise. Immortalisation is frequently associated with the reactivation of telomerase [adapted from ref. 105].

Unlike human cells, the need for disruption of the pRb pathway in bypassing senescence in mouse cells remains a subject of debate. Mouse embryonic fibroblasts (MEFs) lacking pRb undergo growth arrest [14], whereas MEFs derived from p53-null animals maintain their proliferative potential and do not senesce [15]. There appears to be functional redundancy within the pRb and its pocket protein homologues, p107 and p130. Although pRb deficiency or combined p107/p130 deficiency caused a slight S-phase acceleration, the overall growth characteristics of these knockout MEFs did not deviate from those of wild-type cells [16, 17]. Ablation of all three pocket proteins rendered MEFs completely insensitive to any senescence response and immortal clones did not sustain deletions or mutations in p53, thus confirming that pocket proteins are also critical regulators of senescence in mice [18]. This chapter will focus on the role of the pRb pathway in establishing cellular senescence and the cellular signals required for this critical pRb function.

The retinoblastoma pathway and cell cycle control

The function of pRb in the senescence program is tightly coupled with the ability of pRb to maintain cell cycle arrest. pRb regulates cell cycle progression from the G_1-phase through the restriction point and into S-phase. This transition is associated with the sequential expression and activation of cyclin/cyclin-dependent kinase (cdk) complexes [19]. D-type cyclin/cdk4 and cdk6 complexes are activated upon growth factor stimulation of quiescent cells, and promote entry into the G_1 phase of the cell cycle (Figure 2) [20]. Both cdk4 and cdk6 can be inhibited by the INK4 family of cdk inhibitors, including $p15^{INK4b}$, $p16^{INK4a}$, $p18^{INK4c}$ and $p19^{INK4d}$ (Figure 2) [reviewed in ref. 21]. The $p16^{INK4a}$ inhibitor can also induce cell cycle arrest in the G_2 phase by inhibiting the cyclin D3/cdk4 complex (Figure 2) [22].

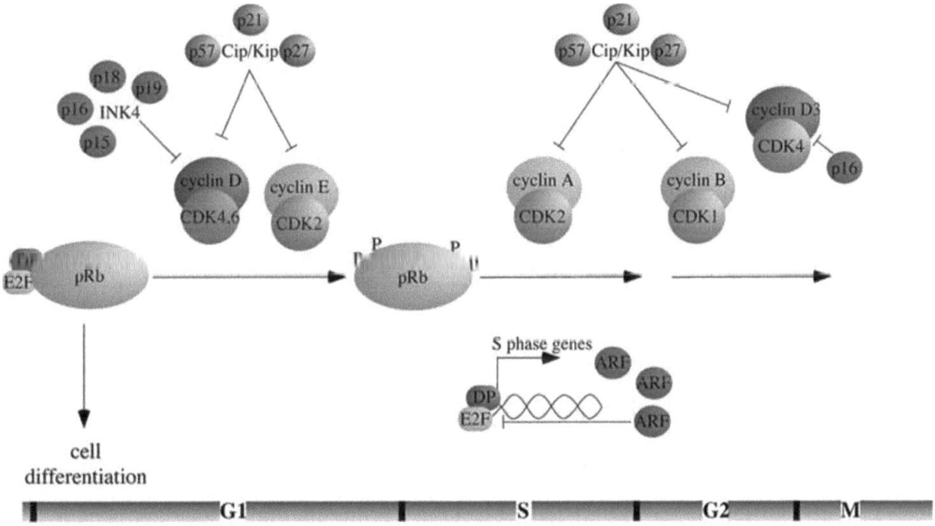

Figure 2. Regulation of the eukaryotic cell cycle. Active, hypophosphorylated pRb binds E2F/DP transcription factors in early G_1 and suppresses the expression of specific target genes. In late G_1, cyclin D/cdk 4,6 and cyclin E/cdk2 complexes initiate and maintain the phosphorylation of pRb. This stimulates the release of E2F/DP factors and the expression of S-phase target genes, including p14ARF. p14ARF acts in a negative feedback loop to destabilise and inhibit the activity of E2F-1, -2 and -3, and also activates the p53 pathway. Cyclin A-associated kinase activity is required for entry and completion of DNA synthesis (S phase) and entry into mitosis (M phase) of the cell cycle. Cyclin/cdk complexes are inhibited by two classes of cyclin dependent kinase inhibitors, INK4 (p15, p16, p18, p19) and Cip/Kip (p21, p27, p57).

The primary substrates of the cyclin D-dependent kinases are the retinoblastoma family of proteins, pRb, p107 and p130 [reviewed in ref. 23]. These functionally and structurally related pocket proteins bind and regulate a large number of cellular proteins, including members of the E2F family of transcription factors (E2F-1 to E2F-6) [24]. E2F factors in combination with a DP family protein (DP-1 to DP-3) control the transcription of many genes that encode proteins involved in cell cycle progression (e.g., p107, E2F-1, E2F-2, cyclin E, cyclin A), DNA metabolism (dihydrofolate reductase, thymidine kinase, thymidylate synthase, histone H2A), the p53 pathway ($p21^{Cip1}$, p14ARF [25]) and proto-oncogenes (Myc, Myb) [reviewed in ref. 23]. Binding of pocket proteins to the E2F factors inhibits the transactivation activity of E2Fs and in some instances converts these transcriptional activators to transcriptional repressors [26].

Phosphorylation of pRb and the other two family members is initiated by the cyclin D-dependent kinases and then accelerated by the cyclin E-cdk2 complex in mid-to-late G_1 (Figure 2) [20]. pRb hyperphosphorylation results in the release of E2F transcription factors and the expression of the above-mentioned E2F-regulated genes. Cyclin A- and cyclin B-dependent kinases maintain pRb in its hyperphosphorylated state as the cycle moves ahead, and pRb is not dephosphorylated until cells complete mitosis and re-enter the G_1 phase (or G_0) (Figure 2). Although all three pocket proteins are likely to be phosphorylated by the cyclin D/cdk complexes, pRb is the critical cell cycle regulator. In almost all cancers either pRb or components of its regulatory pathway are inactivated.

The INK4 family of cdk inhibitors inhibit cyclin D-dependent kinases, while cyclin D-, E- and A-dependent kinases are negatively regulated by the Cip/Kip family of cdk inhibitors, $p21^{Cip1}$, $p27^{Kip1}$ and $p57^{Kip2}$ (Figure 2) [27]. Accumulation of $p21^{Cip1}$ inhibits the activity of cdk2, cdk4 and cdk6 and this prevents the phosphorylation of pRb and passage through the restriction point. $p27^{Kip1}$ may also be involved in restriction point cell cycle control; in quiescent cells, $p27^{Kip1}$ levels are high, but once cells enter the cycle, they rapidly decrease [28].

Role of pRb in cellular senescence

Reintroduction or reactivation of pRb in human tumor cell lines that lack functional pRb often results in senescence, even in cells lacking wild type p53 [29–31]. Cells overexpressing pRb are arrested with G_1 DNA content, and appear enlarged and flattened, a phenotype typical of senescence [30]. The senescence response triggered by pRb is irreversible, but inactivation of pRb in senescent cells leads to the abortive re-entry of these cells into S-phase and their subsequent apoptotic death [31, 32]. Not surprisingly, the senescence response in human diploid fibroblasts is abrogated with the introduction of the pRb-inactivating viral oncoprotein E7 in combination with increased telomerase activity [33]. Similarly, the adenovirus oncoprotein E1A, which inactivates and sequesters pRb, blocks the senescence response triggered by oncogenic Ras in human cells [34]. Finally, MEFs derived from mice homozygous for the constitutively active $cdk4^{R24C}$ mutant displayed decreased doubling times and escaped replicative senescence. $Cdk4^{R24C}$ kinase activity resulted in hyperphosphorylation of all three members of the Rb family [35].

Although pRb mediated E2F repression and subsequent growth arrest may promote senescence, it is not sufficient to maintain permanent growth arrest. The pRb homologues, p107 and p130, also induce cell cycle arrest through E2F repression but induce senescence poorly. Indeed, p107 and p130 may be the primary regulators of E2F-dependent activity. p130 is the main pocket protein bound to E2F target gene promoters in G_0 and early G_1, while p107 dominated in late G_1 and S phases [36]. Further, in synchronized, pRb-deficient MEFs, most E2F target genes are induced at the appropriate time relative to S phase entry [17, 36]. Thus, a feature unique to pRb, possibly $p27^{Kip1}$ acvtivation, induces the senescence response in cells.

pRb exists only in its active (hypophosphorylated) growth inhibitory form in senescent cells. This is because senescent cells accumulate high levels of the cdk inhibitors, $p21^{Cip1}$, $p16^{INK4a}$ and $p27^{Kip1}$. It is not surprising therefore that these cdk inhibitors, which engage the pRb pathway, are critical effectors of cellular senescence.

Role of $p16^{INK4a}$ in pRb-induced replicative and induced senescence

The *CDKN2A/ARF* locus functionally intersects with p53 and pRb and operates a central defence mechanism against hyperproliferative mitogenic and oncogenic signals. Not surprisingly this locus on chromosome band 9p21, is one of the most frequently altered sequences in human cancer [reviewed in ref. 37]. The dual coding capacity of the *INK4a/ARF* locus may account for its high disruption rate in human tumors. This locus encodes two distinct proteins, the $p16^{INK4a}$ cdk inhibitor and the alternative product, p14ARF, which are translated in different reading frames from alternatively spliced transcripts (Figure 3) [38]. Both *INK4a/ARF* encoded proteins maintain cell cycle control: $p16^{INK4a}$ regulates G_1 phase exit by interacting with cdk4 and cdk6 to inhibit the phosphorylation of pRb (see Figure 2), whereas p14ARF interacts with hdm2 to prevent the degradation of p53 [reviewed in ref. 39].

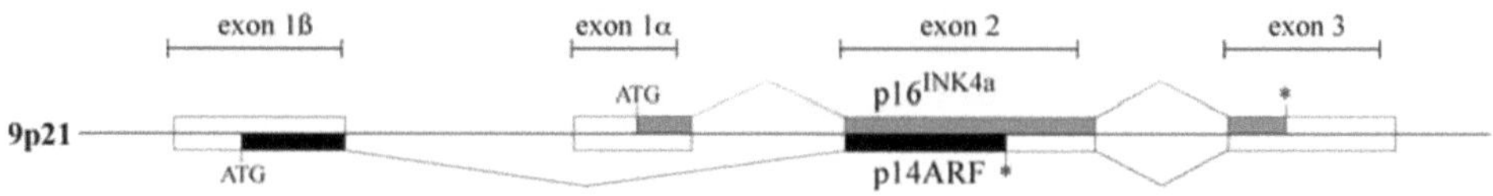

Figure 3. Organization of the INK4a/ARF locus. The coding regions of $p16^{INK4a}$ *are shown in grey, and those of p14ARF are in black. Asterisk (*) denotes the* $p16^{INK4a}$ *and p14ARF termination codons.*

Both $p16^{INK4a}$ and ARF have the potential to act in tumor surveillance and senescence, but their complex genomic organization has made it difficult to resolve their relative importance. In mice, the significance of each gene product has been addressed by elegant gene knock-out experiments. Mice deficient in the mouse p14ARF homologue, p19ARF, alone are highly tumor prone, the resultant MEFs are immortal and also sensitive to transformation by Ras [15, 40]. In contrast, although specific ablation of p16^{INK4a} may contribute to tumor susceptibility, p16^{INK4a}-deficient MEFs undergo senescence and are growth arrested by oncogenic Ras [41, 42]. Thus in mice, ARF is the key regulator of senescence and is invariably disrupted in immortalized MEFs. In human cells, most tumor-associated alterations affecting the *CDKN2A/ARF* locus specifically target and inactivate p16^{INK4a}, and evidence linking p16^{INK4a} to senescence is compelling. High p16^{INK4a} levels are maintained in senescent cells, and human cells that have lost p16^{INK4a} have an extended lifespan [43]. In human diploid fibroblasts, oncogenic Ras elicits the same senescent-like response seen in MEFs [34], but Ras has little impact on p14ARF expression whereas it induces p16^{INK4a} via the Ras/Raf/MEK pathway (Figure 4). Further, cells deficient for p16^{INK4a} immortalise if they also inactivate the p53 pathway and gain the ability to maintain telomere length [44].

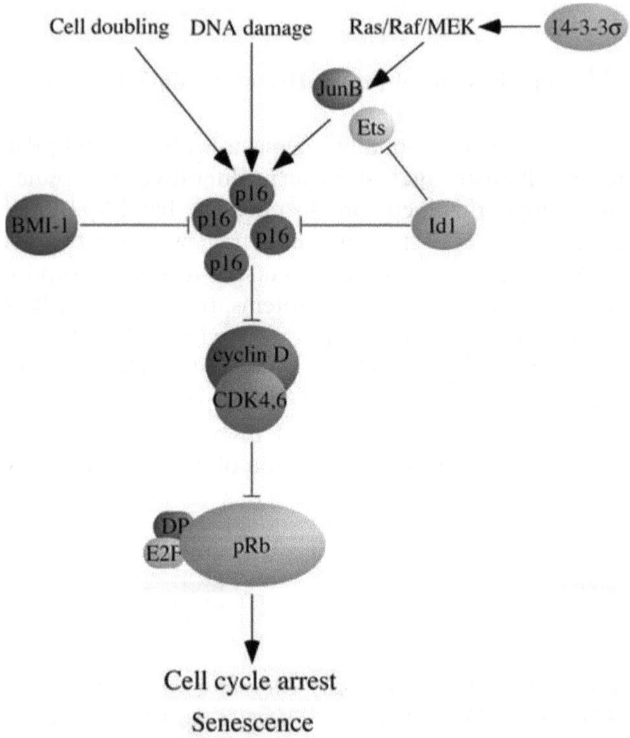

Figure 4. *p16^{INK4a} promotes senescence by maintaining pRb in an active, hypophosphorylated state. p16^{INK4a} accumulation can be triggered via a series of signals, including DNA damage, the Ras/Raf/MEK cascade and continued cell division. Increased levels of p16^{INK4a} inhibit the cyclin D-dependent kinases and prevent the hyperphosphorylation of pRb. Active pRb induces potent G1 cell cycle arrest and can promote both cellular and replicative senescence. The expression of p16^{INK4a} is negatively regulated in early passage cells, where p16^{INK4a} levels are low. The chromatin remodeling protein, BMI-1, and the transcription factor Id-1, repress p16^{INK4a}, and their ectopic expression can bypass the senescence response.*

In normal, early passage cells, such as T-lymphocytes, fibroblasts, keratinocytes and melanocytes, $p16^{INK4a}$ levels are undetectable or low, and they steadily increase with each population doubling until they peak (up to 40-fold compared to early passage cells) at senescence [reviewed in refs. 10, 45]. The increase in $p16^{INK4a}$ as cells enter senescence is due to its transcriptional up-regulation, and its stability; $p16^{INK4a}$ has a half-life of 12–18 h and accumulates with each population doubling (Figure 4) [46–50].

The accumulation of $p16^{INK4a}$ in senescent cells inhibits pRb phosphorylation, and this is regarded as the most important tumor suppressing function of $p16^{INK4a}$. First, $p16^{INK4a}$ promotes rapid pRb-dependent G_1 arrest. Second, $p16^{INK4a}$ maintains pRb in an active state where it induces morphological changes typical of senescence, and independent of E2F binding [51]. Third, $p16^{INK4a}$-mediated senescence in tumor cell lines such as glioma cells, squamous cell carcinoma and melanoma cells correlated with the relative increase of underphosphorylated pRb [52–54].

The importance of $p16^{INK4a}$ and, therefore, pRb in the process of senescence in human cells is probably best demonstrated from studies on melanoma-associated $p16^{INK4a}$ mutations. $p16^{INK4a}$ is frequently inactivated in individuals derived from melanoma-prone kindreds [55], and most of these $p16^{INK4a}$ mutants are unable to inhibit pRb phosphorylation and can not induce potent cell cycle arrest [56–58]. In contrast to wild type $p16^{INK4a}$, these mutants do not produce a typical senescent morphology [58]. More importantly, human fibroblasts derived from an individual who carries mutations in both *CDKN2A* alleles, were resistant to Ras-induced growth arrest and senescence, although mutant $p16^{INK4a}$ levels increased as expected [59]. This supports a model in which the pRb/$p16^{INK4a}$ pathway is critical in establishing and maintaining cellular senescence in human cells, particularly in response to oncogenic insults.

The inverse correlation between $p16^{INK4a}$ expression and proliferation verifies that regulation of $p16^{INK4a}$ is a critical component in establishing senescence. It is known that cells deficient for functional pRb express high levels of $p16^{INK4a}$, and the presence of pRb downregulates $p16^{INK4a}$ at the transcriptional level [47, 60]. pRb-mediated transcriptional repression is complicated, with multiple cellular factors cooperating for transcriptional repression on specific promoters. Transcriptional regulation of $p16^{INK4a}$ via pRb does not depend on E2F transcription factors, but presumably involves other pRb-binding factors. Using cDNA micro-array analysis, some 300 targets were reproducibly repressed by pRb, and many of these were distinct from genes known to be regulated by E2F [61].

The Ras/Raf/MEK signalling cascade is a key regulator of $p16^{INK4a}$, and the activation of this pathway induces cellular senescence accompanied by accumulation of p53 and $p16^{INK4a}$ in several cell types. Members of the Ets family, which includes Ets1, Ets2, PEA3, SAP1 and ELK1, are downstream targets of Ras/Raf/MEK signalling and can be activated by MAPK-mediated phosphorylation. Both Ets-1 and Ets-2 can promote $p16^{INK4a}$ transcription via Ets binding sites in the *CDKN2A* promoter [48], but in early passage human fibroblasts the activity of the predominant Ets protein, Ets-2, is counterbalanced by Id1, and $p16^{INK4a}$ levels remain low (Figure 4). The Id family of helix-loop-helix proteins bind and negatively regulate the Ets family of transcription factors. Furthermore, Id1 directly down-regulates the transcription of the p16^{INK4a} gene, probably via two E-box motifs identified in the *CDKN2A* promoter [62]. Neonatal foreskin keratinocytes with retroviral overexpression of Id1 showed p16^{INK4a} down regulation and delayed senescence entry after an extended lifespan [63] and MEFs lacking functional Id1 entered premature senescence that correlated with increased p16^{INK4a} levels [62, 64]. In senescence cells, however, when Ras/Raf/MEK signalling is attenuated and Ets2 and Id1 levels decline, the upregulation of p16^{INK4a} is driven by increased expression of Ets1, in the absence of any interference by Id1 (Figure 4) [48]. These ideas are consistent with recent reports that ectopic expression of Ets-1 or -2 increased p16^{INK4a} levels, induced growth arrest and senescence in human fibroblasts [48].

Activation of Raf and signalling through the Ras/Raf/MEK cascade requires binding of the dimeric protein, 14-3-3σ, to serine-phosphorylated Raf [65]. 14-3-3 proteins bind a variety of cellular factors, and have been directly associated with the control of the cell cycle. In colorectal cancer cells, exposure to DNA damaging agents results in a p53-dependent induction of 14-3-3σ which, in turn, arrests cells in the G_2/M phase of the cell cycle [66]. 14-3-3σ is abundant in the suprabasal layers of human epidermis and its downregulation allows keratinocytes to escape senescence. The downregulation of 14-3-3σ is associated with p16^{INK4a} repression and maintenance of telomerase activity (Figure 4) [65]. It is not surprising, therefore, that coincident epigenetic inactivation of both p16^{INK4a} and 14-3-3σ has recently been detected in oral and vulval cancers [67, 68].

As with other regulators of p16^{INK4a} expression, JunB, which forms part of the Ap-1 transcription factor complex, is also involved in the process of senescence. JunB activates p16^{INK4a} transcription via interaction with AP-1 binding sites in the *CDKN2A* promoter. JunB dependent upregulation of p16^{INK4a} was associated with growth inhibition and premature senescence in mouse fibroblasts. In normal mouse fibroblasts the levels of JunB increased when cells approached senescence and JunB did not affect cell proliferation or senescence in p16^{INK4a}-deficient mouse fibroblasts [50]. Active AP-1 is a known downstream element of the Ras/Raf/MEK pathway, and is another element that promotes p16^{INK4a} expression and senescence via pRb activation when this cascade is constitutively activated (Figure 4) [69].

Chromatin remodelling can potentially disrupt the expression of many genes and can induce a senescent phenotype in human and mouse cells. The chromatin remodelling protein, BMI-1 is known to immortalise and cooperate with the c-Myc oncogene in transforming MEFs. BMI-1 deficient MEFs overexpressed p16^{INK4a} and p19ARF and underwent premature senescence in culture, which was prevented in MEFs lacking p16^{INK4a} and BMI-1 expression [70]. Conversely, BMI-1 overexpression induced telomerase activity and immortalized p16^{INK4a}-deficient human mammary epithelial cells. The overexpression of BMI-1 also extended the replicative life span of human fibroblasts, but in these cells telomerase was not induced and the cells were not immortalized. Although, the ability of BMI-1 to regulate telomerase appears cell-type specific, the ability of this repressor to suppress p16^{INK4a} transcription occurs in many cell types [71]. As expected, expression of BMI-1 decreases, while p16^{INK4a} steadily accumulates as human fibroblasts near senescence (Figure 4) [49].

Does p14ARF play a role in pRb mediated senescence in human cells?

The ARF tumor suppressor is a key regulator of senescence in mice and is invariably disrupted in immortalized MEFs. In human cells, however, the role of p14ARF in senescence is not so obvious. It is well established that p14ARF activates p53 in response to mitogenic stimuli, including E1A, MYC, oncogenic Ras, v-abl and E2F-1 [reviewed in ref 72]. These stimuli can induce premature senescence in human cells, and it seems likely that ARF may participate in this response.

ARF inhibits cell growth principally by activating the p53 pathway but can also inhibit growth via p53- and hdm2-independent mechanisms [73, 74]. The p53/hdm2-independent functions of ARF are poorly understood but presumably involve additional ARF-binding partners. ARF can interact and destabilise the S-phase inducing transcription factors E2F-1, -2 and -3 (Figure 5) [75–77].

Consistent with a role for E2F degradation in the mechanism of ARF growth suppression, p53-null MEFs engineered to overexpress E2F-1 were at least partially insensitive to the growth inhibitory action of p19ARF [75, 78]. Other proteins also complex with ARF but the extent to which these ARF-complexes modulate the cell cycle inhibitory action of ARF is not well established.

Although p14ARF does not accumulate significantly with prolonged tissue culture or at telomere-based senescence [79, 80], it may still play a role in mediating senescence induced by oncogenic signals. For instance, E2F-1 induces a senescent phenotype when overexpressed in normal human fibroblasts. *In vivo*, inappropriate or excess E2F-1 can result from a number of mechanisms, including DNA damage or upstream oncogenic activation [81, 82]. The ability of unrestricted E2F-1 to induce premature senescence in normal human cells can be accounted for by the induction of ARF by E2F-1; cells with compromised p14ARF or p53 function fail to senesce in response to E2F-1 activation [83]. The regulation of p14ARF by E2F-1 establishes an important feedback loop that acts as a sensor for deregulated pRb function (Figures 2 and 5). Activation of p14ARF in cells overexpressing E2F-1 would promote rapid cell cycle arrest and senescence if the cells maintained an intact p53 pathway.

The ARF/p53/p21^{Cip1} pathway and pRb-mediated senescence

Like p14ARF, the cdk inhibitor, p21^{Cip1}, forms an important link between the pRb and p53 cell cycle regulatory pathways. p21^{Cip1} can interact with all cdk complexes, and depending on its stoichiometry, may inhibit or stimulate cdk4/6 complexes. When overexpressed, p21^{Cip1} inhibits E2F activity indirectly, via cdk inactivation and pRb dephosphorylation (Figures 2 and 5), and directly by binding to E2F. Moreover, p21^{Cip1}, a transcriptional target of the p53 tumor suppressor, accumulates in both quiescent and senescent cells. Like p53, the levels of p21^{Cip1} are transiently elevated during senescence and several weeks after human fibroblast cultures reach senescence, p21^{Cip1} levels gradually decline as p16^{INK4a} levels rise [reviewed in ref. 84].

Induction of p21^{Cip1}, via p53 activation, or other inducers, such as TGF-β and mimosine may contribute to the functional changes that occur in senescent cells. In

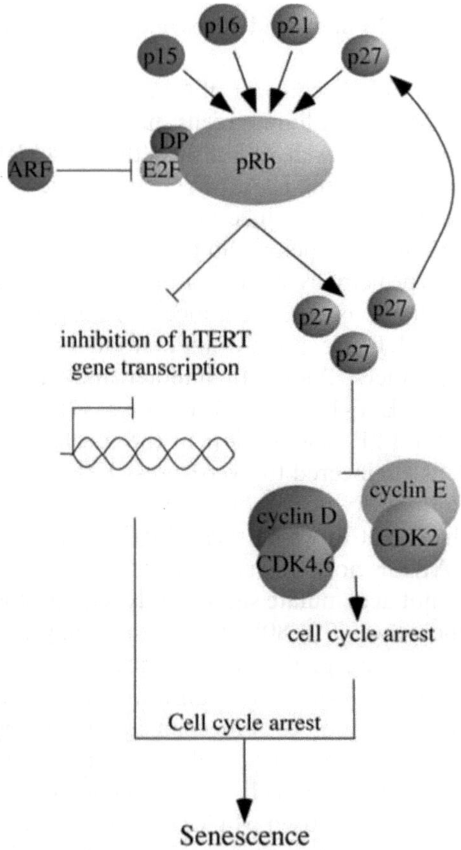

Figure 5. *Model for pRb-mediated senescence. Accumulation of the cyclin dependent kinase inhibitors, p16^{INK4a}, p15^{INK4b}, p27^{Kip1} and p21^{Cip1} maintain pRb in an active hypophosphorylated state. In this state, pRb suppresses the E2F/DP transcription factor complex and the subsequent expression of E2F target genes, which include the telomerase reverse transcriptase (hTERT) gene. Activation of the p14ARF tumour suppressor by abnormal proliferative signals including E1A and c-myc, leads to the repression and degradation of the E2F transcriptional activators. In a non-E2F-dependent manner, hypophosphorylated pRb increases p27^{Kip1} levels postranscriptionally. The inhibition of E2F-mediated transcription along with the accumulation of p27^{Kip1}, induces a potent and persistent G_1 cell cycle arrest that triggers senescence and the associated morphological changes.*

cDNA micro-array analysis, p21^{Cip1} expression was found to upregulate a series of genes, which have been associated with cellular senescence and organismal aging [85]. Conditioned media derived from p21^{Cip1}-induced, but not uninduced cells, stimulated the growth of a slow growing human fibrosarcoma cell line, and protected an apoptosis-prone cell line from serum starvation-induced apoptosis. The transcriptional activating and paracrine activities of p21^{Cip1} are shared by p16^{INK4a}. The consecutive activation of these two cdk inhibitors initiate and maintain senescence-associated arrest, but also contribute to the secretion of mitogenic and anti-apoptotic factors that may promote the growth and survival of carcinoma cells [reviewed in ref 84]

Role of p15^{INK5b} in pRb-associated senescence

The important role of p16^{INK4a} in human cancer development is well established, and yet the contribution of its homologue, p15^{INK4b} in tumorigenesis is unclear. The p15^{INK4b} gene, *CDKN2B*, is frequently deleted, usually along with the p16^{INK4a} gene, in tumor cell lines and primary tumors [86], but only rare cases of p15^{INK4b} inactivation, through gene mutations or hypermethylation have been reported [87, 88]. Moreover, whereas several reports have demonstrated a strong correlation between p16^{INK4a} expression and replicative senescence, the level of p15^{INK4b} does not always change with increasing population doublings [46]. Taken together, there is little evidence that p15^{INK4b} contributes to senescence, and yet overexpression of this cdk inhibitor in primary fibroblasts can induce irreversible growth arrest, the hallmark of senescence. Similarly, glioma cells overexpressing p15^{INK4b} arrested in G_1, and showed the typical senescence-associated changes including cellular enlargement and expression of SA-β-galactosidase [89]. p15^{INK4b}-induced senescence was dependent on the presence of wild type pRb and was indistinguishable from that caused by p16^{INK4a}.

It is possible that p15^{INK4b} may not generally function to induce senescence, but it may do so after exposure of certain cell types to specific stimuli. The expression of p15^{INK4b}, but not p16^{INK4a}, is regulated by exogenous growth-inhibitory factors such as TGF-β and IFN-α [91]. TGF-β induced expression of p15^{INK4b} appears to co-operate with p27^{Kip1} to induce a G_1 phase arrest and quiescence [92]. Further, the TGF-β pathway is involved in the senescence response of primary mouse keratinocytes to oncogenic Ras. The secretion of active TGF-β1, and p15^{INK4b} levels increased in parallel with the growth arrest and senescence of Ras-infected keratinocytes, and preceded the induction of p16^{INK4a}. As with many other cell types the senescence associated growth arrest of TGF-β1 signalling proficient cells occurred in G_1. Thus pRb became unphosphorylated, the total protein levels of pRb decreased, cdk4 and cdk2 kinase activity were reduced as was the expression of cyclins A and B (Figure 5) [93]. Although it is likely that Ras-associated senescence was mediated by the combined action of p15^{INK4b}, p16^{INK4a} and p21^{Waf1}, the likely direct target for the increase in TGF-β secretion is p15^{INK4b}. This may explain the frequent co-deletion and inactivation of the p15^{INK4b} and p16^{INK4a} gene loci in tumors [reviewed in ref 94].

Role of p27^{Kip1} in retinoblastoma protein mediated senescence

The induction of senescence by ectopic pRb in Saos-2 tumor cells, which are defective for both pRb and p53, correlated with the post-transcriptional accumulation of the cdk inhibitor $p27^{Kip1}$, an increase of $p27^{Kip1}$-cyclin E complexes and a concomitant decrease in cyclin E-associated kinase activity (see Figure 2). The accumulation of $p27^{Kip1}$ was maintained by pRb, and not by the p107 and p130 pocket proteins. Given that E2F regulatory functions appear to be adequately controlled by p107 and p130 [36], and that pRb mutants deficient in E2F regulation were still able to induce $p27^{Kip1}$ and senescence, it appears that E2F repression was not required for senescence in these tumor cells. The elimination of $p27^{Kip1}$ expression using specific antisense oligonucleotides, diminished pRb-mediated G_1 arrest and decreased the proportion of pRb-induced senescent cells [30].

Interestingly, whereas $p27^{Kip1}$ produced large rounded Saos-2 cells that were positive for SA-β-galactosidase activity, pRb-induced senescent cells were enlarged and flattened. The morphological differences between pRb and $p27^{Kip1}$ induced senescence involves the reorganization of microtubules in pRb-transfected cells. Microtubule alterations may contribute to pRb-mediated senescence response, but it is clearly not required for senescence and is separable from $p27^{Kip1}$ induction [30].

The general importance of the pRb-$p27^{Kip1}$ pathway in inducing a general senescence response remains to be resolved. Certainly this pathway was activated in a cell line lacking functional p53, and may be subordinated by the existence of a functional p53 pathway. Evidence is accumulating for a $p27^{Kip1}$-dependent, p53-independent mechanism of senescence in human cells. For example, the simultaneous blocking of p53 function and re-establishment of the pRb pathway (using $p16^{INK4a}$) in U20S cells, an osteosarcoma cell line that is wild type for p53 and pRb, resulted in the upregulation of $p27^{Kip1}$ in senescent cells. Also, the inhibition of phosphoinositide 3-kinase in MEFs led to a senescent phenotype that was associated with elevated $p27^{Kip1}$ but not p53, p19ARF, $p16^{INK4a}$ or $p21^{Cip1}$ levels. Finally, iron chelation-induced senescence in hepatocyte cell lines was associated with $p27^{Kip1}$ induction and was p53-independent [96].

A role for the pRb family of proteins in controlling telomere length

Cells lacking telomerase lose up to 200 bp of telomeric DNA per cell division [97]. Human cells with eroded telomeres cease cell division with a senescent phenotype [98], and telomerase re-activation is sufficient to rescue short telomeres and confer unlimited proliferative capacity [99]. The phosphorylation status of pRb is a key regulator of telomerase activity (Figure 5). Restoration of functional pRb in pRb- and p53-defective tumor cells induced growth cessation concurrently with inhibition of telomerase activity [31]. Conversely, overexpression of cdk2, cdk4 or cyclins D1 and E, all of which induce pRb hyperphosphorylation, rescue pRb mediated down-regulation of telomerase activity [97]. E2F-1 is an important target of pRb action, and can specifically bind to two non-canonical E2F consensus sites within the telomerase reverse transcriptase (hTERT) gene promoter [100]. In normal human somatic cells, E2Fs 1-5 activate the hTERT promoter, via these E2F-1 sites [101], and tumor tissues from $E2F^{-/-}$ mice are negative for telomerase activity [97]. In contrast, E2Fs 1-3, repress the hTERT promoter in human tumor cells [101]; human cancer cell lines stably overexpressing E2F-1 exhibit decreased hTERT expression and telomerase activity [100]. The inhibitory action of E2F-1 on hTERT gene transcription occurs by suppressing Sp1-mediated transcriptional activation; Sp1 is strongly associated with the hTERT promoter in human tumor cells [101]. A dual activator and repressor role for the E2F transcription factors is well established; pRb/E2F complexes can actively repress the transcription of E2F-responsive promoters in G_0/G_1 [reviewed in ref. 102], and it is also possible that the regulation of telomerase activity by E2F-1 is cell cycle dependent and influenced by the presence and phosphorylation status of functional pocket-proteins.

In addition to pRb, the retinoblastoma family of proteins, p107 and p130 are also involved in the regulation of telomere length. In particular, MEFs doubly deficient in p107 and p130 or triply deficient in pRb, p107 and p130 have markedly elongated telomeres compared with those of wild type or pRb-deficient cells. The triply deficient cells were immortal, but their long telomere phenotype was probably not responsible for their immortality, as the double knock out cells, which also showed elongated telomeres, were not immortal and showed the typical senescence-like arrest after 5–10 population doublings in culture. It was the additional loss of pRb that permitted triply deficient MEFs with long telomeres to divide immortally. In MEFs deficient in pRb, p107 and p130 the elongation of the telomeres was not associated with any detectable increase in telomerase activity. The lengthening of telomeres may be due to changes in the accessibility of the telomere to telomerase, or to the activation of mechanisms for telomerase-independent telomere elongation, known as alternative lengthening of telomeres [reviewed in ref. 103]. The connection between the regulation of telomere length and cell cycle control could be the basis of lifespan extension exerted by a number of viral oncoproteins that target and inactivate the pRb family [104].

Conclusions

Complex genetic and epigenetic events are associated with cellular aging and senescence. The molecular changes required for the senescence response can vary with cell lineage, tissue type and physiological stimuli. Nevertheless, the same central pathways remain critical for implementing and maintaining senescence growth arrest in mitotic cells. Like p53, pRb is a key regulator of cell cycle progression, growth arrest in response to mitogenic stimuli and cellular senescence. The fact that both these proteins are often inactivated in most human tumors, tell us that cellular senescence is a mechanism for suppressing the development of malignant tumors *in vivo*, and that cells can maintain senescence via multiple pathways in an effort to avoid tumorigenesis.

References

1. Stanulis-Praeger B (1987). Cellular senescence revisited: a review. *Mech Ageing Dev.* 38: 1–48.
2. Campisi J, Dimri GP, Hara E (1996). Control of replicative senescence. In: Schneider E, Rowe J, eds. *Handbook of the Biology of Aging.*New York: Academic Press, pp. 121–49.

3. Cristofalo VJ, Pignolo RJ (1993). Replicative senescence of human fibroblast-like cells in culture. *Physiol Rev.* 73: 617–38.
4. Dimri GP, Lee X, Basile G, *et al.* (1995). A biomarker that identifies senescent human cells in culture and in aging skin *in vivo*. *Proc Natl Acad Sci USA*, 92: 9363–7.
5. Bayreuther K, Francz PI, Gogol J, Kontermann K (1992). Terminal differentiation, aging, apoptosis and spontaneous transformation in fibroblast stem cell systems *in vivo* and *in vitro*. *Ann NY Acad Sci.* 663: 167–79.
6. Haddad MM, Xu W, Medrano EE (1998). Aging in epidermal melanocytes: cell cycle genes and melanins. *J Invest Dermatol Symp Proc.* 3: 36–40.
7. Rohme D (1981). Evidence for a relationship between longevity of mammalian species and life spans of normal fibroblasts *in vitro* and erythrocytes *in vivo*. *Proc Natl Acad Sci USA* 78: 5009–13.
8. Campisi J (1997). Aging and cancer: the double-edged sword of replicative senescence. *J Am Geriatric Soc.* 45: 1–6.
9. Neumeister P, Albanese C, Balent B, Greally J, Pestell RG (2002). Senescence and epigenetic dysregulation in cancer. *Int J Biochem Cell Biol.* 34: 1475–90.
10. Toussaint O, Remacle J, Dierick J-F, *et al.* (2002). From the Hayflick mosaic to the mosaics of aging. Role of stress-induced premature senescence in human aging. *Int J Biochem Cell Biol.* 34: 1415–29.
11. Shay JW, Pereira-Smith OM, Wright WE (1991). A role for both RB and p53 in the regulation of human cellular senescence. *Exp Cell Res.* 196: 33–9.
12. Shay JW, Wright WE, Werbin H (1991). Defining the molecular mechanisms of human cell immortalization. *Biochim Biophys Acta* 1072: 1–7.
13. Hara E, Tsurui H, Shinozaki A, Nakada S, Oda K (1991). Cooperative effect of antisense-Rb and antisense-p53 oligomers on the extension of life span in human diploid fibroblasts, TIG-1. *Biochem Biophys Res Commun.* 179: 528–34.
14. Zindy F, Eischen CM, Randle DH, *et al.* (1998). Myc signaling via the ARF tumor suppressor regulates p53-dependent apoptosis and immortalization. *Genes Dev.* 12: 2424–33.
15. Kamijo T, Zindy F, Roussel MF, *et al.* (1997). Tumor suppression at the mouse INK4a locus mediated by the alternative reading frame product $p19^{ARF}$. *Cell* 91: 649–59.
16. Cobrinik D, Lee MH, Hannon G, *et al.* (1996). Shared role of the pRB-related p130 and p107 proteins in limb development. *Genes Dev.* 10: 1633–44.
17. Herrera RE, Makela TP, Weinberg RA (1996). TGF beta-induced growth inhibition in primary fibroblasts requires the retinoblastoma protein. *Mol Biol Cell* 7: 1335–42.
18. Dannenberg J-H, van Rossum A, Schuijff L, Te Riele H (2000). Ablation of the retinoblastoma gene family deregulates G1 control causing immortalization and increased cell turnover under growth-restricting conditions. *Genes Dev.* 14: 3051–64.
19. Morgan DO (1995). Principles of CDK regulation. *Nature* 374: 131–4.
20. Sherr CJ (1994). G1 phase progression: cyclin on cue. *Cell* 79: 551–5.
21. Sherr CJ, Roberts JM (1995). Inhibitors of mammalian G1 cyclin-dependent kinases. *Genes Dev.* 9: 1149–63.
22. Gabrielli BG, Sarcevic B, Sinnamon J, *et al.* (1999). A cyclin D-CDK4 activity is required for G2 phase cell cycle progression in ultraviolet radiation induced G2 phase delay. *J Biol Chem.* 274: 13961–9.
23. Grana X, Garriga J, Mayol X (1998). Role of the retinoblastoma protein family, pRB, p107 and p130 in the negative control of cell growth. *Oncogene* 17: 3365–83.
24. Johnson DG, Schneider-Broussard R (1998). Role of E2F in cell cycle control and cancer. *Front Biosci.* 27: d447–8.

25. Bates S, Phillips AC, Clark PA, *et al.* (1998). p14ARF links the tumour suppressors RB and p53. *Nature* 395: 124–5.
26. Luo RX, Postigo AA, Dean DC (1998). Rb interacts with histone deacetylase to repress transcription. *Cell* 92: 463–73.
27. Grana X, Reddy P (1995). Cell cycle control in mammalian cells: role of cyclins, cyclin dependent kinases (CDKs), growth suppressor genes and cyclin-dependent kinase inhibitors (CKIs). *Oncogene* 11: 211–19.
28. Kato JY, Matsuoka M, Polyak K, Massague J, Sherr CJ (1994). Cyclic AMP-induced G1 phase arrest mediated by an inhibitor (p27^{Kip1}) of cyclin-dependent kinase 4 activation. *Cell* 79: 487–96.
29. Dai CY, Enders GH (2000). p16^{INK4a} can initiate an autonomous senescence program. *Oncogene* 19: 1613–22.
30. Alexander K, Hinds PW (2001). Requirement for p27^{Kip1} in retinoblastoma protein-mediated senescence. *Mol Cell Biol.* 21: 3616–31.
31. Xu H-J, Zhou Y, Ji W, *et al.* (1997). Reexpression of the retinoblastoma protein in tumor cells induces senescence and telomerase inhibition. *Oncogene* 15: 2589–96.
32. Tiemann F, Hinds PW (1998). Induction of DNA synthesis and apoptosis by regulated inactivation of a temperature-sensitive retinoblastoma protein. *EMBO J.* 17: 104–52.
33. Kiyono T, Foster SA, Koop JI, McDougall JK, Galloway DA, Klingelhutz AJ (1998). Both Rb/p16^{INK4a} inactivation and telomerase activity are required to immortalize human epithelial cells. *Nature* 396: 84–8.
34. Serrano M, Lin AW, McCurrach ME, Beach D, Lowe SW (1997). Oncogenic ras provokes premature cell senescence associated with accumulation of p53 and p16^{INK4a}. *Cell* 88: 593–602.
35. Rane SG, Cosenza SC, Mettus RV, Reddy EP (2002). Germ line transmission of the Cdk4^{R24C} mutation facilitates tumorigenesis and escape from cellular senescence. *Mol Cell Biol.* 22: 644–56.
36. Hurford Jr RK, Cobrinik D, Lee MH, Dyson N (1997). pRB and p107/p130 are required for the regulated expression of different sets of E2F responsive genes. *Genes Dev.* 11: 1447–63.
37. Sherr CJ (1996). Cancer cell cycles. *Science* 274: 1672–7.
38. Quelle DE, Zindy F, Ashmun RA, Sherr CJ (1995). Alternative reading frames of the *INK4a* tumor suppressor gene encode two unrelated proteins capable of inducing cell cycle arrest. *Cell* 83: 993–1000.
39. James MC, Peters G (2000). Alternative product of the p16/CDKN2A locus connects the Rb and p53 tumor suppressors. *Prog Cell Cycle Res.* 4: 71–81.
40. Serrano M, Lee H-W, Chin L, Cordon-Cardo C, Beach D, DePinho RA (1996). Role of the *INK4a* locus in tumor suppression and cell mortality. *Cell* 85: 27–37.
41. Krimpenfort P, Quon KC, Mooi WJ, Loonstra A, Berns A (2001). Loss of p16^{INK4a} confers susceptibility to metastatic melanoma in mice. *Nature* 413: 83–6.
42. Sharpless NE, Bardeesy N, Lee K-H, *et al.* (2001). Loss of p16^{INK4a} with retention of p19^{ARF} predisposes mice to tumorigenesis. *Nature* 413: 86–91.
43. Noble JR, Rogan EM, Neumann AA, Maclean K, Bryan TM, Reddel RR (1996). Association of extended in vitro proliferative potential with loss of p16^{INK4} expression. *Oncogene* 13: 1259–68.
44. Huschtscha LI, Reddel RR (1999). p16^{INK4a} and the control of cellular proliferative life span. *Carcinogenesis* 20: 921–6.

45. Bandyopadhyay D, Medrano EE (2000). Melanin accumulation accelerates melanocyte senescence by a mechanism involving p16^{INK4a}/CDK4/pRB and E2F1. *Ann NY Acad Sci.* 908: 71–84.
46. Palmero I, McConnell B, Parry D, *et al.* (1997). Accumulation of p16^{INK4a} in mouse fibroblasts as a function of replicative senescence and not of retinoblastoma gene status. *Oncogene* 15: 495–503.
47. Hara E, Smith R, Parry D, Tahara H, Steven S, Peters G (1996). Regulation of p16(CDKN2 expression and its implications for cell immortalization and senescence. *Mol Cell Biol.* 16: 859–67.
48. Ohtani N ZZ, Huot TJ, Stinson JA, *et al.* (2001). Opposing effects of Ets and Id proteins on p16INK4a expression during cellular senescence. *Nature* 409: 1067–70.
49. Itahana K, Zou Y, Itahana Y, *et al.* (2003). Control of the replicative life span of human fibroblasts by p16 and the polycomb protein Bmi-1. *Mol Cell Biol.* 23: 389–401.
50. Passegue E, Wagner EF (2000). JunB suppresses cell proliferation by transcriptional activation of p16^{INK4a} expression. *EMBO J.* 19: 2969–79.
51. Sellers WR, Novitch BG, Miyake S, *et al.* (1998). Stable binding to E2F is not required for the retinoblastoma protein to activate transcription, promote differentiation, and suppress tumor cell growth. *Genes Dev.* 12: 95–106.
52. Uhrbom L, Nister M, Westermark B (1997). Induction of senescence in human malignant glioma cells by p16^{INK4a}. *Oncogene* 15: 505–14.
53. Timmermann S, Hinds PW, Munger K (1998). Re-expression of endogenous p16^{INK4a} in oral squamous cell carcinoma lines by 5-aza-2′-deoxycytidine treatment induces a senescence-like state. *Oncogene* 26: 3445–53.
54. Stone S, Dayananth P, Kamb A (1996). Reversible, p16-mediated cell cycle arrest as protection from chemotherapy. *Cancer Res.* 56: 3199–202.
55. Kefford RF, Newton Bishop JA, Bergman W, Tucker MA (1999). Counseling and DNA testing for individuals perceived to be genetically predisposed to melanoma: s consensus dtatement of the melanoma genetics consortium. *J Clin Oncol.* 17: 3245–51.
56. Walker GJ, Gabrielli BG, Castellano M, Hayward NK (1999). Functional reassessment of p16 variants using a transfection-based assay. *Int J Cancer* 82: 305–12.
57. Parry D, Peters G (1996). Temperature-sensitive mutants of p16CDKN2 associated with familial melanoma. *Mol Cell Biol.* 16: 3844–52.
58. Becker TM, Rizos H, Kefford RF, Mann GJ (2001). Functional impairment of melanoma-associated p16^{INK4a} mutants in melanoma cells despite retention of cyclin-dependent kinase 4 binding. *Clin Cancer Res.* 7: 3282–8.
59. Huot TJ, Rowe J, Harland M, *et al.* (2002). Biallelic mutations in p16^{INK4a} confer resistance to Ras- and Ets-induced senescence in human diploid fibroblasts. *Mol Cell Biol.* 22: 8135–43.
60. Li Y, Nichols MA, Shay JW, Xiong Y (1994). Transcriptional repression of the D-type cyclin-dependent kinase inhibitor p16 by the retinoblastoma susceptibility gene product pRb. *Cancer Res.* 54: 6078–82.
61. Markey MP, Angus SP, Strobeck MW, *et al.* (2002). Unbiased analysis of RB-mediated transcriptional repression identifies novel targets and distinctions from E2F action. *Cancer Res.* 62: 6587–97.
62. Alani RM, Young AZ, Shifflett CB (2001). Id1 regulation of cellular senescence through transcriptional repression of p16/INK4a. *Proc Natl Acad Sci USA* 98: 7812–16.
63. Nickoloff BJ, Chaturvedi V, Bacon P, Qin JZ, Denning MF, Diaz MO (2000). Id-1 delays senescence but does not immortalize keratinocytes. *J Biol Chem.* 275: 27501–4.

64. Lyden D, Young AZ, Zagzag D, *et al.*(1999). Id1 and Id3 are required for neurogenesis, angiogenesis and vascularization of tumour xenografts. *Nature* 401: 670–7.
65. Dellambra E, Golisano O, Bondanza S, *et al.* (2000). Downregulation of 14-3-3sigma prevents clonal evolution and leads to immortalization of primary human keratinocytes. *J Cell Biol.* 149: 1117–30.
66. Hermeking H, Lengauer C, Polyak K, *et al.* (1997). 14-3-3 sigma is a p53-regulated inhibitor of G2/M progression. *Mol Cell* 1: 3–11.
67. Gasco M, Bell AK, Heath V, *et al.* (2002). Epigenetic inactivation of 14-3-3 sigma in oral carcinoma: association with $p16^{INK4a}$ silencing and human papillomavirus negativity. *Cancer Res.* 62: 2072–6.
68. Gasco M, Sullivan A, Repellin C, *et al.* (2002). Coincident inactivation of 14-3-3 sigma and $p16^{INK4a}$ is an early event in vulval squamous neoplasia. *Oncogene* 21: 1876–81.
69. Balmanno K, Cook SJ (1999). Sustained MAP kinase activation is required for the expression of cyclin D1, $p21^{Cip1}$ and a subset of AP-1 proteins in CCL39 cells. *Oncogene* 18: 3085–97.
70. Jacobs JJ, Kieboom K, Marino S, DePinho RA, van Lohuizen M (1999). The oncogene and Polycomb-group gene bmi-1 regulates cell proliferation and senescence through the INK4a locus. *Nature* 397: 164–8.
71. Dimri GP, Martinez J-L, Jacobs JJ, *et al.* (2002). The Bmi-1 oncogene induces telomerase activity and immortalizes human mammary epithelial cells. *Cancer Res.* 92: 4736–45.
72. Sherr CJ, Weber JD (2000). The ARF/p53 pathway. *Curr Opin Genet Dev.* 10: 94–9.
73. Sanchez-Cespedes M, Reed AL, Buta M, *et al.* (1999). Inactivation of the INK4A/ARF locus frequently coexists with TP53 mutations in non-small cell lung cancer. *Oncogene* 18: 5843–9.
74. Esteller M, Tortola S, Toyota M, *et al.* (2000). Hypermethylation-associated inactivation of $p14^{ARF}$ is independent of $p16^{INK4a}$ methylation and p53 mutational status. *Cancer Res.* 60: 129–33.
75. Martelli F, Hamilton T, Silver DP, *et al.* (2001). p19ARF targets certain E2F species for degradation. *Proc Natl Acad Sci USA*, 98: 4455–60.
76. Eymin B, Karayan L, Seite P, *et al.* (2001). Human ARF binds E2F1 and inhibits its transcriptional activity. *Oncogene* 20: 1033–41.
77. Datta A, Nag A, Raychaudhuri P (2002). Differential regulation of E2F1, DP1, and the E2F1/DP1 complex by ARF. *Mol Cell Biol.* 22: 8398–408.
78. Carnero A, Hudson JD, Price CM, Beach DH (2000). $p16^{INK4a}$ and p19ARF act in overlapping pathways in cellular immortalization. *Nature Cell Biol.* 2: 148–55.
79. Wei W, Hemmer RM, Sedivy JM (2001). Role of p14ARF in replicative and induced senescence of human fibroblasts. *Mol Cell Biol.* 21: 6748–57.
80. Munro J, Stott FJ, Vousden KH, Peters G, Parkinson EK (1999). Role of the alternative INK4a proteins in human keratinocyte senescence: evidence for the specific inactivation of $p16^{INK4a}$ upon immortalization. *Cancer Res.* 59: 2516–21.
81. Hwang SG, Lee D, Kim J, Seo T, Choe J (2002). Human papillomavirus type 16 E7 binds to E2F1 and activates E2F1-driven transcription in a retinoblastoma protein-independent manner. *J Biol Chem.* 277: 2923–30.
82. Lin WC, Lin FT, Nevins JR (2001). Selective induction of E2F1 in response to DNA damage, mediated by ATM-dependent phosphorylation. *Genes Dev.* 15: 1833–44.
83. Dimri GP, Itahana K, Acosta M, Campisi J (2000). Regulation of a senescence checkpoint response by the E2F1 transcription factor and p14ARF tumor suppressor. *Mol Cell Biol.* 20: 273–85.

84. Roninson IB (2002). Oncogenic functions of tumour suppressor p21Waf1/Cip1/Sdi1: association with cell senescence, tumour-promoting activities of stromal fibroblasts. *Cancer Lett.* 179: 1–4.
85. Chang BD, Watanabe K, Broude EV, *et al.* (2000). Effects of p21Waf1/Cip1/Sdi1 on cellular gene expression: implications for carcinogenesis, senescence and age-related diseases. *Proc Natl Acad Sci USA* 97: 4291–6.
86. Hirama T, Koeffler HP (1995). Role of the cyclin-dependent kinase inhibitors in the development of cancer. *Blood* 86: 841–54.
87. Herman JG, Jen J, Merlo A, Baylin SB (1996). Hypermethylation-associated inactivation indicates a tumor suppressor role for $p15^{INK4B}$. *Cancer Res.* 56: 722–7.
88. Okamoto A, Hussain SP, Hagiwara K, *et al.* (1995). Mutations in the $p16^{INK4}$/MTS1/CDKN2, $p15^{INK4B}$/MTS2, and p18 genes in primary and metastatic lung cancer. *Cancer Res.* 55: 1448–51.
89. Fuxe J, Akusjarvi G, Goike HM, Roos G, Collins VP, Pettersson RF (2000). Adenovirus-mediated overexpression of $p15^{INK4B}$ inhibits human glioma cell growth, induces replicative senescence, and inhibits telomerase activity similarly to $p16^{INK4A}$. *Cell Growth Differ.* 11: 373–84.
90. Hannon GJ, Beach D (1994). p15INK4B is a potential effector of TGF-beta-induced cell cycle arrest. *Nature* 371: 257–61.
91. Sangfelt O, Erickson S, Einhorn S, Grander D (1997). Induction of Cip/Kip and Ink4 cyclin dependent kinase inhibitors by interferon-alpha in hematopoietic cell lines. *Oncogene* 14: 415–23.
92. Reynisdottir I, Polyak K, Iavarone A, Massague J (1995). Kip/Cip and Ink4 Cdk inhibitors cooperate to induce cell cycle arrest in response to TGF-beta. *Genes Dev.* 9: 1831–45.
93. Tremain R, Marko M, Kinnimulki V, Ueno H, Bottinger E, Glick A (2000). Defects in TGFβ signaling overcome senescence of mouse keratinocytes expressing v-rasHa. *Oncogene* 19: 1698–709.
94. Roussel M (1999). The INK4 family of cell cycle inhibitors in cancer. *Oncogene* 18: 5311–17.
95. Collado M, Medema RH, Garcia-Cao I, *et al.* (2000). Inhibition of the phosphoinositide 3-kinase pathway induces a senescence-like arrest mediated by p27Kip1. *J Biol Chem.* 275: 21960–8.
96. Yoon G, Kim HJ, Yoon YS, Cho H, Lim IK, Lee JH (2002). Iron chelation-induced senescence-like growth arrest in hepatocyte cell lines: association of transforming growth factor beta1 (TGF-β1)-mediated $p27^{Kip1}$ expression. *Biochem J* 366: 613–21.
97. Crowe DL, Nguyen DC (2001). Rb, E2F-1 regulate telomerase activity in human cancer cells. *Biochim Biophys Acta* 1518: 1–6.
98. Harley CB, Futcher AB, Greider CW (1990). Telomeres shorten during ageing of human fibroblasts. *Nature* 345: 458–60.
99. Samper E, Flores JM, Blasco MA (2001). Restoration of telomerase activity rescues chromosomal instability and premature aging in $Terc^{-/-}$ mice with short telomeres. *EMBO Rep* 2: 800–7.
100. Crowe DL, Nguyen DC, Tsang KJ, Kyo S (2001). E2F-1 represses transcription of the human telomerase reverse transcriptase gene. *Nucleic Acids Res.* 29: 2789–94.
101. Won J, Yim J, Kim TK (2002). Opposing regulatory roles of E2F in human telomerase reverse transcriptase (hTERT) gene expression in human tumor and normal somatic cells. *FASEB J.* 16: 1943–5.

102. Dyson N (1998). The regulation of E2F by pRB-family proteins. *Genes Dev.* 12: 2245–62.
103. Henson JD, Neumann AA, Yeager TR, Reddel RR (2002). Alternative lengthening of telomeres in mammalian cells. *Oncogene* 21: 598–610.
104. Garcia-Cao M, Gonzalo S, Dean D, Blasco MA (2002). A role for the Rb family of proteins in controlling telomere length. *Nature Genet.* 32: 415–19.
105. Itahana K, Dimri G, Campisi J (2001). Regulation of cellular senescence by p53. *Eur J Biochem.* 268: 2784–91.

Telomeres and Cellular Aging

Christian D. Toouli and Roger R. Reddel
Cancer Research Unit, Children's Medical Research Institute, Sydney, Australia

Cellular senescence

Human somatic cells undergo only a limited number of divisions *in vitro* before they withdraw from the cell division cycle and enter a permanent state of proliferation arrest. This arrest is accompanied by a variety of morphological and biochemical changes in the cells, and the arrest state is referred to as senescence. The *in vitro* senescence process has been widely studied as a model for *in vivo* cellular aging [1].

The phenomenon of senescence was first described by Hayflick and colleagues [2], and the maximum number of population doublings that a particular type of cell can undergo *in vitro* is often referred to as the "Hayflick limit." In addition, various treatments can result in rapid onset of permanent growth arrest and senescence-like morphological and biochemical alterations (Table 1). This is sometimes referred to as "premature senescence" or "stasis" [3], with the term "senescence" or "replicative senescence" being reserved for those cells that have reached the Hayflick limit (Figure 1).

It seems likely, however, that the best available cell culture techniques are an inadequate substitute for *in vivo* growth conditions, so it may be expected that improvements in cell culture will result in an apparent increase in the Hayflick limit. Thus, although the concept that there is an intrinsic limit to the number of times normal human cells can replicate even under perfect culture conditions is likely to be valid, in practice most or all of the senescence observed in *in vitro* experiments may be premature.

Although the evidence is not yet definitive, it seems that senescence is a programmed series of cellular events that may be triggered in response to a variety of stimuli. If this view is correct, then senescence is analogous to apoptosis for which there is a wide range of endogenous and exogenous triggers, with the major difference being that the outcome of apoptosis is destruction of the cell, whereas the senescence program results in a cell that is unable to replicate but remains metabolically active.

S.C. Kaul and R. Wadhwa (eds.), Aging of Cells In and Outside the Body, 171–205.

Table 1. *Stimuli that result in premature senescence – a senescence-like state at a population doubling level below the Hayflick limit*

Reported stimuli	References
Gamma-irradiation	228, 229
Oxidative stress	12, 60, 230, 231
Activated Ha-ras oncogene	232–234
Cancer chemotherapeutic agents	235–238
Chemopreventive agents	239
Cross-linked DNA	240
Ceramide	241
Phosphatidylinositol-3-kinase inhibitor	242
DNA demethylating agents	243, 244
Histone deacetylase inhibitors	245–247
Inhibition of cGMP production	248

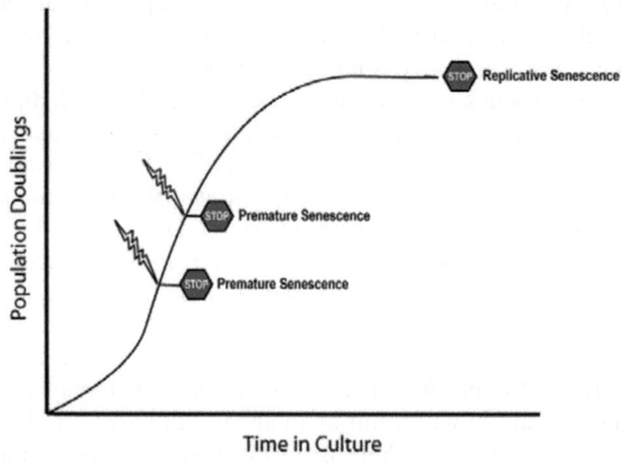

Figure 1. *Senescence of somatic cells in vitro. Cells that are grown under the best available culture conditions will eventually cease replication and exhibit morphological and biochemical changes suggestive of aging; this state of growth arrest is called replicative senescence. Aberrant oncogene expression and various types of cellular insults listed in Table 1 may result in the onset of senescence-like changes regardless of the number of times the culture has replicated; this is sometimes called premature senescence.*

Induction of the senescence program in tumor cells by chemotherapeutic agents may be an important determinant of treatment outcome in cancer therapy [4]. This appears to be true even in a species such as the mouse [5] where it seems that replicative senescence does not occur [3].

The extent to which replicative senescence contributes to human aging *in vivo* is essentially unknown. It is likely to be important in the pathogenesis of various disease states where there is greatly increased proliferation, for example due to cell death and compensatory regeneration in conditions such as the muscular dystrophies [6] or hepatic cirrhosis [7]. Replicative senescence is very likely, however, to play a major role in protection of long-lived organisms, such as humans, against cancer [8].

The mitotic clock concept

The existence of a limit on the number of divisions a cell population can undergo strongly suggests that cells must have some mechanism for keeping track of how many times their predecessors have divided. The term encapsulating this concept that is now entrenched in the literature is "mitotic clock." However, because it is not time, but the number of cell divisions that is more important, the term "replicometer" [9] would be preferable.

There is now a considerable body of evidence indicating that the progressive telomere shortening that occurs with every cell division (see below) acts as a mitotic clock. Although the purpose of this chapter is to summarize information about the relationship between telomeres and cellular aging, it is important to note that other counting mechanisms have been proposed, including progressive loss of DNA methylation with increasing population doublings [10], the progressive decrease in stability of heterochromatin domains [11], accumulation of oxidative DNA damage [12], and accumulation of mitochondrial DNA mutations leading to loss of mitochondrial function [13]. It is possible, especially under non-ideal growth conditions, that there is more than one mitotic clock, with senescence being determined by whichever clock reaches the cutoff point first [14].

Telomeres

Telomeres consist of highly repetitive and conserved DNA at each end of linear chromosomes [15]. Human telomeric DNA is approximately 4–15 kilobases (kb) in length and consists of tandemly repeated copies of the guanine-rich sequence T_2AG_3, orientated in a 5′-3′ direction towards the chromosome end [16]. The telomere ends in a short 3′ G-rich single-stranded tail approximately 30–200 nucleotides in length [17–19]. Telomeres distinguish the ends of chromosomes from broken chromosomes [20–22]. How they accomplish this is incompletely understood, but the mechanism appears to involve binding of specific telomere associated proteins, and formation of three dimensional structures which protect and hide the end of the telomere [23, 24]. In addition to protection of chromosome ends from fusion with other chromosome ends, the functions of telomeres include the organization of chromosomes during meiosis and facilitation of the complete replication of non-telomeric chromosomal DNA [25–27].

Table 2. *Examples of mammalian proteins that affect the telomere*

Protein	Proposed functions at the telomere	References
ATM	Maintenance of telomere stability and telomere length	249
BLM	RecQ helicase; interacts with TRF2 in ALT cells; positive telomere length regulator in ALT cells; together with WRN and Replication Protein A, actively unwind long telomeric duplex regions that are pre-bound by TRF2	250, 251
CDK2	Localizes to the telomeres of chromosomal bivalents throughout meiotic prophase	252
DNA-dependent protein kinase catalytic subunit (DNA-PKcs)	Telomere end capping and telomerase-independent length maintenance; required for end-to-end fusions and apoptosis triggered by critically short telomeres	253, 254
DNA ligase IV	Required for non-homologous end joining-mediated end-to-end fusion of telomeres when there is insufficient TRF2 function	255
Histone H2B variant	Component of double-stranded telomere DNA binding complex in sperm; may have role in attachment of telomere to nuclear membrane	256
hnRNP family proteins	Bind 3′ G-rich telomeric sequence, and telomerase; may regulate telomere length maintenance	42, 257–261
hRAP1	Binds TRF2, negative length regulation	32
KU86:KU70	Binds 3′ G-rich telomere overhang; maintains telomere end integrity; DNA repair	262, 263
MRE11/RAD50/NBS1	Binds TRF2, negative telomere length regulation, recombinational DNA repair	37
Nucleolin	Binds duplex telomeric DNA *in vitro*	264
TP53	Binds telomeric 3′ single-strand overhang and t-loop junction; increases efficiency of TRF2-catalyzed t-loop formation	41

Table 2. (continued)

Protein	Proposed functions at the telomere	References
PINX1	Binds TRF1/Pin2, inhibits telomerase leading to telomere shortening	40
POT1	Binds terminal G strand telomeric DNA	39, 265
Replication Protein A (RPA)	Together with WRN and BLM helicases, actively unwinds long telomeric duplex regions that are pre-bound by TRF2	250
Tankyrase 1	Binds TRF1, positive length regulation; can poly(ADP-ribosyl)ate itself and TRF1	33–35
Tankyrase 2	Binds TRF1, positive length regulation; can poly(ADP-ribosyl)ate itself and TRF1	266, 267
TIN2	Binds TRF1, negative length regulator	36
TRF1/Pin2	Binds telomeric DNA; negative length regulator	29, 268, 269
TRF2	Involved in formation of T-loop structure; negative telomere length regulator; deficiency results in p53- and ATM-dependent apoptosis	30, 269, 270
WRN	RecQ helicase; co-localizes and physically interacts with TRF2 which stimulates helicase activity; together with RPA and BLM, actively unwinds long telomeric duplex regions that are pre-bound by TRF2	250

The term "telosome" is sometimes used to refer to the complex formed by telomeric DNA and telomere binding proteins [28]. There is an expanding list of proteins that have been demonstrated to bind to or associate with single- and/or double-stranded telomeric DNA (Table 2). Telomere-associated proteins are thought to influence telomere length, stability and function [24]. Human proteins that may associate with the telosome protein complex include TRF1 [29], TRF2 [30, 31], hRAP [32], tankyrase 1 and 2 [33–35], TIN2 [36], RAD50, MRE11, and NBS1 [37], the KU heterodimer [38], POT1 [39], PinX1 [40], p53 [41], and heterogeneous nuclear ribonucleoproteins (hnRNPs) [42]. One of the consequences of protein binding to telomeric DNA is the formation of a three dimensional duplex loop structure, called a "telomere loop (t-loop)," that is hypothesized to be critically important for telomere stability and required for its normal function [23, 43]. The t-loop is thought to be generated by the looping of telomeric DNA back onto itself, forming a lariat structure (Figure 2).

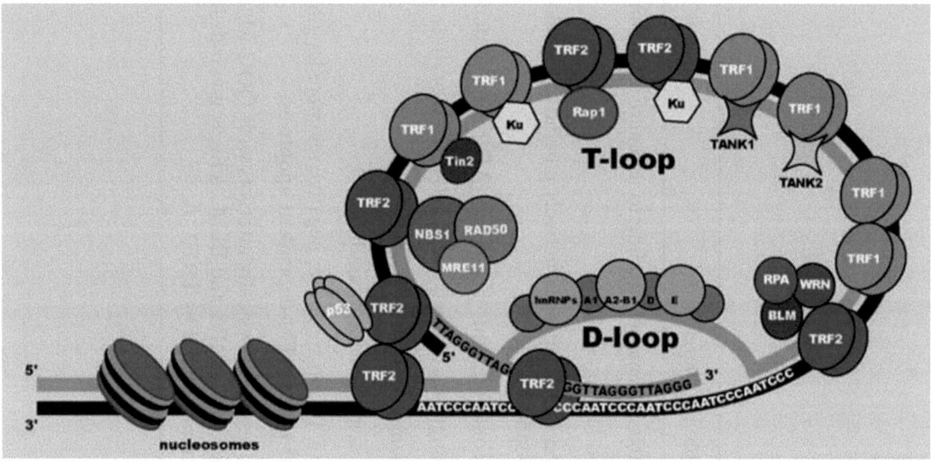

Figure 2. Schematic representation of telomere (t)-loop structure and binding proteins. Proposed functions of the binding proteins are summarized in Table 2. Looping back of the single-stranded telomeric terminus and invasion of a proximal duplex telomeric sequence results in a displacement loop.

T-loop formation is most likely mediated by the binding of telomere binding proteins, especially TRF2, which facilitate the invasion of the 3′ terminal end of the telomere into proximal double-stranded telomeric DNA, forming a heteroduplex structure [23]. The t-loop is thought to protect the G-rich single-stranded terminal DNA from degradation, and to prevent recognition of the telomere end as a double-strand break by DNA repair proteins [24]. The lack of homology between some telomere-associated proteins in different species suggests that there has been an unusually rapid divergence in the proteins that mediate and regulate telomere function, but may also indicate that additional telomere-associated proteins remain to be identified.

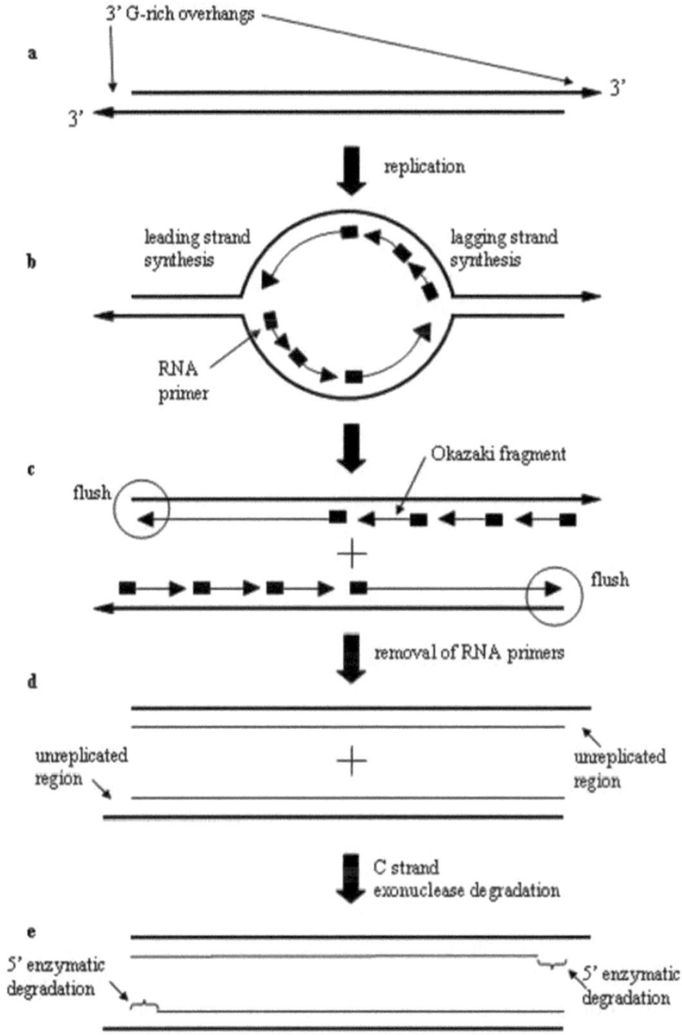

***Figure 3**. Proposed mechanisms of telomere shortening during DNA replication. Schematic representation of the "end replication problem" and exonuclease digestion. (a) The two strands of a linear DNA molecule are shown in black, arrowheads denoting the 3′ G-rich single-strand ends. (b) Leading and lagging strands and RNA primers (boxes) are shown for a single replication bubble; in reality, there are many thousands of replication bubbles per chromosome. (c) Replicated DNA molecules between the RNA primers (Okazaki fragments [53]). Note that the leading strand synthesis continues to the end of the template whereas this is not possible for the lagging strand due to the requirement for a 5′ RNA primer. (d) RNA primers are degraded, the internal gaps are filled in, and the fragments are ligated. The terminal gap is not able to be filled in. (e) Degradation of the 5′ C strand by a putative exonuclease increases the length of the 3′ G-rich strand overhang.*

Why do telomeres shorten?

Average telomere length reduces by approximately 50–200 bp each time a normal human somatic cell divides [44–48]. There are at least two reasons for this loss: incomplete replication, and exonuclease activity. The incomplete replication is thought to be due to the "end replication problem" [49, 50], or "marginotomy" as it was first termed [51, 52]. The end replication problem is the inability of lagging strand DNA synthesis to fully replicate the terminus of a linear DNA molecule (Figure 3).

The RNA molecules that prime DNA synthesis are eventually degraded and replaced by DNA, but this is not possible at the terminus where there is no duplex DNA on the 5′ side of the gap left by the degraded RNA. The result is that leading strand synthesis is complete, but lagging strand synthesis leaves an incompletely replicated strand that is a shortened template in the next round of DNA replication. The end-replication problem has recently been demonstrated in an *in vitro* model system [54]. There is evidence that the C-rich strand may also be shortened by exonuclease activity [17, 18, 55] (Figure 3). This serves to generate a 3′ G-rich telomere overhang which is presumably required for binding of some of the telomere-associated proteins, and for the strand invasion involved in formation of a stable t-loop (Figure 2).

In addition to the putative exonuclease activity that produces G-rich overhangs, there may be other endogenous telomere shortening mechanism(s). MRC-5 embryonal lung fibroblasts held at confluence for three months displayed an accelerated decrease in telomere length on recommencement of proliferation [56]; the mechanism is unknown. Furthermore, telomeres in telomerase-negative immortalized human cells sometimes exhibit a rapid shortening [57], reminiscent of the rapid reduction to wild-type length of over-lengthened yeast telomeres that may occur within a single cell division, referred to as Telomere Rapid Deletion (TRD) [58]. It is unknown whether a TRD-like telomere shortening mechanism occurs in normal human cells.

Telomeres may also shorten as a result of DNA damage. Telomeric DNA is more sensitive to oxidative damage than genomic DNA overall. Exposure of human fibroblasts to oxidative stress caused an accelerated rate of cellular aging and telomere shortening [59–62]. It has been suggested that there is a lower rate of DNA repair at the telomere due to reduced access of DNA repair proteins to DNA packaged in the telosome [63].

The telomere hypothesis of senescence

More than three decades ago, Olovnikov not only predicted that telomere shortening occurs with replication but also hypothesized that this acts as the mitotic clock which is responsible for triggering senescence [51, 52]. This is now known as the telomere hypothesis of senescence [64]. It is supported by a study of cells obtained from 31 individuals in which there was a correlation between telomere length and proliferative capacity [47]. In cattle that were cloned from senescent somatic cells (0–4 population doublings remaining), nuclear transfer resulted in both telomere lengthening and extension of the replicative life span of cells from the cloned cattle to greater than 90 population doublings [65]. This supports an association between telomere length and proliferative capacity and suggests that telomere length was reset by the cloning process in cattle, although cloning of sheep did not appear to result in telomere lengthening [66].

In a more recent refinement of the telomere hypothesis, it has been proposed that senescence is triggered by telomere cap status rather than telomere length *per se* [67]. It might be predicted that cap status would be determined by binding of telomere-associated proteins, and that this in turn would be determined both by the concentration of these proteins and by the length of available telomeric DNA. This concept is supported by the finding that overexpression of TRF2 results in senescence being triggered at a significantly shorter telomere length: TRF2 reduced the senescence setpoint from 7 to 4 kb [68]. Presumably, by permitting formation of a t-loop, the presence of a single-stranded telomeric overhang is also an important determinant of cap status.

As yet there is no direct assay of telomere cap function, so it is mostly telomere length that has been studied in the context of the onset of senescence. It has not been clearly established how telomere shortening results in senescence, and there are many unanswered questions including the following. Is one short/uncapped telomere sufficient to trigger senescence, or is more than one such telomere required? Is there a specific telomere that is usually responsible for onset of senescence? How is telomere shortening or uncapping sensed, and what is the signaling pathway that results in execution of the senescence program?

There is considerable heterogeneity of telomere length within individual human cells. Telomeres on chromosomes 17p, 22q, 19p and 20q were shown to be consistently below average length, while telomeres on chromosomes 4q, 1p, 3p and 5p were observed to be above average length in cells from ten individuals [69]. The rate of telomere shortening at individual telomeres was found to vary between 50 and 150 bp per population doubling and short telomeres with an estimated 1–2 kb of telomere repeats were found to accumulate prior to senescence [70]. In this study, neither the shortest nor the longest chromosomes correlated best with remaining proliferative potential [70]. However, in another study where telomere lengths were artificially increased and homogenized, and then allowed to undergo normal replicative telomere attrition, it was found that the average telomere length at the senescence setpoint was significantly reduced [71], consistent with the notion that it is one or more of the shortest telomeres that triggers senescence.

It has been postulated that the mechanistic link between telomere shortening and senescence might be a shortening-induced change in expression pattern of senescence-related genes located close to the telomere [72]. Telomere position effects (TPE) on gene expression in human cells have been observed [73] but as yet there is no direct support for the involvement of TPE in senescence.

Another proposal is that telomere uncapping or decline in length to the senescence setpoint is detected by the cell as a DNA damage event [62, 74]. In yeast, the loss of a single telomere led to a DNA damage signal and transient growth arrest [22]. In human cells, p53 and $p21^{CIP1/WAF1}$ are usually upregulated both by DNA damage and at the onset of senescence [75–77]. However, the posttranslational changes in p53 induced by UV and ionizing radiation do not completely overlap with those seen at replicative senescence [78]. The signaling generated by telomere attrition thus shares some but not all of the features of DNA double-strand breaks [78]. A recent study has shown that removal of TRF2 from human telomeres induced senescence in a p53-dependent manner, but interestingly in the absence of p53 function senescence could also be induced via the $p16^{INK4A}$/RB pathway [79]. The p53 and $p16^{INK4A}$/RB pathways are activated in a wide variety of situations in which senescence is not the outcome, so many details of the signaling pathways leading from uncapped telomere(s) to senescence remain to be elucidated.

Genetic abnormalities that permit temporary bypass of senescence

As might be expected from what little is known about the mechanisms involved in triggering senescence, impaired function of the p53 and $p110^{Rb}/p16^{INK4A}$ tumor suppressor pathways may permit cells to escape temporarily from the limitation on cellular proliferation that is imposed by senescence [80–84]. When these pathways are inactivated, for example, by the oncoproteins of DNA tumor viruses, exposure to DNA damaging agents, or a combination of inherited and spontaneous mutations, cells may continue dividing even though their telomeres may have become shorter than the length that would be expected to cause senescence [81, 85]. There are several possible explanations for this observation, with the simplest being that the absence of the p53 and $p110^{Rb}/p16^{INK4A}$ tumor suppressor pathways allows the cells to ignore the signals that would otherwise trigger senescence.

The additional proliferation permitted by inactivation of these pathways is not unlimited, however, and the population eventually enters a state referred to as crisis [86, 87]. The characteristics of cell populations at senescence differ substantially from those at crisis, and to denote these differences the states have been called mortality stages 1 and 2 (M1 and M2), respectively [87]. During crisis the cell proliferation rate in the culture is matched and finally exceeded by the cell death rate [86, 88], in contrast to senescence where there is essentially no cell division and the cells are resistant to apoptosis [89]. During crisis the cell morphology changes, leading to the generation of giant multinucleated cells with bizarre lobed nuclei and paranuclear cytoplasmic vacuoles, and the accumulation of cellular debris [86]. Subsets of cells in crisis stain positively for senescence-associated β-galactosidase [90, 91], but other cells undergo apoptotic cell death [91].

At crisis, the telomeres become critically shortened and seriously dysfunctional. Telomeric DNA is lost completely from some chromosome ends [92]. It has been proposed that critical shortening in one or more telomeres leads to chromosomal instability, an inability to undergo cell division, and ultimately to cell death [85, 92]. Complete loss of telomeric DNA will result in the chromosome end resembling a double-strand break and activating DNA repair mechanisms [22]. Cells in crisis may therefore undergo recombination events, end-to-end fusions and bridge-fusion-breakage cycles [92]. This karyotypic instability may result in the random accumulation of genetic alterations, and in rare cells some of these alterations may result in immortalization.

Genetic changes that permit immortalization

Although there is evidence for the existence of dominantly acting genetic changes that may contribute to immortalization [93, 94], somatic cell hybridization experiments have shown that immortalization is essentially a recessive phenotype. Fusion of normal cells and immortal cells resulted in hybrids with a limited proliferative capacity [95–101]. This suggests that immortal cells have lost the function of one or more genes ("senescence genes") that act to suppress the immortal phenotype in the hybrids. Further, when different immortalized cell lines were fused together, in some cases the hybrids were mortal also [102–104], suggesting that the cell lines had become immortalized by losing the function of different senescence genes, so that the set of functional senescence genes contained in the hybrid cells was able to suppress the immortal phenotype. Hybridization of immortalized cell lines has defined four complementation groups, A through D, for immortality [102].

The chromosomal localizations of some senescence genes have been defined by transferring single chromosomes from normal human cells into immortalized cells using microcell-mediated chromosome transfer (MMCT). The genes corresponding to complementation groups B, C and D have been localized to human chromosomes 4, 1, and 7, respectively [105–110], and there is preliminary evidence that the group A gene may be located on chromosome 6 [111]. MMCT has shown, however, that a number of other chromosomes including 2 [112, 113], 3 [114–117], 10 [118], 11 [119], 16 [120,121], 18 [122], X [123], and possibly 17 [124] are also capable of inducing senescence in various types of human cells.

Most of these putative senescence genes have not been identified and their function is not clear. In view of the critical telomere shortening that is characteristic of cells in crisis, it might be predicted that telomere length maintenance would be essential for immortalization and, indeed, in every immortalized cell line examined so far there is evidence that the activation of a telomere maintenance mechanism (TMM) has occurred [125]. Therefore, a further prediction would be that at least some of the putative senescence genes may be involved in repression of telomere length maintenance and this appears to be the case for genes on chromosomes including 3 [116, 117, 126], 4 [127], 6 [128, 129], 7 [130, 131], and 17 [124]. In some cases, however, MMCT of chromosomes from normal cells results in senescence without any detectable effects on the TMM [132].

Senescence genes that are unrelated to telomere length maintenance may be related to other putative mitotic clocks, some of which may also trigger senescence through the p53 and/or $p110^{Rb}/p16^{INK4A}$ tumor suppressor pathways. Even in cells that have an active TMM, restoring expression of genes in these pathways may reimpose a finite proliferative capacity [133, 134]. The putative senescence genes that have been delineated by MMCT and that are apparently unrelated to control of a TMM could therefore be involved in these tumor suppressor pathways, or may be components of other unknown pathways involved in controlling proliferative life span.

Activation of a TMM is nevertheless the most important known change associated with immortalization. This serves both to neutralize the telomere cap-dependent mitotic clock, and to prevent the genetic instability resulting from critically short telomeres and bridge-fusion-breakage cycles. The most intensively studied of the TMMs is telomerase, a ribonucleoprotein (RNP) holoenzyme that synthesizes telomeric DNA by reverse transcription, but there is at least one TMM, referred to as Alternative Lengthening of Telomeres (ALT), that is not dependent on telomerase [135].

Telomerase

Telomerase adds DNA to the telomere ends [136], and thus can maintain telomere length by replacing the DNA that is lost during normal replication. The DNA is added onto the G-rich strand, and it is thought that this then serves as a template for the synthesis of the complementary strand by DNA primase and conventional DNA polymerases. The human telomerase holoenzyme is a complex that contains diverse subunits (Table 3) and is approximately 1000 kDa in size [137, 138]. In the ciliate, *Euplotes crassus*, the size and functionality of the telomerase complex varies according to life cycle stage [139], but it is not clear whether mammalian cells assemble telomerase complexes with different higher order structures to fulfill different roles. The essential subunits of human telomerase include the catalytic subunit, **TE**lomerase **R**everse **T**ranscriptase (hTERT) [140–143]; and an RNA moiety, **TE**lomerase **R**NA (hTER; also referred to as hTR; encoded by the TERC gene) [144]. hTERT has homology to telomerase catalytic subunits in species including *Saccharomyces cerevisiae*, *Schizosaccharomyces pombe*, and *Euplotes aediculatus*, and contains evolutionarily conserved reverse transcriptase motifs that are similar to those of viral reverse transcriptases, including HIV-1 [141]. hTER contains a template region of 11 nucleotides (5′-CUAACCCUAAC), complementary to the human telomere sequence $(TTAGGG)_n$, which is utilized by hTERT to reverse transcribe telomeric DNA. Other proteins with possible roles in telomerase assembly, stabilization and localization include human **TE**lomerase associated **P**rotein 1 (hTEP1) [145], L22 and hStau [146], the chaperone foldosome proteins p23 and hsp90 [147], hnRNPs A1, C1/C2 and D [42], dyskerin [148], Gar1p [149], 14-3-3 [150], and La [151]. Some of the subunits of telomerase might bind only transiently to the RNP complex during specific times in the cell cycle, in particular cellular locations, or when telomerase is involved in specific functions.

Telomerase-positive immortal cells have telomeres of a relatively homogeneous length, with the mean length generally being <10 kb, and shorter than in the cells from which they were derived [152–154]. Telomerase has also been implicated in the healing of broken chromosome ends [155–157] and repair of DNA double-strand breaks [158, 159]. Critically short telomeres are preferentially lengthened or maintained in cells escaping from crisis [92], indicating that telomerase may be directed to those telomeres that are most crucial for karyotypic stability and cellular proliferation.

In addition to its expression in immortal cell lines and tumors, telomerase has been detected in a minority of somatic cells, including human mammary epithelial cells (HMEC) [160], haematopoietic cells [161, 162] including T lymphocytes [163, 164], epidermal keratinocytes [165], hair follicle cells [166], endometrial cells [167] and colonic epithelial cells [168]. Telomerase is also expressed in germ cells [169, 170]. Telomerase activity may play a critical role in embryonic development, as expression

Table 3. *Examples of telomerase-associated molecules*

Protein or RNA	Proposed functions	References
14-3-3	Binds to hTERT; implicated in hTERT/telomerase nuclear localization	150
Dyskerin	Binds hTER; required for stabilization of hTER	reviewed in 148
GARlp	Binds hTER; required for stabilization of hTER	149
hnRNPs A1, C1/C2 and D	Interact with telomerase and telomeres; may recruit telomerase to the telomere and facilitate telomere synthesis	reviewed in 42
hSTAU	Binds to hTER and hTERT; implicated in telomerase biogenesis	146
hTEP1	RNA binding protein that interacts with hTER and hTERT; implicated in p53-mediated regulation of telomerase activity.	137, 145, 271
hTER	RNA template moiety of telomerase; essential for synthesis of telomeric DNA; encoded by hTERC gene	144
hTERT	Catalytic subunit of telomerase; reverse transcribes hTER	140–143
L22	Binds to hTER and hTERT; implicated in telomerase biogenesis	146
La	Binds to hTER; implicated in telomerase biogenesis, function and nuclear localization	151
p23/HSP90 foldosome	Bind to hTERT; implicated in telomerase biogenesis	147

of telomerase has been detected in cells as early as the blastocyst stage, followed by very specific expression in various tissues in the developing embryo [171, 172]. Dyskeratosis congenita (DKC) is an inherited disease associated with diminished telomerase activity in embryonic cells and adult tissues. This results in telomeres that are shorter than in age-matched controls, impaired proliferation of tissues such as the skin and bone marrow *in vivo*, premature *in vitro* proliferative senescence, and genomic instability [148, 173–176]. Expression of telomerase and telomere maintenance therefore appears to be required for the normal function of human tissues. A common feature of normal somatic cells that express telomerase is that they need to divide a large number of times during the human life span. However, telomerase in somatic cells is generally expressed at low levels, and is unable to completely counteract the progressive telomere shortening that occurs with normal cell division [160, 161]. The low telomerase activity seen in somatic cells may decrease the rate of telomere shortening, and/or perform other unknown functions in the cell, that are distinct from telomere maintenance.

The expression of telomerase during specific periods of embryonic development, and only in a minority of somatic adult cells, implies that its activity is tightly regulated, but this regulation is poorly understood at present. The expression of both hTERT and hTER is required for the induction of telomerase activity *in vitro* in rabbit reticulocyte lysates [177, 178]. hTER, but not hTERT, is constitutively expressed in normal human somatic cells. Expression of exogenous hTERT has been found to induce telomerase activity in telomerase negative somatic cells [178, 179], indicating that the amount of available hTERT is limiting and suggesting the hypothesis that telomerase activity may be controlled at the level of hTERT expression. Expression of hTERT mRNA correlates with telomerase activity [142], consistent with the notion that telomerase activity is controlled, at least in part, at the level of hTERT transcription. A number of transcription factors and other proteins have been implicated, but the catalogue of such factors is almost certainly very incomplete, and the relevant *cis*-acting elements in the hTERT promoter/enhancer require further definition [180]. It seems likely that hTERT is also controlled at the post-transcriptional and post-translational levels [140, 181–184].

When exogenous hTERT is expressed under the control of a heterologous promoter in telomerase-negative human cells, the telomerase that is induced [178, 179] is capable of maintaining telomere length [185, 186]. Expression of exogenous hTERT thus allows analyses of the relationship between cellular proliferative potential and maintenance of telomere length. Induction of telomerase in this way resulted in the bypass of crisis in SV40-TAg transformed human post-senescent embryonic kidney cells [187], IMR90 fibroblasts [188], pancreatic cells [189], and bone marrow-derived endothelial cells [190], supporting the importance of critical telomere shortening as a trigger for the induction of crisis, and telomere maintenance as a requirement for continued cellular proliferation and immortalization. However, the relationship between telomere maintenance and senescence (as opposed to crisis) seems more complex. The effect of exogenous hTERT expression on proliferative capacity of normal primary cells depends on the cell type being analyzed (Table 4). The initial studies using BJ foreskin fibroblasts and retinal pigment epithelial cells demonstrated that expression of exogenous telomerase activity permitted bypass of senescence [185, 186]. These cells continued dividing long-term in the absence of any detectable genotypic or phenotypic changes, suggesting that telomere maintenance was sufficient for *in vitro* immortalization [191–193]. This observation supports the role of telomere shortening as a mitotic clock and limit to normal cell proliferation. Similar observations have also been made with other cell types including endothelial and mesothelial cells [194–196]. However, experiments performed in other cell types, including HMECs and keratinocytes [194, 197–199], embryonic lung fibroblasts [200, 201], mammary fibroblasts [202], T-lymphocytes [203] and peripheral blood, microvascular and bone marrow-derived endothelial cells [190, 202] failed to induce immortalization by expression of exogenous hTERT alone. Inactivation of the $p16^{INK4A}/p110^{Rb}$ pathway through spontaneous methylation of the $p16^{INK4A}$ promoter, or expression of the human papillomavirus 16 E7 viral oncogene, was required in addition to telomerase activity to induce immortalization in HMECs and keratinocytes [197–199]. Transfection of exogenous hTERT into MRC-5 embryonic lung fibroblasts [200, 201, 204], T-lymphocyte cultures [203, 205], and endothelial cells [190] led to telomerase activity, but not immortalization in all cells. These results suggest that the requirements for immortalization are cell type specific, and that often hTERT-transfected cells require additional genetic changes to bypass senescence and become immortal.

The requirement for inactivation of the $p16^{INK4A}/p110^{Rb}$ pathway for immortalization of some hTERT-transduced cells suggests that these genes can normally trigger cellular senescence independently of telomere length. It has been hypothesized, however, that the $p16^{INK4A}/p110^{Rb}$ pathway is involved in signaling a form of premature senescence that is induced by the accumulation of changes resulting from inadequate *in vitro* growth conditions [206]. It is well known that conventional *in vitro* tissue culture conditions include a hyperoxic growth environment, unphysiological growth surfaces (e.g., growth on plastic dishes), lack of three-dimensional tissue architecture, physical and enzymatic stresses (e.g., passaging), and possibly inappropriate growth factors and media. Tumor cells *in vivo* often experience stressful growth conditions, and inactivation of the $p16^{INK4A}/p110^{Rb}$ pathway is one of the most common genetic or epigenetic changes of tumors. It seems possible, therefore, that the requirement for loss of this pathway during hTERT-mediated immortalization of some types of human cells under standard *in vitro* growth conditions reflects a process that also occurs in tumors *in vivo*.

It will be very interesting to determine whether cell strains that are capable of hTERT-mediated immortalization in the absence of any detectable additional genetic changes are defective in their ability to undergo telomere-independent senescence, and/or whether they have particularly robust mechanisms for coping with the stresses that cause other cells to undergo premature senescence. The latter explanation appears to apply to one such cell strain, BJ fibroblasts, which is immortalized by hTERT transduction alone [191–193] and has a low intracellular peroxide content, low levels of protein carbonyls, and low steady state lipofuscin content in comparison to other primary fibroblast cell types [207], suggesting that it is unusually resistant to oxidative stress.

Table 4. *In vitro studies of hTERT-induced immortalization of human somatic cells*

Cell type/ tissue	Bypass of senescence	Requirements for bypass of senescence[b]	References
Fibroblasts			
Lung	Yes	hTERT	204, 272
Lung	No	hTERT and additional genetic changes	200, 201
Foreskin	Yes	hTERT	192–194, 198, 206
Skin	Yes	hTERT	273
HCA2 strain	Yes	hTERT	274
Mammary	No	hTERT and SV40-TAg	202
Bloom Syndrome	Yes	hTERT	272
Roberts Syndrome	Yes	hTERT	272
Xeroderma Pigmentosum[a]	Yes	hTERT	272
Werner Syndrome	Yes	hTERT	272, 274, 275
Hutchinson-Gilford Progeria Syndrome	Yes	hTERT	272
Ataxia telangiectasia	Yes	hTERT	273, 276
Nijmegen breakage syndrome	Yes	hTERT	277
Endothelial			
Umbilical	Yes	hTERT	195
Aortic	Yes	hTERT	195, 278
Dermal microvascular	Yes	hTERT	195
Saphenous vein	Yes	hTERT	195
Coronary arterial	Yes	hTERT	195
Microvascular	No	hTERT and SV40-TAg	202
Peripheral blood	No	No lifespan extension	190
Bone marrow	No	hTERT and additional genetic changes, or SV40-TAg	190

Table 4. *(continued)*

Cell type/ tissue	Bypass of senescence	Requirements for bypass of senescence[b]	References
Epithelial			
Retinal pigment	Yes	hTERT	185
Adenoid	No	hTERT and additional genetic changes (including inactivation of $p16^{INK4A}$ or $p14^{ARF}$)	197
Mesothelial	Yes	hTERT	194
Mammary	No	hTERT and HPV-16 E6 and E7, or $p16^{INK4A}$ inactivation	198, 199, 279, 280
Mammary	Yes	hTERT	281
Thyroid	No	No lifespan extension	282
Tracheobronchial	No	hTERT and SV40-TAg	283
Melanocytes	Yes	hTERT	284
Keratinocytes			
Foreskin	Yes	hTERT	206
Foreskin	No	hTERT and HPV-16 E7	198
Epidermal	No	hTERT and escape from senescence (involving inactivation of $p16^{INK4A}$)	194
Oral mucosal	No	hTERT and escape from senescence (involving inactivation of $p16^{INK4A}$)	194
Lymphocytes	Yes	hTERT	205
	No	No lifespan extension	203
Myometrial	Yes	hTERT	285, 286
Osteoblasts	Yes	hTERT	287
Pancreatic islet	No	No lifespan extension	288
Uterine leiomyoma	Yes	hTERT	286

[a]Complementation groups E and V; [b]where additional genetic changes or oncogenes are listed, expression of exogenous hTERT was insufficient to induce immortalization

Alternative lengthening of telomeres

Some cell lines and cancers that are telomerase-negative maintain the length of their telomeres by one or more mechanisms referred to as Alternative Lengthening of Telomeres (ALT) [208]. Rapid changes in telomere length that were observed in a telomerase-negative cell line were suggestive of a recombinational mechanism of telomere maintenance [57]. In at least some ALT cells, it appears that the ALT mechanism involves a DNA strand from one telomere annealing with the complementary strand of another telomere, thereby priming synthesis of new telomeric DNA using the complementary strand as a copy template [209] (Figure 4).

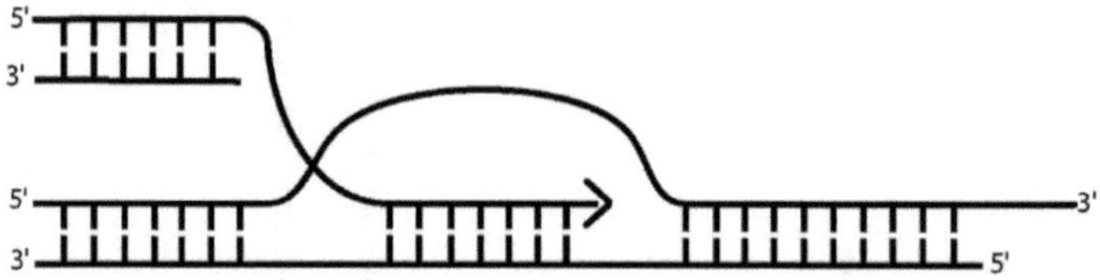

***Figure 4**. Proposed Alternative Lenthening of Telomeres (ALT) mechanism. In ALT cells, one telomere may use another telomere (or extrachromosomal telomeric DNA) as a copy template [209]. The single-stranded telomeric DNA may invade duplex telomeric DNA and anneal with the complementary strand. Extension of the invading strand can then occur using the invaded telomere as a copy template for DNA synthesis, after which the other strand can be filled in. The net amount of telomeric DNA thus increases.*

Evidence for such a "recombination/copy switching" mechanism was obtained by inserting DNA tags into the telomeres of ALT cells and showing that the tags were copied to other telomeres. In telomerase-positive cells there was no detectable copying of an intratelomeric tag to other telomeres [209].

Telomerase-independent telomere maintenance has been found in a number of other species. In the mosquito, *Anopheles gambiae*, the predominant mechanism involves recombination [210]. In the fruitfly, *Drosophila melanogaster*, telomere length maintenance normally involves retrotransposition events [211, 212]. In several species of yeast including *S. cerevisiae* [213, 214] and *Kluyveromyces lactis* [215, 216] where telomerase is normally used for telomere maintenance, telomerase-null mutants can survive by activating a mechanism that is dependent on genes involved in recombination. The telomerase-null *S. cerevisiae* survivors fall into two categories (types I and II) by telomere structure and genetic analyses [214, 217], and it will also be interesting to determine whether there is more than one ALT mechanism in mammalian cells.

All of the human ALT cell lines studied so far have telomeres that are very heterogeneous in length – ranging from extremely short to greater than 50 kb, with the mean length being approximately 20–25 kb [135, 152, 218]. Presumably the heterogeneity is generated by the combined effects of the gradual telomere shortening that accompanies DNA replication and large increases in telomere length due to the

ALT mechanism. The increases in telomere length appear to be stochastic, but are more likely to involve short telomeres [57].

A second characteristic of ALT cells is the presence of PML nuclear bodies (PNBs) with specific contents [219]. PML nuclear bodies (PNBs) are subnuclear complexes that are found in almost all cells, and contain PML and other proteins. They are dynamic structures that have been associated with numerous processes including tumor suppression, cell cycle regulation, senescence and apoptosis [220]. In ALT cells, a subset of the PNBs contain telomere-related components including the telomere binding proteins TRF1 and TRF2 and extrachromosomal telomeric DNA. Because PNBs with these components have not been detected in telomerase-positive cell lines or in non-immortalized cells, they are referred to as ALT-associated PNBs (APBs) [219]. APBs also contain proteins involved in DNA replication and recombination: RAD51, RAD52, RPA, MRE11, RAD50, NBS1, BLM and WRN [reviewed by 221]. In cells that escape from crisis by activating ALT, APBs are detectable as soon as the characteristic heterogeneous telomere length pattern is detectable [219]. The function of APBs is unknown, but the co-localization in APBs of telomeric DNA with telomere binding proteins and proteins affiliated with recombination suggests that these structures may be directly involved in ALT telomere maintenance, possibly as sites for recombination and telomere synthesis, or as storage locations for proteins involved in this process. Alternatively, APBs may act to remove by-products of the ALT process, in accord with the proposal that PNBs may act as sites of intranuclear proteolysis [222]. Although APBs have been found in every ALT cell line examined, within an exponentially dividing ALT population APBs are present in only about 5% of the cells [219]. They are found primarily during the late S/G2/M phases of the cell cycle, suggesting that they appear and disappear in a cell cycle-dependent manner [223, 224].

Although human tumors have not yet been surveyed extensively for ALT, it seems that ALT is more common in sarcomas than in carcinomas, i.e., more common in tumors of mesenchymal origin than in those of epithelial origin [208, 221, 225]. The reasons for this are not yet clear, but it is possible that telomerase is more tightly repressed in mesenchymal cells than in epithelial cells, and thus there is a higher relative probability that ALT will be activated in tumors arising from mesenchyme [221].

As yet, there is no clear evidence for an ALT-like mechanism in normal mammalian cells [221]. However, preferential maintenance of short telomeres compared to longer telomeres has been observed in near-senescent fibroblasts: longer telomeres were observed to shorten rapidly, while shorter telomeres were maintained [70]. It was suggested that non-reciprocal recombination events between short and long telomeres may help maintain the length of critically short telomeres at the expense of the longer telomeres, but when the average telomere length shortens to a level unable to support recombination, the shorter telomeres continue to shorten, and senescence occurs [70]. Late passage telomerase knockout ($TERC^{-/-}$) mouse embryonic stem cells activated a telomerase-independent telomere length maintenance mechanism [226]. ALT-like activity in mice *in vivo* has been inferred from the observation that in late generation telomerase knockout mice, the telomeres in proliferating splenocytes of immunized mice were longer than in non-immunized mice [227]. It is unknown whether such activity occurs in normal mammalian cells.

Even if there is some ALT-like activity in normal somatic cells, it is clear that they do not exhibit the full-blown ALT process associated with telomere length maintenance and immortalization of cell lines and tumors. Cell hybridization experiments in which normal cells and ALT cells were fused have shown that ALT was repressed in the hybrids, indicating that normal cells contain repressors of ALT [101]. The identity of the repressors is currently unknown.

Concluding comments

There is a very substantial body of evidence that telomere status, especially telomere length and binding of specific binding proteins, is related to the proliferative capacity of human cells. In at least some cells, senescence can be bypassed by telomerase activity that is induced by the expression of exogenous hTERT. It is currently not possible to test whether ALT has the same effect because there is no known technique for inducing ALT activity. Nevertheless, it is clear that activation of a TMM, telomerase or ALT, is always associated with immortalization in cells that become immortalized spontaneously. For some types of cells grown under standard cell culture conditions expression of exogenous hTERT is not sufficient for immortalization, and other changes such as loss of function of the $p16^{INK4A}$/RB pathway are also required. The extent to which this is cell-type specific or a consequence of environmental stress needs further investigation. At present, the possibility that there are mitotic clocks in addition to telomere shortening cannot be definitively excluded. The exact nature of the altered telomere state that triggers senescence and the nature of the ensuing signaling pathway also require elucidation.

Acknowledgments

The authors thank Axel Neumann and Elizabeth Collins for illustrations. Work in the authors' laboratory was supported by the Carcinogenesis Fellowship of the Cancer Council NSW, project grants from the National Health and Medical Research Council, Australia, and the Nippon Boehringer Ingelheim Virtual Research Institute of Aging.

References

1. Goldstein S (1990). Replicative senescence: the human fibroblast comes of age. *Science* 249: 1129–33.
2. Hayflick L, Moorhead PS (1961). The serial cultivation of human diploid cell strains. *Exp Cell Res.* 25: 585–621.
3. Wright WE, Shay JW (2002). Historical claims and current interpretations of replicative aging. *Nat Biotechnol.* 20: 682–88.
4. Chang B-D, Broude EV, Dokmanovic M, *et al.* (1999). A senescence-like phenotype distinguishes tumor cells that undergo terminal proliferation arrest after exposure to anticancer agents. *Cancer Res.* 59: 3761–7.

5. Schmitt CA, Fridman JS, Yang M, *et al.* (2002). A senescence program controlled by p53 and p16(INK4a) contributes to the outcome of cancer therapy. *Cell.* 109: 335–46.
6. Decary S, Ben Hamida C, Mouly V, Barbet JP, Hentati F, Butler-Browne GS (2000). Shorter telomeres in dystrophic muscle consistent with extensive regeneration in young children. *Neuromusc Disord.* 10: 113–20.
7. Kitada T, Seki S, Kawakita N, Kuroki T, Monna T (1995). Telomere shortening in chronic liver diseases. *Biochem Biophys Res Commun.* 211: 33–9.
8. Reddel RR (2000). The role of senescence and immortalization in carcinogenesis. *Carcinogenesis* 21: 477–84.
9. Hayflick L (1997). Mortality and immortality at the cellular level. A review. *Biochemistry (Mosc)* 62: 1180–90.
10. Matsumura T, Malik F, Holliday R (1989). Levels of DNA methylation in diploid and SV40 transformed human fibroblasts. *Exp Gerontol.* 24: 477–81.
11. Howard BH (1996). Replicative senescence: considerations relating to the stability of heterochromatin domains. *Exp Gerontol.* 31: 281–93.
12. Chen Q, Fischer A, Reagan JD, Yan L-J, Ames BN (1995). Oxidative DNA damage and senescence of human diploid fibroblast cells. *Proc Natl Acad Sci USA* 92: 4337–41.
13. Gershon D (1997). More on mitochondria and senescence. *BioEssays* 19: 533–4.
14. Reddel RR (1998). A reassessment of the telomere hypothesis of senescence. *BioEssays* 20: 977–84.
15. Blackburn EH (1991). Structure and function of telomeres. *Nature* 350: 569–73.
16. Moyzis RK, Buckingham JM, Cram LS, *et al.* (1988). A highly conserved repetitive DNA sequence, $(TTAGGG)_n$, present at the telomeres of human chromosomes. *Proc Natl Acad Sci USA* 85: 6622–6.
17. Makarov VL, Hirose Y, Langmore JP (1997). Long G tails at both ends of human chromosomes suggest a C strand degradation mechanism for telomere shortening. *Cell* 88: 657–66.
18. McElligott R, Wellinger RJ (1997). The terminal DNA structure of mammalian chromosomes. *EMBO J.* 16: 3705–14.
19. Wright WE, Tesmer VM, Huffman KE, Levene SD, Shay JW (1997). Normal human chromosomes have long G-rich telomeric overhangs at one end. *Genes Dev.* 11: 2801–9.
20. McClintock B (1941). The stability of broken ends of chromosomes in *Zea mays*. *Genetics* 26: 234–82.
21. Muller HJ (1938). The remaking of chromosomes. *Collecting Net* 13: 181–98.
22. Sandell LL, Zakian VA (1993). Loss of a yeast telomere: arrest, recovery, and chromosome loss. *Cell* 75: 729–39.
23. Griffith JD, Comeau L, Rosenfield S, *et al.* (1999). Mammalian telomeres end in a large duplex loop. *Cell* 97: 503–14.
24. de Lange T (2002). Protection of mammalian telomeres. *Oncogene* 21: 532–40.
25. Blackburn EH (1994). Telomeres: No end in sight. *Cell* 77: 621–3.
26. Greider CW (1996). Telomere length regulation. *Annu Rev Biochem.* 65: 337–65.
27. Zakian VA (1995). Telomeres: beginning to understand the end. *Science* 270: 1601–7.
28. Wright JH, Gottschling DE, Zakian VA (1992) *Saccharomyces* telomeres assume a non-nucleosomal chromatin structure. *Genes Dev.* 6: 197–210.
29. Chong L, van Steensel B, Broccoli D, *et al.* (1995). A human telomeric protein. *Science* 270: 1663–7.
30. Bilaud T, Brun C, Ancelin K, Koering CE, Laroche T, Gilson E (1997). Telomeric localization of TRF2, a novel human telobox protein. *Nat Genet.* 17: 236–9.

31. Broccoli D, Smogorzewska A, Chong L, de Lange T (1997). Human telomeres contain two distinct Myb-related proteins, TRF1 and TRF2. *Nat Genet.* 17: 231–5.
32. Li B, Oestreich S, de Lange T (2000). Identification of human Rap1: implications for telomere evolution. *Cell* 101: 471–83.
33. Smith S, Giriat I, Schmitt A, de Lange T (1998). Tankyrase, a poly(ADP-ribose) polymerase at human telomeres. *Science* 282: 1484–7.
34. Smith S, de Lange T (2000). Tankyrase promotes telomere elongation in human cells. *Curr Biol.* 10: 1299–302.
35. Seimiya H, Smith S (2002). The telomeric poly(ADP-ribose) polymerase, tankyrase 1, contains multiple binding sites for telomeric repeat binding factor 1 (TRF1) and a novel acceptor, 182-kDa tankyrase-binding protein (TAB182). *J Biol Chem.* 277: 14116–26.
36. Kim S, Kaminker P, Campisi J (1999). TIN2, a new regulator of telomere length in human cells. *Nat Genet.* 23: 405–12.
37. Zhu X-D, Küster B, Mann M, Petrini JHJ, de Lange T (2000). Cell-cycle-regulated association of RAD50/MRE11/NBS1 with TRF2 and human telomeres. *Nat Genet.* 25: 347–52.
38. Hsu H-L, Gilley D, Galande SA, *et al.* (2000). Ku acts in a unique way at the mammalian telomere to prevent end joining. *Genes Dev.* 14: 2807–12.
39. Baumann P, Cech TR (2001). Pot1, the putative telomere end-binding protein in fission yeast and humans. *Science* 292: 1171–5.
40. Zhou XZ, Lu KP (2001). The Pin2/TRF1-interacting protein PinX1 is a potent telomerase inhibitor. *Cell* 107: 347–59.
41. Stansel RM, Subramanian D, Griffith JD (2002). p53 binds telomeric single strand overhangs and t-loop junctions *in vitro. J Biol Chem.* 277: 11625–8.
42. Ford LP, Wright WE, Shay JW (2002). A model for heterogeneous nuclear ribonucleoproteins in telomere and telomerase regulation. *Oncogene* 21: 580–3.
43. Stansel RM, de Lange T, Griffith JD (2001). T-loop assembly *in vitro* involves binding of TRF2 near the 3′ telomeric overhang. *EMBO J.* 20: 5532–40.
44. Harley CB, Futcher AB, Greider CW (1990). Telomeres shorten during ageing of human fibroblasts. *Nature* 345: 458–60.
45. Hastie ND, Dempster M, Dunlop MG, Thompson AM, Green DK, Allshire RC (1990). Telomere reduction in human colorectal carcinoma and with ageing. *Nature* 346: 866–8.
46. Lindsey J, McGill NI, Lindsey LA, Green DK, Cooke HJ (1991). *In vivo* loss of telomeric repeats with age in humans. Mutat Res, 256: 45–8.
47. Allsopp RC, Vaziri H, Patterson C, *et al.* (1992). Telomere length predicts replicative capacity of human fibroblasts. *Proc Natl Acad Sci USA* 89: 10114–8.
48. Allsopp RC, Harley CB (1995). Evidence for a critical telomere length in senescent human fibroblasts. *Exp Cell Res.* 219: 130–6.
49. Watson JD (1972). Origin of concatemeric T7 DNA. *Nat New Biol.* 239: 197–201.
50. Levy MZ, Allsopp RC, Futcher AB, Greider CW, Harley CB (1992). Telomere end-replication problem and cell aging. *J Mol Biol.* 225: 951–60.
51. Olovnikov AM (1971). [Principle of marginotomy in template synthesis of polynucleotides]. *Doklady Akademii Nauk SSR* 201: 1496–9.
52. Olovnikov AM (1973). A theory of marginotomy. The incomplete copying of template margin in enzymic synthesis of polynucleotides, biological significance of the phenomenon. *J Theor Biol.* 41: 181–90.
53. Okazaki R, Okazaki T, Sakabe K, Sugimoto K, Sugino A (1968). Mechanism of DNA chain growth. I. Possible discontinuity and unusual secondary structure of newly synthesized chains. *Proc Natl Acad Sci USA* 59: 598–605.

54. Ohki R, Tsurimoto T, Ishikawa F (2001). *In vitro* reconstitution of the end replication problem. *Mol Cell Biol.* 21: 5753–66.
55. Wellinger RJ, Ethier K, Labrecque P, Zakian VA (1996). Evidence for a new step in telomere maintenance. *Cell* 85: 423–33.
56. Sitte N, Saretzki G, Von Zglinicki T (1998). Accelerated telomere shortening in fibroblasts after extended periods of confluency. *Free Radic Biol Med.* 24: 885–93.
57. Murnane JP, Sabatier L, Marder BA, Morgan WF (1994). Telomere dynamics in an immortal human cell line. *EMBO J.* 13: 4953–62.
58. Li B, Lustig AJ (1996). A novel mechanism for telomere size control in *Saccharomyces cerevisiae. Genes Dev.* 10: 1310–26.
59. Petersen S, Saretzki G, Von Zglinicki T (1998). Preferential accumulation of single-stranded regions in telomeres of human fibroblasts. *Exp Cell Res.* 239: 152–60.
60. Von Zglinicki T, Saretzki G, Döcke W, Lotze C (1995). Mild hyperoxia shortens telomeres and inhibits proliferation of fibroblasts: A model for senescence? *Exp Cell Res.* 220: 186–93.
61. Von Zglinicki T, Pilger R, Sitte N (2000). Accumulation of single-strand breaks is the major cause of telomere shortening in human fibroblasts. *Free Radic Biol Med.* 28: 64–74.
62. Vaziri H, West MD, Allsopp RC, *et al.* (1997). ATM-dependent telomere loss in aging human diploid fibroblasts and DNA damage lead to the post-translational activation of p53 protein involving poly(ADP-ribose) polymerase. *EMBO J.* 16: 6018–33.
63. Von Zglinicki T (2000). Role of oxidative stress in telomere length regulation and replicative senescence. *Ann NY Acad Sci.* 908: 99–110.
64. Harley CB, Vaziri H, Counter CM, Allsopp RC (1992). The telomere hypothesis of cellular aging. *Exp Gerontol.* 27: 375–82.
65. Lanza RP, Cibelli JB, Blackwell C, *et al.* (2000). Extension of cell life-span and telomere length in animals cloned from senescent somatic cells. *Science* 288: 665–9.
66. Shiels PG, Kind AJ, Campbell KHS, Waddington D, Wilmut I, Colman A, Schnieke AE (1999). Analysis of telomere lengths in cloned sheep. *Nature* 399: 316–17.
67. Blackburn EH (2000). Telomere states and cell fates. *Nature* 408: 53–6.
68. Karlseder J, Smogorzewska A, de Lange T (2002). Senescence induced by altered telomere state, not telomere loss. *Science* 295: 2446–9.
69. Martens UM, Zijlmans JM, Poon SS, *et al.* (1998). Short telomeres on human chromosome 17p. *Nat Genet.* 18: 76–80.
70. Martens UM, Chavez EA, Poon SS, Schmoor C, Lansdorp PM (2000). Accumulation of short telomeres in human fibroblasts prior to replicative senescence. *Exp Cell Res.* 256: 291–9.
71. Steinert S, Shay JW, Wright WE (2000). Transient expression of human telomerase extends the life span of normal human fibroblasts. *Biochem Biophys Res Commun.* 273: 1095–8.
72. Wright WE, Shay JW (1992). Telomere positional effects and the regulation of cellular senescence. *Trends Genet.* 8: 193–7.
73. Baur JA, Zou Y, Shay JW, Wright WE (2001). Telomere position effect in human cells. *Science* 292: 2075–7.
74. Allsopp RC (1996). Models of initiation of replicative senescence by loss of telomeric DNA. *Exp Gerontol.* 31: 235–43.
75. Tahara H, Sato E, Noda A, Ide T (1995). Increase in expression level of $p21^{sdi1/cip1/waf1}$ with increasing division age in both normal and SV40-transformed human fibroblasts. *Oncogene* 10: 835–40.

76. Kulju KS, Lehman JM (1995). Increased p53 protein associated with aging in human diploid fibroblasts. *Exp Cell Res.* 217: 336–45.
77. Noda A, Ning Y, Venable SF, Pereira-Smith OM, Smith JR (1994). Cloning of senescent cell-derived inhibitors of DNA synthesis using an expression screen. *Exp Cell Res.* 211: 90–8.
78. Webley K, Bond JA, Jones CJ, *et al.* (2000). Posttranslational modifications of p53 in replicative senescence overlapping but distinct from those induced by DNA damage. *Mol Cell Biol.* 20: 2803–8.
79. Smogorzewska A, de Lange T (2002). Different telomere damage signaling pathways in human and mouse cells. *EMBO J.* 21: 4338–48.
80. Bond JA, Wyllie FS, Wynford-Thomas D (1994). Escape from senescence in human diploid fibroblasts induced directly by mutant p53. *Oncogene* 9: 1885–9.
81. Rogan EM, Bryan TM, Hukku B, *et al.* (1995). Alterations in p53 and $p16^{INK4}$ expression and telomere length during spontaneous immortalization of Li-Fraumeni syndrome fibroblasts. *Mol Cell Biol.* 15: 4745–53.
82. Noble JR, Rogan EM, Neumann AA, Maclean K, Bryan TM, Reddel RR (1996). Association of extended *in vitro* proliferative potential with loss of $p16^{INK4}$ expression. *Oncogene* 13: 1259–68.
83. Brown JP, Wei W, Sedivy JM (1997). Bypass of senescence after disruption of $p21^{CIP1/WAF1}$ gene in normal diploid human fibroblasts. *Science* 277: 831–4.
84. Jarrard DF, Sarkar S, Shi Y, *et al.* (1999). p16/pRb pathway alterations are required for bypassing senescence in human prostate epithelial cells. *Cancer Res.* 59: 2957–64.
85. Counter CM, Avilion AA, LeFeuvre CE, *et al.* (1992). Telomere shortening associated with chromosome instability is arrested in immortal cells which express telomerase activity. *EMBO J.* 11: 1921–9.
86. Girardi AJ, Jensen FC, Koprowski H (1965). SV40-induced transformation of human diploid cells: crisis and recovery. *J Cell Comp Physiol.* 65: 69–84.
87. Wright WE, Pereira-Smith OM, Shay JW (1989). Reversible cellular senescence: Implications for immortalization of normal human diploid fibroblasts. *Mol Cell Biol.* 9: 3088–92.
88. Rubelj I, Pereira-Smith OM (1994). SV40-transformed human cells in crisis exhibit changes that occur in normal cellular senescence. *Exp Cell Res.* 211: 82–9.
89. Wang E (1995). Senescent human fibroblasts resist programmed cell death, and failure to suppress *bcl2* is involved. *Cancer Res.* 55: 2284–92.
90. Dimri GP, Lee X, Basile G, *et al.* (1995). A biomarker that identifies senescent human cells in culture and in aging skin *in vivo. Proc Natl Acad Sci USA* 92: 9363–7.
91. Macera-Bloch L, Houghton J, Lenahan M, Jha KK, Ozer HL (2002). Termination of lifespan of SV40-transformed human fibroblasts in crisis is due to apoptosis. *J Cell Physiol.* 190: 332–44.
92. Ducray C, Pommier J-P, Martins L, Boussin FD, Sabatier L (1999). Telomere dynamics, end-to-end fusions and telomerase activation during the human fibroblast immortalization process. *Oncogene* 18: 4211–23.
93. Cuthill S, Agarwal P, Sarkar S, Savelieva E, Reznikoff CA (1999). Dominant genetic alterations in immortalization: role for 20q gain. *Genes Chromosomes Cancer* 26: 304–11.
94. Nonet GH, Stampfer MR, Chin K, Gray JW, Collins CC, Yaswen P (2001). The *ZNF217* gene amplified in breast cancers promotes immortalization of human mammary epithelial cells. *Cancer Res.* 61: 1250–4.

95. Muggleton-Harris AL, DeSimone DW (1980). Replicative potentials of various fusion products between WI-38 and SV40 transformed WI-38 cells and their components. *Somatic Cell Genet.* 6: 689–98.
96. Bunn CL, Tarrant GM (1980). Limited lifespan in somatic cell hybrids and cybrids. *Exp Cell Res.* 127: 385–96.
97. Pereira-Smith OM, Smith JR (1981). Expression of SV40 T antigen in finite life-span hybrids of normal and SV40-transformed fibroblasts. *Somatic Cell Genet.* 7: 411–21.
98. Pereira-Smith OM, Smith JR (1983). Evidence for the recessive nature of cellular immortality. *Science* 221: 964–6.
99. Pereira-Smith OM, Robetorye S, Ning Y, Orson FM (1990). Hybrids from fusion of normal human T lymphocytes with immortal human cells exhibit limited life span. *J Cell Physiol.* 144: 546–9.
100. Pereira-Smith OM, Stein GH, Robetorye S, Meyer-Demarest S (1990). Immortal phenotype of the HeLa variant D98 is recessive in hybrids formed with normal human fibroblasts. *J Cell Physiol.* 143: 222–5.
101. Perrem K, Bryan TM, Englezou A, Hackl T, Moy EL, Reddel RR (1999). Repression of an alternative mechanism for lengthening of telomeres in somatic cell hybrids. *Oncogene* 18: 3383–90.
102. Pereira-Smith OM, Smith JR (1988). Genetic analysis of indefinite division in human cells: identification of four complementation groups. *Proc Natl Acad Sci USA* 85: 6042–6.
103. Whitaker NJ, Kidston EL, Reddel RR (1992). Finite life span of hybrids formed by fusion of different simian virus 40-immortalized human cell lines. *J Virol.* 66: 1202–6.
104. Duncan EL, Whitaker NJ, Moy EL, Reddel RR (1993). Assignment of SV40-immortalized cells to more than one complementation group for immortalization. *Exp Cell Res.* 205: 337–44.
105. Ning Y, Weber JL, Killary AM, Ledbetter DH, Smith JR, Pereira-Smith OM (1991). Genetic analysis of indefinite division in human cells: evidence for a cell senescence-related gene(s) on human chromosome 4. *Proc Natl Acad Sci USA* 88. 5635–9.
106. Bertram MJ, Bérubé NG, Hang-Swanson X, *et al.* (1999). Identification of a gene that reverses the immortal phenotype of a subset of cells and is a member of a novel family of transcription factor-like genes. *Mol Cell Biol.* 19: 1479–85.
107. Sugawara O, Oshimura M, Koi M, Annab LA, Barrett JC (1990). Induction of cellular senescence in immortalized cells by human chromosome 1. *Science* 247: 707–10.
108. Hensler PJ, Annab LA, Barrett JC, Pereira-Smith OM (1994). A gene involved in control of human cellular senescence on human chromosome 1q. *Mol Cell Biol.* 14: 2291–7.
109. Ogata T, Ayusawa D, Namba M, Takahashi E, Oshimura M, Oishi M (1993). Chromosome 7 suppresses indefinite division of nontumorigenic immortalized human fibroblast cell lines KMST-6 and SUSM-1. *Mol Cell Biol.* 13: 6036–43.
110. Ogata T, Oshimura M, Namba M, Fujii M, Oishi M, Ayusawa D (1995). Genetic complementation of the immortal phenotype in group D cell lines by introduction of chromosome 7. *Jpn J Cancer Res.* 86: 35–40.
111. Sandhu AK, Hubbard K, Kaur GP, Jha KK, Ozer HL, Athwal RS (1994). Senescence of immortal human fibroblasts by the introduction of normal human chromosome 6. *Proc Natl Acad Sci USA* 91: 5498–502.
112. Uejima H, Mitsuya K, Kugoh H, Horikawa I, Oshimura M (1995). Normal human chromosome 2 induces cellular senescence in the human cervical carcinoma cell line SiHa. *Genes Chrom Cancer,* 14: 120–7.

113. Uejima H, Shinohara T, Nakayama Y, Kugoh H, Oshimura M (1998). Mapping a novel cellular-senescence gene to human chromosome 2q37 by irradiation microcell-mediated chromosome transfer. *Mol Carcinog*. 22: 34–45.
114. Ohmura H, Tahara H, Suzuki M, *et al.* (1995). Restoration of the cellular senescence program and repression of telomerase by human chromosome 3. *Jpn J Cancer Res.* 86: 899–904.
115. Horikawa I, Oshimura M, Barrett JC (1998). Repression of the telomerase catalytic subunit by a gene on human chromosome 3 that induces cellular senescence. *Mol Carcinog.* 22: 65–72.
116. Tanaka H, Shimizu M, Horikawa I, *et al.* (1998). Evidence for a putative telomerase repressor gene on the 3p14.2-p21.1 region. *Genes Chrom Cancer* 23: 123–33.
117. Cuthbert AP, Bond J, Trott DA, *et al.* (1999). Telomerase repressor sequences on chromosome 3 and induction of permanent growth arrest in human breast cancer cells. *J Natl Cancer Inst.* 91: 37–45.
118. Poignée M, Backsch C, Beer K, *et al.* (2001). Evidence for a putative senescence gene locus within the chromosomal region 10p14-p15. *Cancer Res.* 61: 7118–21.
119. Kugoh H, Fujiwara M, Kihara K, *et al.* (2000). Cellular senescence of a human bladder carcinoma cell line (JTC-32) induced by a normal chromosome 11. *Cancer Genet Cytogenet.* 116: 158–63.
120. Reddy DE, Sandhu AK, DeRiel JK, Athwal RS, Kaur GP (1999). Identification of a gene at 16q24.3 that restores cellular senescence in immortal mammary tumor cells. *Oncogene* 18: 5100–7.
121. Reddy DE, Keck CL, Popescu N, Athwal RS, Kaur GP (2000). Identification of a YAC from 16q24 carrying a senescence gene for breast cancer cells. *Oncogene* 19: 217–22.
122. Sasaki M, Honda T, Yamada H, Wake N, Barrett JC, Oshimura M (1994). Evidence for multiple pathways to cellular senescence. *Cancer Res.* 54: 6090–3.
123. Klein CB, Conway K, Wang XW, *et al.* (1991). Senescence of nickel-transformed cells by an X chromosome: possible epigenetic control. *Science* 251: 796–9.
124. Yang X, Tahin Q, Hu YF, *et al.* (1999). Functional roles of chromosomes 11 and 17 in the transformation of human breast epithelial cells *in vitro*. *Int J Oncol.* 15: 629–38.
125. Colgin LM, Reddel RR (1999). Telomere maintenance mechanisms and cellular immortalization. *Curr Opin Genet Dev.* 9: 97–103.
126. Ducrest A-L, Amacker M, Mathieu YD, *et al.* (2001). Regulation of human telomerase activity: repression by normal chromosome 3 abolishes nuclear telomerase reverse transcriptase transcripts but does not affect c-myc activity. *Cancer Res.* 61: 7594–602.
127. Backsch C, Wagenbach N, Nonn M, *et al.* (2001). Microcell-mediated transfer of chromosome 4 into HeLa cells suppresses telomerase activity. *Genes Chromosomes Cancer* 31: 196–8.
128. Steenbergen RDM, Kramer D, Meijer CJLM, *et al.* (2001). Telomerase suppression by chromosome 6 in a human papillomavirus type 16-immortalized keratinocyte cell line and in a cervical cancer cell line. *J Natl Cancer Inst.* 93: 865–72.
129. Kumata M, Shimizu M, Oshimura M, Uchida M, Tsutsui T (2002). Induction of cellular senescence in a telomerase negative human immortal fibroblast cell line, LCS-AF.1-3, by human chromosome 6. *Int J Oncol.* 21: 851–6.
130. Nakabayashi K, Ogata T, Fujii M, *et al.* (1997). Decrease in amplified telomeric sequences and induction of senescence markers by introduction of human chromosome 7 or its segments in SUSM-1. *Exp Cell Res.* 235: 345–53.

131. Nakabayashi K, Ogino H, Michishita E, Satoh N, Ayusawa D (1999). Introduction of chromosome 7 suppresses telomerase with shortening of telomeres in a human mesothelial cell line. *Exp Cell Res.* 252: 376–82.
132. Tanaka H, Horikawa I, Kugoh H, Shimizu M, Barrett JC, Oshimura M (1999). Telomerase-independent senescence of human immortal cells induced by microcell-mediated chromosome transfer. *Mol Carcinog*. 25: 249–55.
133. Uhrbom L, Nistér M, Westermark B (1997). Induction of senescence in human malignant glioma cells by $p16^{INK4A}$. *Oncogene* 15: 505–14.
134. Vogt M, Haggblom C, Yeargin J, Christiansen-Weber T, Haas M (1998). Independent induction of senescence by $p16^{INK4a}$ and $p21^{CIP1}$ in spontaneously immortalized human fibroblasts. *Cell Growth Differ*. 9: 139–46.
135. Bryan TM, Reddel RR (1997). Telomere dynamics and telomerase activity in *in vitro* immortalised human cells. *Eur J Cancer* 33: 767–73.
136. Greider CW, Blackburn EH (1985). Identification of a specific telomere terminal transferase activity in *Tetrahymena* extracts. *Cell* 43: 405–13.
137. Nakayama J, Saito M, Nakamura H, Matsuura A, Ishikawa F (1997) *TLP1*: a gene encoding a protein component of mammalian telomerase is a novel member of WD repeats family. *Cell* 88: 875–84.
138. Schnapp G, Rodi H-P, Rettig WJ, Schnapp A, Damm K (1998). One-step affinity purification protocol for human telomerase. *Nucleic Acids Res.* 26: 3311–13.
139. Greene EC, Shippen DE (1998). Developmentally programmed assembly of higher order telomerase complexes with distinct biochemical and structural properties. *Genes Dev.* 12: 2921–31.
140. Kilian A, Bowtell DDL, Abud HE, *et al.* (1997). Isolation of a candidate human telomerase catalytic subunit gene, which reveals complex splicing patterns in different cell types. *Hum Mol Genet*. 6: 2011–19.
141. Nakamura TM, Morin GB, Chapman KB, *et al.* (1997). Telomerase catalytic subunit homologs from fission yeast and human. *Science* 277: 955–9.
142. Meyerson M, Counter CM, Eaton EN, *et al.* (1997) *hEST2*, the putative human telomerase catalytic subunit gene, is up-regulated in tumor cells and during immortalization. *Cell* 90: 785–95.
143. Harrington L, Zhou W, McPhail T, *et al.* (1997). Human telomerase contains evolutionarily conserved catalytic and structural subunits. *Genes Dev.* 11: 3109–15.
144. Feng J, Funk WD, Wang S-S, *et al.* (1995). The RNA component of human telomerase. *Science* 269: 1236–41.
145. Harrington L, McPhail T, Mar V, *et al.* (1997). A mammalian telomerase-associated protein. *Science* 275: 973–7.
146. Le S, Sternglanz R, Greider CW (2000). Identification of two RNA-binding proteins associated with human telomerase RNA. *Mol Biol Cell* 11: 999–1010.
147. Holt SE, Aisner DL, Baur J, *et al.* (1999). Functional requirement of p23 and Hsp90 in telomerase complexes. *Genes Dev.* 13: 817–26.
148. Collins K, Mitchell JR (2002). Telomerase in the human organism. *Oncogene* 21: 564–79.
149. Dragon F, Pogacic V, Filipowicz W (2000). *In vitro* assembly of human H/ACA small nucleolar RNPs reveals unique features of U17 and telomerase RNAs. *Mol Cell Biol.* 20: 3037–48.
150. Seimiya H, Sawada H, Muramatsu Y, *et al.* (2000). Involvement of 14-3-3 proteins in nuclear localization of telomerase. *EMBO J.* 19: 2652–61.

151. Ford LP, Shay JW, Wright WE (2001). The La antigen associates with the human telomerase ribonucleoprotein and influences telomere length *in vivo*. RNA, 7: 1068–75.
152. Bryan TM, Englezou A, Gupta J, Bacchetti S, Reddel RR (1995). Telomere elongation in immortal human cells without detectable telomerase activity. *EMBO J.* 14: 4240–8.
153. Counter CM, Botelho FM, Wang P, Harley CB, Bacchetti S (1994). Stabilization of short telomeres and telomerase activity accompany immortalization of Epstein-Barr virus-transformed human B lymphocytes. *J Virol.* 68: 3410–14.
154. Bacchetti S, Counter CM (1995). Telomeres and telomerase in human cancer. *Int J Oncol.* 7: 423–32.
155. Harrington LA, Greider CW (1991). Telomerase primer specificity and chromosome healing. *Nature* 353: 451–4.
156. Morin GB (1991). Recognition of a chromosome truncation site associated with α-thalassaemia by human telomerase. *Nature* 353: 454–6.
157. Wilkie AO, Lamb J, Harris PC, Finney RD, Higgs DR (1990). A truncated human chromosome 16 associated with alpha thalassaemia is stabilized by addition of telomeric repeat $(TTAGGG)_n$. *Nature* 346: 868–71.
158. Bryan TM, Cech TR (1999). Telomerase and the maintenance of chromosome ends. *Curr Opin Cell Biol.* 11: 318–24.
159. Slijepcevic P, Bryant PE (1998). Chromosome healing, telomere capture and mechanisms of radiation-induced chromosome breakage. *Int J Radiat Biol.* 73: 1–13.
160. Sun W, Kang K-S, Morita I, Trosko JE, Chang C-C (1999). High susceptibility of a human breast epithelial cell type with stem cell characteristics to telomerase activation and immortalization. *Cancer Res.* 59: 6118–23.
161. Broccoli D, Young JW, de Lange T (1995). Telomerase activity in normal and malignant hematopoietic cells. *Proc Natl Acad Sci USA* 92: 9082–6.
162. Hiyama K, Hirai Y, Kyoizumi S, *et al.* (1995). Activation of telomerase in human lymphocytes and hematopoietic progenitor cells. *J Immunol.* 155: 3711–15.
163. Weng N-P, Levine BL, June CH, Hodes RJ (1996). Regulated expression of telomerase activity in human T lymphocyte development and activation. *J Exp Med.* 183: 2471–9.
164. Liu K, Schoonmaker MM, Levine BL, June CH, Hodes RJ, Weng N-P (1999). Constitutive and regulated expression of telomerase reverse transcriptase (hTERT) in human lymphocytes. *Proc Natl Acad Sci USA* 96: 5147–52.
165. Yasumoto S, Kunimura C, Kikuchi K, *et al.* (1996). Telomerase activity in normal human epithelial cells. *Oncogene* 13: 433–9.
166. Ramirez RD, Wright WE, Shay JW, Taylor RS (1997). Telomerase activity concentrates in the mitotically active segments of human hair follicles. *J Invest Dermatol.* 108: 113–17.
167. Kyo S, Takakura M, Kohama T, Inoue M (1997). Telomerase activity in human endometrium. *Cancer Res.* 57: 610–14.
168. Hiyama E, Hiyama K, Tatsumoto N, Kodama T, Shay JW, Yokoyama T (1996). Telomerase activity in human intestine. *Int J Oncol.* 9: 453–8.
169. Wright WE, Piatyszek MA, Rainey WE, Byrd W, Shay JW (1996). Telomerase activity in human germline and embryonic tissues and cells. *Dev Genet.* 18: 173–9.
170. Kim NW, Piatyszek MA, Prowse KR, *et al.* (1994). Specific association of human telomerase activity with immortal cells and cancer. *Science* 266: 2011–15.
171. Ulaner GA, Giudice LC (1997). Developmental regulation of telomerase activity in human fetal tissues during gestation. *Mol Hum Reprod.* 3: 769–73.
172. Thomson JA, Itskovitz-Eldor J, Shapiro SS, *et al.* (1998). Embryonic stem cell lines derived from human blastocysts. *Science* 282: 1145–7.

173. Vulliamy T, Marrone A, Goldman F, *et al.* (2001). The RNA component of telomerase is mutated in autosomal dominant dyskeratosis congenita. *Nature* 413: 432–5.
174. Mitchell JR, Wood E, Collins K (1999). A telomerase component is defective in the human disease dyskeratosis congenita. *Nature* 402: 551–5.
175. Vulliamy TJ, Knight SW, Mason PJ, Dokal I (2001). Very short telomeres in the peripheral blood of patients with X-linked and autosomal dyskeratosis congenita. *Blood Cells Mol Dis.* 27: 353–7.
176. Dokal I, Bungey J, Williamson P, Oscier D, Hows J, Luzzatto L (1992). Dyskeratosis congenita fibroblasts are abnormal and have unbalanced chromosomal rearrangements. *Blood* 80: 3090–6.
177. Beattie TL, Zhou W, Robinson MO, Harrington L (1998). Reconstitution of human telomerase activity *in vitro. Curr Biol.* 8: 177–80.
178. Weinrich SL, Pruzan R, Ma L, *et al.* (1997). Reconstitution of human telomerase with the template RNA component hTR and the catalytic protein subunit hTRT. *Nat Genet.* 17: 498–502.
179. Counter CM, Meyerson M, Eaton EN, *et al.* (1998). Telomerase activity is restored in human cells by ectopic expression of hTERT (hEST2), the catalytic subunit of telomerase. *Oncogene* 16: 1217–22.
180. Ducrest AL, Szutorisz H, Lingner J, Nabholz M (2002). Regulation of the human telomerase reverse transcriptase gene. *Oncogene* 21: 541–52.
181. Liu J-P (1999). Studies of the molecular mechanisms in the regulation of telomerase activity. *FASEB J.* 13: 2091–104.
182. Colgin LM, Wilkinson C, Englezou A, Kilian A, Robinson MO, Reddel RR (2000). The hTERTα splice variant is a dominant negative inhibitor of telomerase activity. *Neoplasia* 2: 426–32.
183. Yi X, White DM, Aisner DL, Baur JA, Wright WE, Shay JW (2000). An alternate splicing variant of the human telomerase catalytic subunit inhibits telomerase activity. *Neoplasia* 2: 433–40.
184. Kyo S, Inoue M (2002). Complex regulatory mechanisms of telomerase activity in normal and cancer cells: how can we apply them for cancer therapy? *Oncogene* 21: 688–97.
185. Bodnar AG, Ouellette M, Frolkis M, *et al.* (1998). Extension of life-span by introduction of telomerase into normal human cells. *Science* 279: 349–52.
186. Vaziri H, Benchimol S (1998). Reconstitution of telomerase activity in normal human cells leads to elongation of telomeres and extended replicative life span. *Curr Biol.* 8: 279–82.
187. Counter CM, Hahn WC, Wei W, *et al.* (1998). Dissociation among *in vitro* telomerase activity, telomere maintenance, and cellular immortalization. *Proc Natl Acad Sci USA* 95: 14723–8.
188. Zhu J, Wang H, Bishop JM, Blackburn EH (1999). Telomerase extends the lifespan of virus-transformed human cells without net telomere lengthening. *Proc Natl Acad Sci USA* 96: 3723–8.
189. Halvorsen TL, Leibowitz G, Levine F (1999). Telomerase activity is sufficient to allow transformed cells to escape from crisis. *Mol Cell Biol.* 19: 1864–70.
190. MacKenzie KL, Franco S, Naiyer AJ, *et al.* (2002). Multiple stages of malignant transformation of human endothelial cells modelled by co-expression of telomerase reverse transcriptase, SV40 T antigen and oncogenic N-ras. *Oncogene* 21: 4200–11.
191. Morales CP, Holt SE, Ouellette M, *et al.* (1999). Absence of cancer-associated changes in human fibroblasts immortalized with telomerase. *Nat Genet.* 21: 115–18.

192. Jiang XR, Jimenez G, Chang E, *et al.* (1999). Telomerase expression in human somatic cells does not induce changes associated with a transformed phenotype. *Nat Genet.* 21: 111–14.
193. Vaziri H, Squire JA, Pandita TK, *et al.* (1999). Analysis of genomic integrity and p53-dependent G_1 checkpoint in telomerase-induced extended-life-span human fibroblasts. *Mol Cell Biol.* 19: 2373–9.
194. Dickson MA, Hahn WC, Ino Y, *et al.* (2000). Human keratinocytes that express hTERT and also bypass a $p16^{INK4a}$-enforced mechanism that limits life span become immortal yet retain normal growth and differentiation characteristics. *Mol Cell Biol.* 20: 1436–47.
195. Yang J, Chang E, Cherry AM, *et al.* (1999). Human endothelial cell life extension by telomerase expression. *J Biol Chem.* 274: 26141–8.
196. Venetsanakos E, Mirza A, Fanton C, Romanov SR, Tlsty T, McMahon M (2002). Induction of tubulogenesis in telomerase-immortalized human microvascular endothelial cells by glioblastoma cells. *Exp Cell Res.* 273: 21–33.
197. Farwell DG, Shera KA, Koop JI, *et al.* (2000). Genetic and epigenetic changes in human epithelial cells immortalized by telomerase. *Am J Pathol.* 156: 1537–47.
198. Kiyono T, Foster SA, Koop JI, McDougall JK, Galloway DA, Klingelhutz AJ (1998). Both $Rb/p16^{INK4a}$ inactivation and telomerase activity are required to immortalize human epithelial cells. *Nature* 396: 84–8.
199. Toouli CD, Huschtscha LI, Neumann AA, *et al.* (2002). Comparison of human mammary epithelial cells immortalized by simian virus 40 T-Antigen or by the telomerase catalytic subunit. *Oncogene* 21: 128–39.
200. Franco S, MacKenzie KL, Dias S, Alvarez S, Rafii S, Moore MAS (2001). Clonal variation in phenotype and life span of human embryonic fibroblasts (MRC-5) transduced with the catalytic component of telomerase (hTERT). *Exp Cell Res.* 268: 14–25.
201. MacKenzie KL, Franco S, May C, Sadelain M, Moore MAS (2000). Mass cultured human fibroblasts overexpressing hTERT encounter a growth crisis following an extended period of proliferation. *Exp Cell Res.* 259: 336–50.
202. O'Hare MJ, Bond J, Clarke C, *et al.* (2001). Conditional immortalization of freshly isolated human mammary fibroblasts and endothelial cells. *Proc Natl Acad Sci USA* 98: 646–51.
203. Migliaccio M, Amacker M, Just T, *et al.* (2000). Ectopic human telomerase catalytic subunit expression maintains telomere length but is not sufficient for $CD8^+$ T lymphocyte immortalization. *J Immunol.* 165: 4978–84.
204. McSharry BP, Jones CJ, Skinner JW, Kipling D, Wilkinson GW (2001). Human telomerase reverse transcriptase-immortalized MRC-5 and HCA2 human fibroblasts are fully permissive for human cytomegalovirus. *J Gen Virol.* 82: 855–63.
205. Hooijberg E, Ruizendaal JJ, Snijders PJF, Kueter EWM, Walboomers JMM, Spits H (2000). Immortalization of human $CD8^+$ T cell clones by ectopic expression of telomerase reverse transcriptase. *J Immunol.* 165: 4239–45.
206. Ramirez RD, Morales CP, Herbert B-S, *et al.* (2001). Putative telomere-independent mechanisms of replicative aging reflect inadequate growth conditions. *Genes Dev.* 15: 398–403.
207. Lorenz M, Saretzki G, Sitte N, Metzkow S, Von Zglinicki T (2001). BJ fibroblasts display high antioxidant capacity and slow telomere shortening independent of hTERT transfection. *Free Radic Biol Med.* 31: 824–31.

208. Bryan TM, Englezou A, Dalla-Pozza L, Dunham MA, Reddel RR (1997). Evidence for an alternative mechanism for maintaining telomere length in human tumors and tumor-derived cell lines. *Nat Med.* 3: 1271–4.
209. Dunham MA, Neumann AA, Fasching CL, Reddel RR (2000). Telomere maintenance by recombination in human cells. *Nat Genet.* 26: 447–50.
210. Roth CW, Kobeski F, Walter MF, Biessmann H (1997). Chromosome end elongation by recombination in the mosquito *Anopheles gambiae. Mol Cell Biol.* 17: 5176–83.
211. Biessmann H, Champion LE, O'Hair M, Ikenaga K, Kasravi B, Mason JM (1992). Frequent transpositions of *Drosophila melanogaster* HeT-A transposable elements to receding chromosome ends. *EMBO J.* 11: 4459–69.
212. Mason JM, Biessmann H (1995). The unusual telomeres of *Drosophila. Trends Genet.* 11: 58–62.
213. Lundblad V, Blackburn EH (1993). An alternative pathway for yeast telomere maintenance rescues *est1*$^{-}$ senescence. *Cell* 73: 347–60.
214. Teng S-C, Zakian VA (1999). Telomere-telomere recombination is an efficient bypass pathway for telomere maintenance in *Saccharomyces cerevisiae. Mol Cell Biol.* 19: 8083–93.
215. McEachern MJ, Blackburn EH (1996). Cap-prevented recombination between terminal telomeric repeat arrays (telomere CPR) maintains telomeres in *Kluyveromyces lactis* lacking telomerase. *Genes Dev.* 10: 1822–34.
216. McEachern MJ, Iyer S (2001). Short telomeres in yeast are highly recombinogenic. *Mol Cell* 7: 695–704.
217. Chen Q, Ijpma A, Greider CW (2001). Two survivor pathways that allow growth in the absence of telomerase are generated by distinct telomere recombination events. *Mol Cell Biol.* 21: 1819–27.
218. Opitz OG, Suliman Y, Hahn WC, Harada H, Blum HE, Rustgi AK (2001). Cyclin D1 overexpression and p53 inactivation immortalize primary oral keratinocytes by a telomerase-independent mechanism. *J Clin Invest.* 108: 725–32.
219. Yeager TR, Neumann AA, Englezou A, Huschtscha LI, Noble JR, Reddel RR (1999). Telomerase-negative immortalized human cells contain a novel type of promyelocytic leukemia (PML) body. *Cancer Res.* 59: 4175–9.
220. Zhong S, Salomoni P, Pandolfi PP (2000). The transcriptional role of PML and the nuclear body. *Nat Cell Biol.* 2: E85–90.
221. Henson JD, Neumann AA, Yeager TR, Reddel RR (2002). Alternative lengthening of telomeres in mammalian cells. *Oncogene* 21: 598–610.
222. Lallemand-Breitenbach V, Zhu J, Puvion F, *et al.* (2001). Role of promyelocytic leukemia (PML) sumolation in nuclear body formation, 11S proteasome recruitment, and As_2O_3-induced PML or PML/retinoic acid receptor α degradation. *J Exp Med.* 193: 1361–72.
223. Grobelny JV, Godwin AK, Broccoli D (2000). ALT-associated PML bodies are present in viable cells and are enriched in cells in the G_2/M phase of the cell cycle. *J Cell Sci.* 113: 4577–85.
224. Wu G, Lee W-H, Chen P-L (2000). NBS1 and TRF1 colocalize at promyelocytic leukemia bodies bodies during late S/G_2 phases in immortalized telomerase-negative cells: Implication of NBS1 in alternative lengthening of telomeres. *J Biol Chem.* 275: 30618–22.
225. Scheel C, Schaefer K-L, Jauch A, *et al.* (2001). Alternative lengthening of telomeres is associated with chromosomal instability in osteosarcomas. *Oncogene* 20: 3835–44.

226. Niida H, Shinkai Y, Hande MP, *et al.* (2000). Telomere maintenance in telomerase-deficient mouse embryonic stem cells: characterization of an amplified telomeric DNA. *Mol Cell Biol.* 20: 4115–27.
227. Herrera E, Martínez C, Blasco MA (2000). Impaired germinal center reaction in mice with short telomeres. *EMBO J.* 19: 472–81.
228. Di Leonardo A, Linke SP, Clarkin K, Wahl GM (1994). DNA damage triggers a prolonged p53-dependent G1 arrest and long-term induction of Cip1 in normal human fibroblasts. *Genes Dev.* 8: 2540–51.
229. Suzuki K, Mori I, Nakayama Y, Miyakoda M, Kodama S, Watanabe M (2001). Radiation-induced senescence-like growth arrest requires TP53 function but not telomere shortening. *Radiat Res.* 155: 248–53.
230. Chen Q, Ames BN (1994). Senescence-like growth arrest induced by hydrogen peroxide in human diploid fibroblast F65 cells. *Proc Natl Acad Sci USA* 91: 4130–4.
231. Bladier C, Wolvetang EJ, Hutchinson P, De Haan JB, Kola I (1997). Response of a primary human fibroblast cell line to H_2O_2: Senescence-like growth arrest or apoptosis? *Cell Growth Differ.* 8: 589–98.
232. Hicks GG, Egan SE, Greenberg AH, Mowat M (1991). Mutant p53 tumor suppressor alleles release *ras*-induced cell cycle growth arrest. *Mol Cell Biol.* 11: 1344–52.
233. Serrano M, Lin AW, McCurrach ME, Beach D, Lowe SW (1997). Oncogenic *ras* provokes premature cell senescence associated with accumulation of p53 and $p16^{INK4a}$. *Cell* 88: 593–602.
234. Wei S, Wei W, Sedivy JM (1999). Expression of catalytically active telomerase does not prevent premature senescence caused by overexpression of oncogenic Ha-Ras in normal human fibroblasts. *Cancer Res.* 59: 1539–43.
235. Wang X, Wong SCH, Pan J, *et al.* (1998). Evidence of cisplatin-induced senescent-like growth arrest in nasopharyngeal carcinoma cells. *Cancer Res.* 58: 5019–22.
236. Robles SJ, Adami GR (1998). Agents that cause DNA double strand breaks lead to $p16^{INK4a}$ enrichment and the premature senescence of normal fibrolasts. *Oncogene* 16: 1113–23.
237. Chang B-D, Xuan Y, Broude EV, *et al.* (1999). Role of p53 and $p21^{waf1/cip1}$ in senescence-like terminal proliferation arrest induced in human tumor cells by chemotherapeutic drugs. *Oncogene* 18: 4808–18.
238. Michishita E, Nakabayashi K, Ogino H, Suzuki T, Fujii M, Ayusawa D (1998). DNA topoisomerase inhibitors induce reversible senescence in normal human fibroblasts. *Biochem Biophys Res Commun.* 253: 667–71.
239. Christov KT, Shilkaitis AL, Kim ES, Steele VE, Lubet RA (2003). Chemopreventive agents induce a senescence-like phenotype in rat mammary tumours. *Eur J Cancer* 39: 230–9.
240. Weeda G, Donker I, De Wit J, *et al.* (1997). Disruption of mouse *ERCC1* results in a novel repair syndrome with growth failure, nuclear abnormalities and senescence. *Curr Biol.* 7: 427–39.
241. Venable ME, Lee JY, Smyth MJ, Bielawska A, Obeid LM (1995). Role of ceramide in cellular senescence. *J Biol Chem.* 270: 30701–8.
242. Tresini M, Mawal-Dewan M, Cristofalo VJ, Sell C (1998). A phosphatidylinositol 3-kinase inhibitor induces a senescent-like growth arrest in human diploid fibroblasts. *Cancer Res.* 58: 1–4.
243. Fairweather DS, Fox M, Margison GP (1987). The *in vitro* lifespan of MRC-5 cells is shortened by 5-azacytidine-induced demethylation. *Exp Cell Res.* 168: 153–9.

244. Holliday R (1986). Strong effects of 5-azacytidine on the *in vitro* lifespan of human diploid fibroblasts. *Exp Cell Res.* 166: 543–52.
245. Ogryzko VV, Hirai TH, Russanova VR, Barbie DA, Howard BH (1996). Human fibroblast commitment to a senescence-like state in response to histone deacetylase inhibitors is cell cycle dependent. *Mol Cell Biol.* 16: 5210–18.
246. Xiao H, Hasegawa T, Miyaishi O, Ohkusu K, Isobe K (1997). Sodium butyrate induces NIH3T3 cells to senescence-like state and enhances promoter activity of $p21^{WAF/CIP1}$ in p53-independent manner. *Biochem Biophys Res Commun.* 237: 457–60.
247. Terao Y, Nishida J, Horiuchi S, *et al.* (2001). Sodium butyrate induces growth arrest and senescence-like phenotypes in gynecologic cancer cells. *Int J Cancer* 94: 257–67.
248. Lodygin D, Menssen A, Hermeking H (2002). Induction of the Cdk inhibitor p21 by LY83583 inhibits tumor cell proliferation in a p53-independent manner. *J Clin Invest.* 110: 1717–27.
249. Pandita TK (2002). ATM function and telomere stability. *Oncogene* 21: 611–18.
250. Opresko PL, Von Kobbe C, Laine JP, Harrigan J, Hickson ID, Bohr VA (2002). Telomere binding protein TRF2 binds to and stimulates the Werner and Bloom syndrome helicases. *J Biol Chem.* 277: 41110–19.
251. Stavropoulos DJ, Bradshaw PS, Li X, *et al.* (2002). The Bloom syndrome helicase BLM interacts with TRF2 in ALT cells and promotes telomeric DNA synthesis. *Hum Mol Genet.* 11: 3135–44.
252. Ashley T, Walpita D, de Rooij DG (2001). Localization of two mammalian cyclin dependent kinases during mammalian meiosis. *J Cell Sci.* 114: 685–93.
253. Gilley D, Tanaka H, Hande MP, *et al.* (2001). DNA-PKcs is critical for telomere capping. *Proc Natl Acad Sci USA* 98: 15084–8.
254. Espejel S, Franco S, Sgura A, *et al.* (2002). Functional interaction between DNA-PKcs and telomerase in telomere length maintenance. *EMBO J.* 21: 6275–87.
255. Smogorzewska A, Karlseder J, Holtgreve-Grez H, Jauch A, de Lange T (2002). DNA ligase IV-dependent NHEJ of deprotected mammalian telomeres in G1 and G2. *Curr Biol.* 12: 1635–44.
256. Gineitis AA, Zalenskaya IA, Yau PM, Bradbury EM, Zalensky AO (2000). Human sperm telomere-binding complex involves histone H2B and secures telomere membrane attachment. *J Cell Biol.* 151: 1591–8.
257. Ishikawa F, Matunis MJ, Dreyfuss G, Cech TR (1993). Nuclear proteins that bind the pre-mRNA 3′ splice site sequence r(UUAG/G) and the human telomeric DNA sequence $d(TTAGGG)_n$. *Mol Cell Biol.* 13: 4301–10.
258. McKay SJ, Cooke H (1992). hnRNP A2/B1 binds specifically to single stranded vertebrate telomeric repeat $TTAGGG_n$. *Nucleic Acids Res.* 20: 6461–4.
259. LaBranche H, Dupuis S, Ben-David Y, Bani M-R, Wellinger RJ, Chabot B (1998). Telomere elongation by hnRNP A1 and a derivative that interacts with telomeric repeats and telomerase. *Nat Genet.* 19: 199–202.
260. Dallaire F, Dupuis S, Fiset S, Chabot B (2000). Heterogeneous nuclear ribonucleoprotein A1 and UP1 protect mammalian telomeric repeats and modulate telomere replication *in vitro*. *J Biol Chem.* 275: 14509–16.
261. Fiset S, Chabot B (2001). hnRNP A1 may interact simultaneously with telomeric DNA and the human telomerase RNA *in vitro*. *Nucleic Acids Res.* 29: 2268–75.
262. Hsu H-L, Gilley D, Blackburn EH, Chen DJ (1999). Ku is associated with the telomere in mammals. *Proc Natl Acad Sci USA* 96: 12454–8.
263. Bianchi A, de Lange T (1999). Ku binds telomeric DNA *in vitro* [published erratum appears in *J Biol Chem.* 1999 274(49):35284]. *J Biol Chem.* 274: 21223–7.

264. Pollice A, Zibella MP, Bilaud T, Laroche T, Pulitzer JF, Gilson E (2000) *In vitro* binding of nucleolin to double-stranded telomeric DNA. *Biochem Biophys Res Commun.* 268: 909–15.
265. Baumann P, Podell E, Cech TR (2002). Human pot1 (protection of telomeres) protein: cytolocalization, gene structure, and alternative splicing. *Mol Cell Biol.* 22: 8079–87.
266. Cook BD, Dynek JN, Chang W, Shostak G, Smith S (2002). Role for the related poly(ADP-Ribose) polymerases tankyrase 1 and 2 at human telomeres. *Mol Cell Biol.* 22: 332–42.
267. Lyons RJ, Deane R, Lynch DK, *et al.* (2001). Identification of a novel human tankyrase through its interaction with the adaptor protein Grb14. *J Biol Chem.* 276: 17172–80.
268. Shen M, Haggblom C, Vogt M, Hunter T, Lu KP (1997). Characterization and cell cycle regulation of the related human telomeric proteins Pin2 and TRF1 suggest a role in mitosis. *Proc Natl Acad Sci USA* 94: 13618–23.
269. Smogorzewska A, van Steensel B, Bianchi A, *et al.* (2000). Control of human telomere length by TRF1 and TRF2. *Mol Cell Biol.* 20: 1659–68.
270. Karlseder J, Broccoli D, Dai Y, Hardy S, de Lange T (1999). p53- and ATM-dependent apoptosis induced by telomeres lacking TRF2. *Science* 283: 1321–5.
271. Li H, Cao Y, Berndt MC, Funder JW, Liu J-P (1999). Molecular interactions between telomerase and the tumor suppressor protein p53 *in vitro. Oncogene* 18: 6785–94.
272. Ouellette MM, McDaniel LD, Wright WE, Shay JW, Schultz RA (2000). The establishment of telomerase-immortalized cell lines representing human chromosome instability syndromes. *Hum Mol Genet.* 9: 403–11.
273. Nakamura H, Fukami H, Hayashi Y, *et al.* (2002). Establishment of immortal normal and ataxia telangiectasia fibroblast cell lines by introduction of the hTERT gene. *J Radiat Res (Tokyo)* 43: 167–74.
274. Wyllie FS, Jones CJ, Skinner JW, *et al.* (2000). Telomerase prevents the accelerated cell ageing of Werner syndrome fibroblasts. *Nat Genet.* 24: 16–17.
275. Choi D, Whittier PS, Oshima J, Funk WD (2001). Telomerase expression prevents replicative senescence but does not fully reset mRNA expression patterns in Werner syndrome cell strains. *FASEB J.* 15: 1014–20.
276. Wood LD, Halvorsen TL, Dhar S, *et al.* (2001). Characterization of ataxia telangiectasia fibroblasts with extended life-span through telomerase expression. *Oncogene* 20: 278–88.
277. Ranganathan V, Heine WF, Ciccone DN, *et al.* (2001). Rescue of a telomere length defect of Nijmegen breakage syndrome cells requires NBS and telomerase catalytic subunit. *Curr Biol.* 11: 962–6.
278. Matsushita H, Chang E, Glassford AJ, Cooke JP, Chiu CP, Tsao PS (2001). eNOS activity is reduced in senescent human endothelial cells: preservation by hTERT immortalization. *Circ Res.* 89: 793–8.
279. Kim H, Farris J, Christman SA, *et al.* (2002). Events in the immortalizing process of primary human mammary epithelial cells by the catalytic subunit of human telomerase. *Biochem J.* 365: 765–72.
280. Elenbaas B, Spirio L, Koerner F, *et al.* (2001). Human breast cancer cells generated by oncogenic transformation of primary mammary epithelial cells. *Genes Dev.* 15: 50–65.
281. Herbert BS, Wright WE, Shay JW (2002). p16(INK4a) inactivation is not required to immortalize human mammary epithelial cells. *Oncogene* 21: 7897–900.
282. Jones CJ, Kipling D, Morris M, *et al.* (2000). Evidence for a telomere-independent "clock" limiting *RAS* oncogene-driven proliferation of human thyroid epithelial cells. *Mol Cell Biol.* 20: 5690–9.

283. Lundberg AS, Randell SH, Stewart SA, *et al.* (2002). Immortalization and transformation of primary human airway epithelial cells by gene transfer. *Oncogene* 21: 4577–86.
284. Bandyopadhyay D, Timchenko N, Suwa T, Hornsby PJ, Campisi J, Medrano EE (2001). The human melanocyte: a model system to study the complexity of cellular aging and transformation in non-fibroblastic cells. *Exp Gerontol.* 36: 1265–75.
285. Condon J, Yin S, Mayhew B, *et al.* (2002). Telomerase immortalization of human myometrial cells. *Biol Reprod.* 67: 506–14.
286. Carney SA, Tahara H, Swartz CD, *et al.* (2002). Immortalization of human uterine leiomyoma and myometrial cell lines after induction of telomerase activity: molecular and phenotypic characteristics. *Lab Invest.* 82: 719–28.
287. Yudoh K, Matsuno H, Nakazawa F, Katayama R, Kimura T (2001). Reconstituting telomerase activity using the telomerase catalytic subunit prevents the telomere shorting and replicative senescence in human osteoblasts. *J Bone Miner Res.* 16: 1453–64.
288. Halvorsen TL, Beattie GM, Lopez AD, Hayek A, Levine F (2000). Accelerated telomere shortening and senescence in human pancreatic islet cells stimulated to divide *in vitro. J Endocrinol.* 166: 103–9.

Molecular Chaperones and Cellular Aging

Geneviève Morrow and Robert M. Tanguay
Laboratory of Cellular & Developmental Genetics, CREFSIP and Dept Medicine, Pav. Marchand, Université Laval, Ste-Foy, Québec, Canada G1K 7P4

Summary

Chaperones are phylogenetically conserved proteins involved in the proper folding of nascent proteins and repair of damaged or misfolded proteins. They are ubiquitously expressed and have been shown to play important roles in many cellular processes in addition to protein synthesis and folding. Chaperones are important for intracellular protein transport, membrane translocation and cytoprotection against many environmental and physiological insults. Their decrease in expression during aging could account for numerous features of cellular senescence. As many proteins have been shown to have chaperone activity, we focus here on heat shock proteins (Hsp), a major class of chaperones, and on their involvement in the aging process.

Introduction

According to the free radical theory, lifespan is determined by the capacity of an organism to cope with random damages induced by the toxic effects of reactive oxygen species (ROS) which include superoxide anion ($O_2^{\cdot -}$), hydrogen peroxide (H_2O_2), and hydroxyl radicals (OH·) [1]. Features of aging include cumulative damages to macromolecules with further decline of cell and organ functions. In attempts to understand the mechanisms underlying the aging process, multiple mutants with increased longevity of the fruitfly *Drosophila melanogaster* and the nematode *Caenorhabditis elegans*, have been obtained and analyzed for differences in gene expression at the genetic level [2–8]. Manipulating metabolic rate through calorie restriction or feeding agents with antioxidant properties have also been under investigation [9–13]. An increase in thermotolerance and resistance to stress has been observed in several model organisms featuring extended longevity phenotypes [2, 3, 6]. In many organisms, Hsps, which are for the majority unspecific molecular

Sunil C. Kaul and Renu Wadhwa (eds.), Aging of Cells In and Outside the Body, 207–223.

chaperones, have been shown to be major determinants in the acquisition of thermotolerance and to confer resistance to oxidative and chemical stresses [14, 15].

Heat shock proteins are ubiquitously expressed in all living organisms and in addition to their main function in protein folding and trafficking, also insure cell protection and survival during whole lifetime, especially in stressed conditions. These molecular chaperones are divided into different families on the basis of their molecular weights (Table 1). Hsps have been shown to be involved in numerous cellular processes; thermotolerance (small Hsp (sHsps), Hsp70, Hsp90, Hsp100) [16–23], inhibition of apoptosis (Hsp27, Hsp70, Hsp90) [24–34] activation of apoptosis (Hsp10, Hsp60) [35–37], reservoirs of misfolded proteins (sHsps) [33, 38], refolding of

Table 1*. The HSP families*

Families	Weight (kDa)	Functions	References
Small Hsp	10–40	Thermotolerance	16, 17, 20
		Inhibition of apoptosis	24–26, 29, 30, 33, 34
		Molecular chaperone	33, 38, 137
		Maintenance of cytoskeleton	33, 138
		Preservation of membrane integrity	139
Hsp60	60	Activation of apoptosis	35–37
		Folding of nascent or denatured proteins	39–44
		Protection of Fe/S proteins and of iron release	140
Hsp70	70	Thermotolerance	15
		Inhibition of apoptosis	15, 27–29, 31–32
		Refolding of nascent or denatured proteins	39–42, 44, 45
		Targeting of proteins to organelles and driving force for membrane translocation	39, 44, 46
		Assembly of mitochondrial Fe/S clusters	44
Hsp90/Grp94	83–100	Thermotolerance	15
		Inhibition of apoptosis	15
		Capacitor for morphological evolution	55, 56
		Cell proliferation	141, 142
		Regulation and stabilization of proteins involved in cell cycle control and signal transduction	142–148
		Protein assembly, folding	136, 149
Hsp100/Clp		Dissagregation of proteins for refolding by HSP70	150, 151
		Chaperone	44, 152
		Degradation of protein	44, 48, 51, 53, 54
		Activation of DNA replication factor	151
		Thermotolerance	18, 19, 21–23

nascent or denatured proteins (Hsp60, Hsp70) [39–45], targeting of nascent proteins to proper organelles and passage of proteins through membranes (Hsp70) [39, 44, 46], regulating receptor endocytosis and activity, targeting of proteins to proteolytic pathways (Hsp70, Hsp90, Hsp100) [44, 47–54], stabilizing damaged proteins and genetics variations (Hsp90) [55, 56]. Many chaperones like Hsp60, Hsp70 and Hsp90 are assisted by co-chaperones in their function [49, 50, 57, 58]. In general, chaperones of the Hsp family do not show substrate specificity although there are some exceptions like Hsp47 whose chaperone activity is restricted to collagen [59]. Altogether, chaperone's basal functions are to fold nascent proteins correctly and to repair altered ones in order to maintain the pool of active proteins needed for cell viability.

During aging the amount of damaged proteins tends to increase in cells [reviewed in refs 60, 61]. Some of the age-associated post-translational modifications that occur are stable and thereby alter protein function. This is the case for deamination of asparaginyl and glutamyl residues, protein glycation, methionine oxidation, among others [62, 63]. Some proteins seem to be more susceptible to oxidative damages (Aconitase [64]). Such modified proteins can aggregate and impair the cellular proteolytic activity of the proteasome [65, 66]. The accumulation of protein aggregates can thus lead to detrimental effects on cell function [67, 68].

Because they are ubiquitous and multifunctional, molecular chaperones are essential to cell viability and good health. A down regulation or a default in one or multiple chaperones would result in increased level of damaged proteins hence leading to possible dramatic effects at the whole organism level [67, 69]. In this chapter we take a closer look at chaperone expression during lifespan.

The stress response changes during aging

One feature of aging is the accumulation of damaged proteins and the increased sensitivity of organisms to environmental stresses. Flemings *et al.* [70] were the first to demonstrate a difference in the set of thermally induced-proteins between young and old *Drosophila*. Feeding young flies with canavanine, an arginine analogue used to mimic accumulation of damaged proteins, resulted in a protein expression pattern similar to that of old flies. Many authors have also documented a decrease in Hsp induction during aging [71]. For example, Locke and Tanguay [72] demonstrated a decreased capacity of Hsp70 induction by heat in aged rats; this decreased capacity was not due to the absence of the heat shock factor 1 (Hsf1), a general stress transcription factor responsible for Hsp induction, but rather to a lack of activation of Hsf1. Moreover, Garigan *et al.* [73] isolated a mutant of *C. elegans* by RNA interference (RNAi) with a rapid-aging phenotype. They further showed that this mutant was encoding Hsf and suggested that Hsp could function to slow the rate of aging. Changes in heat-induction of Hsp during aging have also been reported for Hsp27 [74], Hsp32 [75], Hsp60 [74] and Hsp70 [76–78].

Chaperones and senescence

Hsp27, Hsp70 and Hsp90 have been reported to be involved in the regulation of cellular senescence. For example, over expression of Hsp27 in bovine endothelial cells lead to accelerated growth and senescence [79]. Expression of Mortalin, an Hsp70-homologue, has been shown to confer cellular senescence in transfected NIH 3T3 cells [80, 81]. Hsp90 has also been reported to play an indirect role in senescence by folding and stabilizing the telomerase, which is responsible for the maintenance of telomeres length [82].

Age modulation of chaperones

In addition to being coordinately expressed following exposure to different stresses, Hsps can display differential expression pattern during normal development and aging. Although the heat shock response is often down regulated during aging, some Hsps clearly show an increased level of expression in aged animals [83, reviewed in refs 60, 61]. Interestingly, the age-associated post-translational modification of αB-crystallin, a member of the small Hsp family, results in a decreased chaperone activity [84, 85]. In *Drosophila melanogaster,* a muscle-specific increased expression of Hsp70 has been observed in aged flies [86]. Northern blot analysis data showed that the induction was not correlated with an increase in mRNA level suggesting that the induction of Hsp70 levels was regulated post-transcriptionally. Fly mutants for either Cu/ZnSOD or catalase showed a similar muscle-specific induction of an Hsp70-β-galactosidase reporter gene suggesting that Hsp70 expression during aging might result from oxidative damage. Whether expression of Hsp70 simply represents a stress response or whether Hsp70 might have a role in protecting muscles cells from oxidative damage remains unanswered at that time. Thermally conditioned flies or flies engineered to over express Hsp70 show an extended longevity at normal growth temperatures [87]. Hsp70 is up regulated with age in rat kidneys [88] but remains unchanged in aged rat skeletal muscle [89]. Hsc70 is elevated in very old rats [90], unchanged in hepatocytes [91] and decreased in testis [92]. This apparent discrepancy may reflect the different needs of cells or the cell-specific response of Hsf members [93].

Up regulation of the mRNA levels of the small Hsp22 and Hsp23 of *Drosophila* has also been observed in aged flies [86, 94]. Induction of Hsp22 mRNA during aging showed a 60-fold increase in the head comparatively to 16-fold in abdomen and 2.5-fold in thorax. Hsp23 RNA was induced 5-fold in the head [94]. No age-related induction of the mRNAs encoding the other two main sHsps of *Drosophila*, Hsp26 and Hsp27, was observed. This preferential induction of Hsp22 could reflect the sensitivity of mitochondrial proteins to aging since this sHsp localizes to the mitochondrial matrix [95] while Hsp23 and Hsp26 are cytoplasmic and Hsp27 nuclear [96, reviewed in refs 97, 98]. A potentially beneficial effect of Hsp22 and Hsp23 expression in the aging process is suggested by the observation that flies genetically selected for increased longevity display an earlier onset of Hsp22 and Hsp23 expression [99]. Preliminary data obtained in our laboratory also argue for a beneficial effect of sHsp in *Drosophila* aging (Morrow *et al.*, in preparation).

A genome wide microarray study performed on aged *Drosophila* showed that Hsp26 mRNA was down regulated during aging, and that a gene encoding an Hsp60-like protein was up regulated [4]. This gene was also up regulated in paraquat fed flies. Thus a high level of expression of Hsp22 [86, 94] and Hsp60-homolog [4] is correlated with aging, a finding in agreement with the hypothesis that mitochondrial proteins may be particularly sensitive during aging. This is further supported by recent RNAi data on the nematode *C. elegans* identifying a critical role for genes involved in mitochondrial function in the aging process [100, 101].

Over expression of Hsps increases lifespan

Organisms showing extended lifespan are often more resistant to environmental stress than normal ones at the same age [3, 6, 8, 102]. Since some Hsps are up regulated during aging, over expression of Hsp could have a beneficial effect in lifespan determination of these organisms. While there are few direct experiments addressing that question, some published data tend to support this idea. Early experiments using mild pre-conditioning heat stresses demonstrated that Hsps could promote lifespan extension. Organisms submitted to such mild heat stress showed an increased capacity to survive otherwise lethal conditions. This phenomenon was called acquired thermotolerance. Thus repeated mild stresses allowed the enhancement of cellular defense mechanisms and hence were favorable to survival under drastic conditions and to lifespan extension [15, 102–107] a phenomenon termed hormesis. Butov *et al.* [106] proposed a model where a balance between macromolecular damages and the level of Hsp production could explain the hormetic effect of mild heat shock.

In *C. elegans*, the *Age-1* mutation has been shown to confer an extended longevity phenotype and increased thermotolerance. Walker *et al.* [108] suggested that this phenotype was the result of elevated molecular chaperones and demonstrated that *age-1* mutants displayed increased levels of the small heat shock protein Hsp16 following thermal stress throughout the adult lifespan. However although expression of a single Hsp, Hsp16, was high in these mutants, other proteins, including other chaperones and/or antioxidant enzymes, may account for the long survival phenotype. Indeed flies over expressing antioxidant enzymes in specific tissues have a considerably higher longevity [109–112, reviewed in refs 113, 114].

In *Drosophila melanogaster*, over expression of Hsp70 during a short time laps (10 to 30 min) has been shown to increase lifespan by 10% to 30% [87] in the two weeks following the treatment. In these experiments, the authors used transgenic fly strains carrying different copy numbers of *hsp70*. Exceeding by 10–12% the normal content of Hsp70 was needed to observe a subsequent increase in survival to stress. Moreover, flies carrying extra copies had greater effects on their survival. This suggested that over expression of Hsp70 could act by delaying senescence of critical and specific pathways or tissues due to the increased chaperone activity for renaturing and properly refolding denatured or modified proteins. However there was no increase in the mean lifespan in these experiments. Minois *et al.* [105] did not observe any beneficial effects in flies over expressing Hsp70 but Chong *et al.* [115] have observed protection against oxidative stress.

Seong *et al.* [111] elaborated a gene search strategy using the GAL4/UAS system in *Drosophila melanogaster.* In their experiment, GAL4 expression is driven by *hsp70* promoter and activates UAS sequence inserted in another P-element directed to express sequences flanking the P-element. Using this strategy to search for genes influencing longevity, they found six genes related to stress resistance and redox balance including Hsp26 [111]. They observed an approximate 15% increase in longevity by directing ubiquitous expression of this sole Hsp using this strategy. Similar to lifespan extension of flies, nematodes and human cells with overexpression of the Hsp70 family protein, mortalin, were seen to have longer lifespan as compared to the controls [116, 117]. A listing summarizing the main observations arguing for a beneficial effect of HSP in aging is presented in Table 2.

***Table 2**. Chaperones involvement in aging*

Findings	References
Aging is associated with accumulation of inactive enzymes	153
Aging is accompanied by a different expression pattern of Hsp	70
Increased stress resistance is often associated with increased lifespan	102, 154
The heat shock response tends to decrease in aging animals	72
Hsp70 protein is up regulated during aging	86
Increased lifespan in *Drosophila* transgenic lines over expressing Hsp70	87, 105
Thermally conditioned *Drosophila* exhibit greater longevity	103, 107
Hsp22 and Hsp23 RNA are up regulated during aging	86, 94
Hsp22 and Hsp23 RNA are increased in *Drosophila* lines genetically selected for increased longevity	99
Hsp26 is a candidate gene which might be critical for longevity	111

Chaperones and age-related cellular processes

Aging has been associated with oxidative stress due to mitochondrial alteration. Aging cells display a decrease in their adaptive response to environmental stress resulting in an overall increase of proteotoxic condition within cells. The cellular decrease in chaperone level leads to a decrease of damaged proteins recognition, repair or degradation. The resulting increase of modified proteins may act by inhibiting the proteasome machinery [118, 119]. The accumulation of modified macromolecules leads inevitably to impairment of cellular functions. Hence a decrease in stress response and/or chaperone expression could account for various features of aging namely decline of nervous and cardiovascular system and organs functions.

The role of sick chaperones in cellular functions during aging has been reviewed extensively [69]. Chaperones keep key proteins in active form and should therefore be important during aging and in response to stress. For example, in yeast Hsp70 and Hsp104 are involved in maintenance of functional snRNPs and a default in one or the other can result in inactivation of snRNPs and compromise mRNA splicing [120]. Chaperones p23 and Hsp90 are required for telomerase assembly and hence to telomeres length maintenance. The absence of these chaperones or defects in one could result in a rapid decrease of telomeres length and cell division arrest [82]. Another example is Hsp47, a chaperone responsible for proper pro-collagen folding. A default in this chaperone would lead to collagen deterioration, another feature of the aging process [59]. Hsp90 is needed for ligand-dependent transcriptional activity of steroid receptors. Some Hsps make heterocomplexes with transcription factors, interact with components of the cell cycle [121] or are involved in centrosome assembly [121, 122]. Any default in these proteins could lead to bad signaling and would have generalized consequences typical of aging due to the extent of tissues affected [123, 124]. Other specialized functions of Hsps explaining some aging features have recently been discussed [15].

A decline in systemic functions is also observed during aging. In the cardiovascular system of young animals, Hsp70 is induced by acute hypertension. As the animal become older their stress response is reduced resulting in damages to blood vessels and heart function [72, 125]. Aging is also associated with a decline in nervous system function. Hsc70 as recently been described to be part of a complex important for synapse function [126]. Hsc70 is also involved in clathrin-coated vesicles recycling which is an important feature of neurotransmitter efficiency [127]. As can be seen, many cellular processes in which Hsp have been shown to participate, can also have an incidence on aging and longevity.

Chaperones in aging diseases

Many late-life diseases are associated with misfolding of proteins and formation of aggregates. This is the case of cataracts and many neurodegenerative disorders such as Alzheimer and Poly(Q) repeat diseases [128, 129]. Cataracts result from protein aggregates formation in the lens due to inactivation of scavenging enzymes and to a decreased amount of αB-crystallin, a member of the sHsp family with chaperone activity [85].

Neurons are highly metabolically active post-mitotic cells and hence more sensitive to oxidative damage due to accumulation of damaged proteins [114]. Early proteins aggregates have been shown to be toxic [67, 69] and chaperones can prevent formation of these aggregates [128, 130–132]. In *C. elegans*, yeast Hsp104 transgenes have been shown to prevent protein aggregation in a Poly(Q) disease model [133]. Hsp70 has also been demonstrated to suppress neurodegeneration induced by a Poly(Q) transgene in *Drosophila* [134]. The beneficial effect of Hsps in neurodegenerative diseases has also been reported following dietary restriction. In fact, the neuroprotective effects of dietary restriction have been associated with the production of neurotrophic factors and an increase in protein chaperones [135].

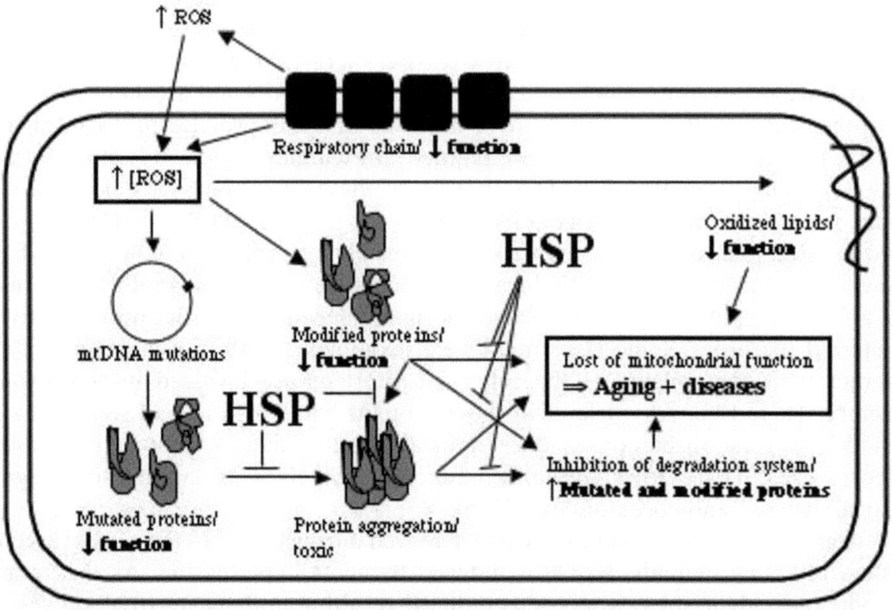

Figure 1. *Aging is associated with oxidative damages. Mitochondria are responsible for producing the energy needed by cells. The respiratory chain localized across the two mitochondrial membranes, is responsible for the transformation of NADH and $FADH_2$ in ATP. This reaction leads to the production of oxygen radicals converted to H_2O and O_2 by detoxifying enzymes (superoxyde dismutase, catalase and glutathione peroxydase). A small concentration of OH· radical is also produced by the Fenton reaction during detoxification of oxygen radicals leading to increasing ROS in the mitochondrial matrix. Mitochondrial DNA is particularly sensitive to ROS and damages result in production of mutated proteins less active and which tend to aggregate. Proteins can also be post-transcriptionally modified by ROS and this contributes to increase the pool of aggregated and/or inactive proteins. All these modified proteins inhibit the protein degradation system and contribute to increase ROS concentration in mitochondria since major functions cannot be accomplished (i.e., impairment of respiratory chain and detoxifying system). This results in a feedback loop where increasing amount of ROS are produced with no mean to degrade them. Lipids are also sensitive to ROS and membranes functions can be altered. Altogether, these damages lead to the loss of mitochondrial functions and the corresponding lack of energy in the cell may be responsible for aging and disease.*

Perspectives

The free radical theory of aging is gaining increasing support. Many proteins of the anti-oxidant family have been shown to be important in the determination of lifespan [109–112]. Recent results obtained by RNAi in *C. elegans* have shown that mitochondria play a critical role in longevity [100, 101]. Three genes of the mitochondrial respiratory-chain have been particularly shown to increase lifespan following down regulation [100]. Maintaining mitochondrial protein integrity in order to prevent excessive oxidative damages thus seems to be particularly important in the determination of longevity. Mitochondrial chaperones could play an important role in maintaining protein homeostasis in mitochondria as shown in the model illustrated in Figure 1. It has been suggested that the aging rates and the acquisition of thermotolerance are determined by overlapping sets of genes including molecular chaperones [136]. Chaperones may modulate the aging rate by preventing accumulation of misfolded protein and their subsequent interferences with multiple cellular processes. Since chaperones are ubiquitous and multifunctional, the impact of their decline during aging could be devastating and account for many features of cellular senescence.

Acknowledgments

We thank the Canadian Institutes of Health Research for grant and fellowship support to RMT and GM respectively.

References

1. Sohal RS (2002). Role of oxidative stress and protein oxidation in the aging process. *Free Rad Biol Med.* 33: 37–44.
2. Lin YJ, Seroude L, Benzer S (1998). Extended life-span and stress resistance in the *Drosophila* mutant methuselah. *Science* 282: 943–6.
3. Rogina B, Reenan RA, Nilsen SP, Helfand SL (2000). Extended life-span conferred by cotransporter gene mutations in *Drosophila. Science* 290: 2137–40.
4. Zoo S, Meadows S, Sharp L, Jan LY, Nung Jan Y (2000). Genome-wide study of aging and oxidative stress response in *Drosophila melanogaster. Proc Natl Acad Sci USA* 97: 13726–31.
5. Chavous DA, Jackson FR, O'Connor CM (2001). Extension of the *Drosophila* lifespan by overexpression of a protein repair methyltransferase. *Proc Natl Acad Sci USA* 98: 14814–18.
6. Ekengren S, Tryselius Y, Dushay MS, Liu G, Steiner H, Hultmark D (2001). A humoral stress response in *Drosophila. Curr Biol.* 11:714–18.
7. Seong KH, Matsuo T, Fuyama Y, Aigaki T (2001). Neural-specific overexpression of *Drosophila* plenty of SH3s (DPOSH) extends the longevity of adult flies. *Biogerontology* 2: 271–81.
8. Tatar M, Kopelman A, Epstein D, Tu MP, Yin CM, Garofalo RS (2001). A mutant *Drosophila* insulin receptor homolog that extends life-span and impairs neuroendocrine function. *Science* 292: 107–10.
9. Brack C, Bechter-Thuring E, Labuhn M (1997). N-acetylcysteine slows down ageing and increases the life span of *Drosophila melanogaster. Cell Mol Life Sci.* 53: 960–6.
10. Anisimov VN, Mylnikov SV, Khavinson VK (1998). Pineal peptide preparation epithalamin increases the lifespan of fruit flies, mice and rats. *Mech Ageing Dev.* 103: 123–32.

11. Khavinson VK, Izmaylov DM, Obukhova LK, Malinin VV (2000). Effect of epitalon on the lifespan increase in *Drosophila melanogaster*. *Mech Ageing Dev.* 120: 141–9.
12. Bonilla E, Medina-Leendertz S, Diaz S (2002). Extension of life span and stress resistance of *Drosophila melanogaster* by long-term supplementation with melatonin. *Exp Gerontol.* 37: 629–38.
13. Kang HL, Benzer S, Min KT (2002). Life extension in *Drosophila* by feeding a drug. *Proc Natl Acad Sci USA* 99: 838–43.
14. Feder ME, Hofmann GE (1999). Heat-shock proteins, molecular chaperones, and the stress response: evolutionary and ecological physiology. *Annu Rev Physiol.* 61: 243–82.
15. Verbeke P, Fonager J, Clark BFC, Rattan SIS (2001). Heat shock response and ageing: mechanisms and applications. *Cell Biol Int.* 25: 845–57.
16. Berger EM, Woodward MP (1983). Small heat shock proteins in *Drosophila* may confer thermal tolerance. *Exp Cell Res.* 147: 437–42.
17. Landry J, Chrétien P, Lambert H, Hickey E, Weber LA (1989). Heat shock resistance conferred by expression of the human HSP27 gene in rodent cells. *J Cell Biol.* 109: 7–15.
18. Squires C, Pedersen S, Ross BM, Squires C (1991). ClpB is the *Escherichia coli* heat shock protein F84.1. *J Bacteriol.* 173: 4254–64.
19. Kitagawa M, Wada C, Yoshioka S, Yura T (1991). Expression of ClpB, an analog of the ATP-dependent protease regulatory subunit in *Escherichia coli*, is controlled by a heat shock sigma factor (sigma 32). *J Bacteriol.* 173L 4247–53.
20. Rollet E, Lavoie JN, Landry J, Tanguay RM (1992). Expression of *Drosophila's* 27 kDa heat shock protein into rodent cells confers thermal resistance. *Biochem Biophys Res Commun.* 185: 116–20.
21. Laxkowska E, Kuczynska-Wisnik D, Skorko-Glonek J, Taylor A (1996). Degradation by proteases Lon, Clp and HtrA, of *Escherichia coli* proteins aggregated *in vivo* by heat shock; HtrA protease action *in vivo* and *in vitro*. *Mol Microbiol.* 22: 555–71.
22. Mogk A, Tomoyasu T, Goloubinoff P, *et al.* (1999). Identification of thermolabile *Escherichia coli* proteins: prevention and reversion of aggregation by DnaK and ClpB. *EMBO J.* 18: 6934–49.
23. Keeler SJ, Boettger CM, Haynes JG, *et al.* (2000). Acquired thermotolerance and expression of the HSP100/ClpB genes of lima bean. *Plant Physiol.* 123: 1121–32.
24. Mehlen P, Briolay J, Smith L, *et al.* (1993). Analysis of the resistance to heat and hydrogen peroxide stresses in COS cells transiently expressing wild type or deletion mutants of the *Drosophila* 27-kDa heat-shock protein. *Eur J Biochem.* 215: 277–84.
25. Mehlen P, Preville X, Chareyron P, Briolay J, Klemenz R, Arrigo AP (1995). Constitutive expression of human hsp27, *Drosophila* hsp27, or human alpha B-crystallin confers resistance to TNF- and oxidative stress-induced cytotoxicity in stably transfected murine L929 fibroblasts. *J Immunol.* 154: 363–74.
26. Mehlen P, Schulze-Osthoff K, Arrigo AP (1996). Small stress proteins as novel regulators of apoptosis. Heat shock protein 27 blocks Fas/APO-1- and staurosporine-induced cell death. *J Biol Chem.* 271: 16510–14.
27. Brar BK, Stephanou A, Wagstaff MJ, *et al.* (1999). Heat shock proteins delivered with a virus vector can protect cardiac cells against apoptosis as well as against thermal or hypoxic stress. *J Mol Cell Cardiol.* 31: 135–46.
28. Garrido C, Bruey JM, Fromentin A, Hammann A, Arrigo AP, Solary E (1999). HSP27 inhibits cytochrome c-dependent activation of procaspase-9. *FASEB J.* 13: 2061–70.
29. Bruey JM, Ducasse C, Bonniaud P, *et al.* (2000). Hsp27. negatively regulates cell death by interacting with cytochrome c. *Nat Cell Biol.* 2: 645–52.

30. Charette SJ, Lavoie JN, Lambert H, Landry J (2000). Inhibition of Daxx-mediated apoptosis by heat shock protein 27. *Mol Cell Biol.* 20: 7602–12.
31. Creagh EM, Carmody RJ, Cotter TG (2000). Heat shock protein 70 inhibits caspase-dependent and independent apoptosis in Jurkat T cells. *Exp Cell Res.* 257: 58–66.
32. Mosser DD, Caron AW, Bourget L, *et al.* (2000). The chaperone function of hsp70 is required for protection against stress-induced apoptosis. *Mol Cell Biol.* 20: 7146–59.
33. Haslbeck M (2002). sHsps and their role in the chaperone network. *Cell Mol Life Sci.* 59: 1649–57.
34. Concannon CG, Gorman AM, Samali A (2003). On the role of Hsp27 in regulating apoptosis. *Apoptosis* 8: 61–70.
35. Samali A, Cai J, Zhivotovsky B, Jones DP, Orrenius S (1999). Presence of a pre-apoptotic complex of pro-caspase-3, Hsp60 and Hsp10 in the mitochondrial fraction of Jurkat cells. *EMBO J.* 18: 2040–8.
36. Xanthoudakis S, Nicholson DW (2000). Heat-shock proteins as death determinants. *Nat Cell Biol.* 2: E163–5.
37. Gupta S, Knowlton AA (2002). Cytosolic heat shock protein 60, hypoxia, and apoptosis. *Circulation* 106: 2727–33
38. Ehrnsperger M, Gaestel M, Buchner J (2000). Analysis of chaperone properties of small Hsp's. *Methods Mol Biol.* 99: 421–9.
39. Neupert W (1997). Protein import into mitochondria. *Annu Rev Biochem.* 66: 863–917.
40. Bukau B, Horwich AL (1998). The Hsp70 and Hsp60 chaperone machines. *Cell* 92: 351–66.
41. Beck FX, Neuhofer W, Muller W (2000). Molecular chaperones in the kidney: distribution, putative roles, and regulation. *Am J Physiol Renal Physiol.* 279: F203–15.
42. Hartl FU, Hayer-Hartl M (2002). Molecular chaperones in the cytosol: from nascent chain to folded protein. *Science* 295: 1852–8.
43. Lee KH, Kim HS, Jeong HS, Lee YS (2002). Chaperonin GroESL mediates the protein folding of human liver mitochondrial aldehyde dehydrogenase in *Escherichia coli*. *Biochem Biophys Res Commun.* 298: 216–24.
44. Voos W, Rottgers K (2002). Molecular chaperones as essential mediators of mitochondrial biogenesis. *Biochim Biophys Acta* 1592: 51–62.
45. Zavilgelsky GB, Kotova VY, Mazhul' MM, Manukhov IV (2002). Role of Hsp70 (DnaK-DnaJ-GrpE) and Hsp100 (ClpA and ClpB) chaperones in refolding and increased thermal stability of bacterial luciferase in *Escherichia coli* cells. *Biochemistry* 67: 986–92
46. Ngosuwan J, Wang NM, Fung KL, Chirico WJ (2002). Roles of cytosolic Hsp70 and Hsp40 molecular chaperones in post-translational translocation of presecretory proteins into the endoplasmic reticulum. *J Biol Chem.* 278: 7034–42.
47. Imamura T, Haruta T, Takata Y, *et al.* (1998). Involvement of heat shock protein 90 in the degradation of mutant insulin receptors by the proteasome. *J Biol Chem.* 273: 11183–8.
48. Hoskins JR, Yanagihara K, Mizuuchi K, Wickner S (2002). ClpAP and ClpXP degrade proteins with tags located in the interior of the primary sequence. *Proc Natl Acad Sci USA* 99: 11037–42.
49. Jin T, Gu Y, Zanusso G, *et al.* (2000). The chaperone protein BiP binds to a mutant prion protein and mediates its degradation by the proteasome. *J Biol Chem.* 275: 38699–704.
50. Luders J, Demand J, Hohfeld J (2000). The ubiquitin-related BAG-1 provides a link between the molecular chaperones Hsc70/Hsp70 and the proteasome. *J Biol Chem.* 275: 4613–17.

51. Burton BM, Williams TL, Baker TA (2001). ClpX-mediated remodelling of mu transpososomes: selective unfolding of subunits destabilizes the entire complex. *Mol Cell* 8: 449–54.
52. Connell P, Ballinger CA, Jiang J, *et al.* (2001). The co-chaperone CHIP regulates protein triage decisions mediated by heat-shock proteins. *Nat Cell Biol.* 3: 93–6.
53. Hoskins JR, Singh SK, Maurizi MR, Wickner S (2000). Protein binding and unfolding by the chaperone ClpA and degradation by the protease ClpAP. *Proc Natl Acad Sci USA* 97: 8892–7.
54. Rottgers K, Zufall N, Guiard B, Voos W (2002). The ClpB homolog Hsp78 is required for the efficient degradation of proteins in the mitochondrial matrix. *J Biol Chem.* 277: 45829–37.
55. Rutherford SL, Lindquist S (1998). Hsp90 as a capacitor for morphological evolution. *Nature* 396: 336–42.
56. Solars V, Lu X, Xiao L, Wang X, Garfinkel MD, Ruden DM (2003). Evidence for an epigenetic mechanism by which Hsp90 acts as a capacitor for morphological evolution. *Nat Genet.* 33: 70–4.
57. Abbas-Terki T, Briand PA, Donze O, Picard D (2002). The Hsp90 co-chaperones Cdc37 and Sti1 interact physically and genetically. *Biol Chem.* 383: 1335–42.
58. Angeletti PC, Walker D, Panganiban AT (2002). Small glutamine-rich protein/viral protein U-binding protein is a novel cochaperone that affects heat shock protein 70 activity. *Cell Stress Chaperones* 7: 258–68.
59. Tasab M, Batten MR, Bulleid NJ (2000). Hsp47: a molecular chaperone that interacts with and stabilizes correctly-folded procollagen. *EMBO J.* 19: 2204–11.
60. Söti C, Csermely P (2000). Molecular chaperones and the aging process. *Biogerontology* 1: 225–33.
61. Söti C, Csermely P (2002). Chaperones come of age. *Cell Stress Chaperones* 7: 186–90.
62. Wright HT (1991). Nonenzymatic deamination of asparaginyl and glutaminyl redidues in proteins. *Crit Rev Biochem Mol Biol.* 26: 1–52.
63. Sun H, Gao J, Ferrington DA Biesiada H, Williams TD, Squier TC (1999). Repair of oxidized calmodulin by methionine sulfoxide reductase restores ability to activate the plasma membrane Ca-ATPase. *Biochemistry* 38: 105–12.
64. Das N, Levine RL, Orr WC, Sohal RS (2001). Selectivity of protein oxidative damage during aging in *Drosophila melanogaster. Biochem J.* 360: 209–16.
65. Bence NF, Sampat RM, Kopito RR (2001). Impairment of the ubiquitin-proteasome system by protein aggregation. *Science* 292: 1552–5.
66. Bulteau A-L, Verbeke P, Petropoulos I, Chaffotte A-F, Friguet B (2001). Proteasome inhibition in glyoxal-treated fibroblasts and resistance of glycated glucose-6-phosphate dehydrogenase to 20S proteasome degradation *in vitro. J Biol Chem.* 274: 662–8.
67. Bucciantini M, Giannoni E, Chiti F, *et al.* (2002). Inherent toxicity of aggregates implies a common mechanism for protein misfolding diseases. *Nature* 416: 507–11.
68. Ellis RJ, Pinheiro TJ (2002). Danger misfolding proteins. *Nature* 416: 483–4.
69. Macario AJL, de Macario EC (2002). Sick chaperones and ageing: a perspective. *Ageing Res Rev.* 1: 295–311.
70. Fleming JE, Walton JK, Dubitski R, Bensch KG (1988). Aging results in an unusual expression of *Drosophila* heat shock proteins. *Proc Natl Acad Sci USA* 85: 4099–103.
71. Finkel T, Holbrook NJ (2000). Oxidants, oxidative stress and the biology of ageing. *Nature* 408: 239–47.
72. Locke M, Tanguay RM (1996). Diminished heat shock response in the aged myocardium. *Cell Stress Chaperones* 1: 251–60.

73. Garigan D, Hsu A-L, Fraser AG, Kamath TS, Ahringer J, Kenyon C (2002). Genetic analysis of tissue aging in *Caenorhabditis elegans*: a role for heat-shock factor and bacterial proliferation. *Genetics* 161: 1101–12.
74. Rao DV, Watson K, Jones GL (1999). Age-related attenuation in the expression of the major heat shock proteins in human peripheral lymphocytes. *Mech Ageing Dev.* 107: 105–18.
75. Nakanishi Y, Yasumoto K (1997). Induction after administering paraquat of heme oxidgenase-1 and heat shock protein in the liver of senescence-accelerated mice. *Biosci Biotechnol Biochem.* 61: 1302–6.
76. Fargnoli J, Kunisada T, Fornace A J Jr, Schneider EL, Holbrook NJ (1990). Decreased expression of heat shock protein 70 mRNA and protein after heat treatment in cells of aged rats. *Proc Natl Acad Sci USA* 87: 846–50.
77. Blake MJ, Udelsman R, Feulner GJ, Norton DD, Holbrook NJ (1991). Stress-induced heat shock protein 70 expression in adrenal cortex: an adrenocorticotropic hormone-sensitive, age-dependent response. *Proc Natl Acad Sci USA* 88: 9873–7.
78. Kregel KC, Moseley PL, Skidmore R, Gutierrez JA, Guerriero V (1995). HSP70 accumulation in tissues of heat-stressed rats in blunted with advancing age. *J Appl Physiol.* 79: 1673–8.
79. Piotrowicz RS, Weber LA, Hickey E, Levin EG (1995). Accelerated growth and senescence of arterial cells expressing the small molecular weight heat-shock protein Hsp27. *FASEB J.* 9: 1079–84.
80. Wadhwa R, Kaul SC, Ikawa Y, Sugimoto Y (1993). Identification of a novel member of mouse hsp70 family. Its association with cellular mortal phenotype. *J Biol Chem.* 268: 6615–21.
81. Wadhwa R, Kaul SC, Sugimoto Y, Mitsui Y (1993). Induction of cellular senescence by transfection of cytosolic mortalin cDNA in NIH 3T3 cells. *J Biol Chem.* 268: 22239–42.
82. Holt SE, Aisner DL, Baur J, *et al.* (1999). Functional requirement of p23 and Hsp90 in telomerase complexes. *Genes Dev.* 13: 817–26.
83. Cherkasova V, Ayyadevara S, Egilmez N, Shmookler Reis R (2000). Diverse *Caenorhabditis elegans* genes that are upregulated in dauer larvae also show elevated transcript levels in long-lived, aged, or starved adults. *J Mol Biol.* 300: 433–48.
84. Cherian M, Abraham WC (1995). Decreased molecular chaperone property of alpha-crystallin due to posttranslational modifications. *Biochem Biophys Res Commun.* 208: 675–9.
85. Derham BK, Harding JJ (1997). Effect of aging on the chaperone-like function of human α-crystallin assessed by three methods. *Biochem J.* 328: 763–8.
86. Wheeler JC, Bieschke ET, Tower J (1995). Muscle-specific expression of *Drosophila* hsp70 in response to aging and oxidative stress. *Proc Natl Acad Sci USA* 92: 10408–12.
87. Tatar M, Khazaeli AA, Curtsinger JW (1997). Chaperoning extended life. *Nature* 390: 30.
88. Maiello M, Boeri D, Sampietro L, Pronzato MA, Odetti P, Marinari UM (1998). Basal synthesis of heat shock protein 70 increases with age in rat kidneys. *Gerontology* 44: 15–20.
89. Locke M (2000). Heat shock transcription factor activation and Hsp72 accumulation in aged skeletal muscle. *Cell Stress Chaperones* 5: 45–51.
90. Cuervo AM, Dice JF (2000). Age-related decline in chaperone-mediated autophagy. *J Biol Chem.* 275: 31505–13.

91. Wu B, Gu MJ, Heydari AR, Richardson A (1993). Protein oxidation associated with aging is reduced by dietary restriction of protein or calories. *Proc Natl Acad Sci USA* 89: 9112–16.
92. Krawczyk Z, Szymik N (1989). Effect of age and busulphan treatment of the hsp70 gene-related transcript level in rat testes. *Int J Androl.* 12: 72–9.
93. Morimoto RI, Sarge KD, Abravaya K (1992). Transcriptional regulation of heat shock genes. A paradigm for inducible genomic responses. *J Biol Chem.* 267: 21987–90.
94. King V, Tower J (1999). Aging-specific expression of *Drosophila* hsp22. *Dev Biol.* 207: 107–18.
95. Morrow G, Inaguma Y, Kato K, Tanguay RM (2000). The small heat shock protein Hsp22 of *Drosophila melanogaster* is a mitochondrial protein displaying oligomeric organization. *J Biol Chem.* 275: 31204–10.
96. Beaulieu JF, Arrigo AP, Tanguay RM (1989). Interaction of *Drosophila* 27,000 Mr heat-shock protein with the nucleus of heat-shocked and ecdysone-stimulated culture cells. *J Cell Sci.* 92: 29–36.
97. Tanguay RM, Joanisse DR, Inaguma Y, Michaud S (1999). Small heat shock proteins: in search of functions *in vivo.* In: Storey KB, ed. *Environmental Stress and Gene Regulation.* Oxford: Bios Scientific Publishers, pp. 125–38.
98. Michaud S, Morrow G, Marchand J, Tanguay RM (2002) *Drosophila* small heat shock proteins: cell and organelle-specific chaperones? *Prog Mol Subcell Biol.* 28: 79–101.
99. Kurapati R, Passananti HB, Rose MR, Tower J (2000). Increased hsp22 RNA levels in *Drosophila* lines genetically selected for increased longevity. *J Gerontol.* 55A: B552–9.
100. Dillin A, Hsu A-L, Arantes-Oliveira N, *et al.* (2002). Rates of behaviour and aging specified by mitochondrial function during development. *Science* 298: 2398–401.
101. Lee SS, Lee RYN, Fraser AG, Kamath RS, Ahringer J, Ruvkun G (2003). A systematic RNAi screen identifies a critical role for mitochondria in *C. elegans* longevity. *Nat Genet.* 33: 40–8.
102. Lithgow GJ, White TM, Melov S, Johnson TE (1995). Thermotolerance and extended life-span conferred by single-gene mutations and induced by thermal stress. *Proc Natl Acad Sci USA* 92: 754–4.
103. Khazaeli AA, Tatar M, Pletcher SD, Curtsinger JW (1997). Heat-induced longevity extension in Drosophila. I. Heat treatment, mortality, and thermotolerance. *J Gerontol A Biol Sci Med Sci.* 52: B48–52.
104. Rattan SI (1998). Repeated mild heat shock delays ageing in cultured human skin fibroblasts. *Biochem Mol Biol Int.* 45: 753–9.
105. Minois N, Khazaeli AA, Curtsinger JW (2001). Locomotor activity as a function of age and life span in *Drosophila melanogaster* overexpressing hsp70. *Exp Gerontol.* 36: 1137–53.
106. Butov A, Johnson T, Cypser J, *et al.* (2001). Hormesis and debilitation effects in stress experiments using the nematode worm *Caenorhabditis elegans*: the model of balance between cell damage and HSP levels. *Exp Gerontol.* 37: 57–66.
107. Le Bourg E, Valenti P, Lucchetta P, Payre F (2001). Effects of mild heat shocks at young age on aging and longevity in *Drosophila melanogaster. Biogerontology* 2: 155–64.
108. Walker GA, White TM, McColl G, *et al.* (2001). Heat shock protein accumulation is upregulated in a long-lived mutant of *Caenorhabditis elegans. J Gerontol.* 56A, B281–7.
109. Parkes TL, Elia AJ, Dickinson D, Hilliker AJ, Phillips JP, Boulianne GL (1998). Extension of *Drosophila* lifespan by overexpression of human SOD1 in motorneurons. *Nat Genet.* 19: 171–4.

110 Sun J, Tower J (1999). FLP recombinase-mediated induction of Cu/Zn-superoxide dismutase transgene expression can extend the life span of adult *Drosophila melanogaster* flies. *Mol Cell Biol.* 19: 216–28.
111. Seong K-H, Ogashiwa T, Matsuo T, Fuyama Y, Aigaki T (2001). Application of the gene search system to screen for longevity genes in *Drosophila. Biogerontology* 2: 209–17.
112. Sun J, Folk D, Bradley TJ, Tower J (2002). Induced overexpression of mitochondrial Mn-superoxide dismutase extends the life span of adult *Drosophila melanogaster. Genetics* 161: 661–72.
113. Tower J. (2000). Transgenic methods for increasing *Drosophila* life span. *Mech Ageing Dev.* 118: 1–14.
114. Boulianne GL (2001). Neuronal regulation of lifespan: clues from flies and worms. *Mech Ageing Dev.* 122: 883–94.
115. Chong KY, Lai CC, Lille S, Chang C, Su CY (1998). Stable overexpression of the constitutive form of heat shock protein 70 confers oxidative protection. *J Mol Cell Cardiol.* 30: 599–608.
116. Yokoyama K, Fukumoto K, Murakami T, *et al.* (2002). Extended longevity of *Caenorhabditis elegans* by knocking in extra copies of hsp70F, a homolog of mot-2 (mortalin)/mthsp70/Grp75. *FEBS Lett.* 516: 53–7.
117. Kaul S, Reddel RR, Sugihara T, Mitsui Y, Wadhwa R (2000). Inactivation of p53 and life span extension of human diploid fibroblasts by mot-2. *FEBS Lett.* 474: 159–64.
118. Friguet B, Stadman ER, Szweda LI (1994). Modification of glucose-6-phosphate dehydrogenase by 4-hydroxy-2-nonenal. Formation of cross-linked protein that inhibits the multicatalytic protease. *J Biol Chem.* 269: 21639–43.
119. Lee H-C, Wei Y-H (2001). Mitochondrial alterations, cellular response to oxidative stress and defective degradation of proteins in aging. *Biogerontology* 2: 231–44.
120. Bracken AP, Bond U (1999). Reassembly and protection of small nuclear ribonucleoprotein particles by heat shock proteins in yeast cells. *RNA.* 5: 1586–96.
121. Helmbrecht K, Zeise E, Rensing L (2000). Chaperones in cell cycle regulation and mitogenic signal transduction: a review. *Cell Prolif.* 33: 341–65.
122. Laszlo A (1992). The effects of hyperthermia on mammalian cell structure and function. *Cell Prolif.* 25: 59–87.
123. Buchner J (1999). Hsp90 and Co a holding for folding. *Trends Biochem Sci.* 24: 136–41.
124. Galigniana MD, Radanyi C, Renoir J-M, Housley PR, Pratt WB (2001). Evidence that the peptidylprolyl isomerase domain of the Hsp90-binding immunophilin FKBP52 is involved in both dynein interaction and glucocorticoid receptor movement to the nucleus. *J Biol Chem.* 276, 14884–9.
125. Udelsman R, Blake MJ, Stagg CA, Li DG, Putney DJ, Holbrook NJ (1993). Vascular heat shock protein expression in response to stress. Endocrine and autonomic regulation of this age-dependent response. *J Clin Invest.* 91: 464–73.
126. Tobaden S, Thakur P, Fernandez-Chacon R, Suedhof TC, Rettig J, Stahl B (2001). A trimeric protein complex functions as a synaptic chaperone machine. *Neuron* 31: 987–99.
127. Morgan JR, Prasad K, Jin S, Augustine GJ, Lafer EM (2001). Uncoating of clathrin-coated vesicles in presynaptic terminals: roles for Hsp70 and auxilin. *Neuron* 32: 289–300.
128. Muchowski PJ (2002). Protein misfolding, amyloid formation, and neurodegeneration: a critical role for molecular chaperones. *Neuron* 35: 9–12.
129. Harding JJ (2002). Viewing molecular mechanisms of ageing through a lens. *Ageing Res Rev.* 1: 465–79.

130. Link CD (2001). Transgenic invertebrate models of age-associated neurodegenerative diseases. *Mech Ageing Dev.* 122: 1639–49.
131. Sakahira H, Breuer P, Hayer-Hartl MK, Hartl FU (2002). Molecular chaperones as modulators of polyglutamine protein aggregation and toxicity. *Proc Natl Acad Sci USA* 99: 16412–18.
132. Söti C, Csermely P (2002). Chaperones and aging: role in neurodegeneration and in other civilizational diseases. *Neurochem Int.* 41: 383–9.
133. Satyal SH, Schmidt E, Kitagawa K, *et al.* (2000). Polyglutamine aggregates alter protein folding homeostasis in *Caenorhabditis elegans. Proc Natl Acad Sci USA* 97: 5750–5.
134. Chan HY, Warrick JM, Gray-Board BL, Paulson JL, Bonini NM (2000). Mechanisms of chaperone suppression of polyglutamine disease: selectivity, synergy and modulation of protein solubility in *Drosophila. Hum Mol Genet.* 9: 2811–20.
135. Prolla TA, Mattson MP (2001). Molecular mechanisms of brain aging and neurodegenerative disorders: lessons from dietary restriction. *Trends Neurosci.* 24: S21–31.
136. Walker GA, Walker DW, Lithgow JL (1998). Genes that determines both thermotolerance and rate of aging in *Caenorhabditis elegans. Ann NY Acad Sci.* 851: 444–9.
137. Jakob U, Gaestel M, Engel K, Buchner J (1993). Small heat shock proteins are molecular chaperones. *J Biol Chem.* 1993 268: 1517–20.
138. Mounier N, Arrigo AP (2002). Actin cytoskeleton and small heat shock proteins: how do they interact? *Cell Stress Chaperones* 7: 167–76.
139. Tsvetkova NM, Horvath I, Torok Z, *et al.* (2002). Small heat-shock proteins regulate membrane lipid polymorphism. *Proc Natl Acad Sci USA* 99: 13504–9.
140. Cabiscol E, Belli G, Tamarit J, Echave P, Herrero E, Ros J (2002). Mitochondrial Hsp60, resistance to oxidative stress, and the labile iron pool are closely connected in *Saccharomyces cerevisiae. J Biol Chem.* 277: 44531–8.
141. Ou J, Ou Z, Ackerman AW, Oldham KT, Pritchard KA (2003). Inhibition of heat shock protein 90 (hsp90) in proliferating endothelial cells uncouples endothelial nitric oxide synthase activity. *Free Radic Biol Med.* 34: 269–76.
142. Takahashi S, Mendelsohn ME (2003). Calmodulin-dependent and independent activation of endothelial nitric oxide synthase by heat shock protein 90. *J Biol Chem.* 278: 9339–44.
143. Boudeau J, Deak M, Lawlor MA, Morrice NA, Alessi DR (2002). Hsp90/Cdc37 interact with LKB1 and regulate its stability. *Biochem J.* 370: 849–57.
144. Joseph K, Tholanikunnel BG, Kaplan AP (2002). Activation of the bradykinin-forming cascade on endothelial cells: a role for heat shock protein 90. *Int Immunopharmacol.* 2: 1851–9.
145. Marcu MG, Doyle M, Bertolotti A, Ron D, Hendershot L, Neckers L (2002). Heat shock protein 90 modulated the unfolded protein response by stabilizing IRE1alpha. *Mol Cell Biol.* 22: 8506–13.
146. Picard D (2002). Heat-shock protein 90, a chaperone for folding and regulation. *Cell Mol Life Sci.* 59: 1640–8.
147. Vanden Berghe T, Kalai M, Van Loo G, Declercq W, Vandenabeele P (2002). Disruption of HSP90 function reverts TNF-induced necrosis to apoptosis. *J Biol Chem.* 278:5622–9.
148. Wang C, Chen J (2003). Phosphorylation and hsp90 binding mediate heat shock stabilization of p53. *J Biol Chem.* 278: 2066–71.
149. Gavrilovich Zgoda V, Arison B, Mkrtchian S, Igelman-Sundberg M, Almira Coreeia M (2002). Hemin-mediated restoration of allylisopropylacetamide-inactivated CYP2B1: a role for glutathione and GRP94 in the heme-protein assembly. *Arch Biochem Biophys.* 408: 58–68.

150. Krzewska J, Langer T, Liberek K (2001). Mitochondrial Hsp78, a member of the Clp/Hsp100 family in *Saccharomyces cerevisiae*, cooperates with Hsp70 in protein refolding. *FEBS Lett*. 489: 92–6.
151. Koniesczny I, Liberek K (2002). Cooperative action of *Escherichia coli* ClpB protein and DnaK chaperone in the activation of a replication initiation protein. *J Biol Chem*. 277: 18483–8.
152. Beinker P, Schlee S, Groempin Y, Seidel R, Reinstein J (2002). The N terminus of ClpB from *Thermus thermophilus* is not essential for the chaperone activity. *J Biol Chem*. 277: 47160–6.
153. Gershon H, Gershon D (1970). Detection of inactive enzyme molecules in ageing organisms. *Nature* 227: 1214–17.
154. Lithgow GJ, Walker GA (2002). Stress resistance as a determinate of *C. elegans* lifespan. *Mech Ageing Dev*. 123: 765–71.

Cellular Senescence Pathways in Mouse and Human

Renu Wadhwa, Md. Kamrul Hasan and Sunil C. Kaul
National Institute of Advanced Industrial Science and Technology (AIST), 1-1-1 Higashi, Tsukuba, Ibaraki 305-8566, Japan

Introduction

Cell proliferation is essential for the survival of multi-cellular organisms. While assuring the renovation of the tissues, transfer of genetic material to daughter cells, it puts genetic material to inherent risk of mutations that might occur during the process of replication and failure of genetic fidelity. Such mutations may cause genomic instability, neoplastic transformation and development of cancer [1]. Perhaps to avoid this risky journey, normal somatic cells are programmed to divide only for a limited number of times in culture first demonstrated in human fibroblasts by Hayflick and Moorhead [2, 3] and has been subsequently documented for a variety of cells from different organisms [4–13]. This phenomenon is called replicative senescence, a most consistent manifestation of cellular senescence (functional decline and inability of cells to maintain themselves forever). The senescent cells from different organisms show similar biochemical and morphological features collectively referred to as senescent/aging phenotype, most consistently characterized by functional decline. The senescent cells are irreversibly arrested in G_1 stage of the cell cycle but remain viable and metabolically active. By entailing continuous propagation of cells, senescence thus acts as a barrier to carcinogenesis and it is therefore referred to as a tumor suppression mechanism [1, 14, 15]. Two other mechanisms that operate in parallel to control cell number during development and eliminate cancer cells are apoptosis and premature senescence. Although monitored by same components of the cellular machinery, yet these operate through distinct pathways and have peculiarities from one another [1, 16–22].

In last two decades, molecular and genetics studies have unraveled many aspects of cellular senescence. These include understanding the role of tumor suppressor genes and of the telomeres in the control of normal cell proliferation. Emerging understanding of aging has following components: (i) it is an inevitable outcome of gene

Sunil C. Kaul and Renu Wadhwa (eds.), Aging of Cells In and Outside the Body, 225–238.

activities (longevity assuring genes) that are essential for maintaining the order, integrity and normal function of a living organism. (ii) since living systems are not equipped with perfect maintenance and repair mechanisms, these confront accumulation of genetic, cellular and extra-cellular damage. (iii) longevity assuring functions, but not their by-product accumulation of damage, are genetically programmed and are monitored by gene activities including tumor suppression mechanism(s) such as activities of negative regulators of cell cycle and telomere shortening. (iv) abrogation of one or more tumor suppressor gene functions and maintenance of repair mechanisms is sufficient to escape senescence at least *in vitro* [1, 23–30].

A number of organismic models ranging from yeast, nematode, insect to mouse have been developed to study the genetics and mechanisms relevant to longevity and aging. Genes that affect longevity in multiple models might be expected to be the best candidates for longevity-associated genes. Although mouse model is most popular among the other widely studied genetically malleable species, it has limitations because of its specific peculiarities and divergence from human.

Why use mouse for aging studies?

- Availability of wild, inbred and mixed inbred strains make the genetic studies feasible.
- Evolutionarily close relationship to humans.
- Short lifespan of 2–3 years allows to perform aging studies within a reasonable length of time.
- Development, genetics, biology, patho-biology and aging of various strains of mice have been well characterized.
- Availability of spontaneous and engineered mouse mutants.
- Technical ability to manipulate the mouse germline through transgenic and knockout methods.
- Unlike worms and flies, whose somatic cells are virtually all post-mitotic as adults, mice have mitotic compartments of self-renewing tissue stem cells that play an important role in organismic aging by maintaining organ homeostasis.
- Technical ability to culture mouse cells from a variety of lineages.
- Mouse genome has been totally sequenced predicting the presence of around 30 000 genes (very close to human) with only lacking 300 genes indicating that 99% of human genes are present in mice.

Genetic manipulations both resulting in lifespan extension and reduction can serve as a powerful tool to understand the mechanisms of aging and its intervention [31, 32]. Since the reduction in lifespan can also occur by non-aging related pathology, the mutant models resulting in lifespan shortening have inherit difficulty in its acceptance. On the other hand, mutant mice that have enhanced longevity are easily conceived to document a bonafide alteration in one or more aging or aging regulatory pathways. Representative examples of single gene mutations resulting in lifespan extension in mouse include mutations in genes affecting growth hormone production. The Ames Dwarf, Snell dwaft and Little mouse have autosomal recessive mutation in the Prop-1, Pit-1 and GHRH genes, respectively [33]. These mutant mice experience pituitary failure to produce growth hormone, prolactin and thyroid stimulating hormone, are resistant to oxidative stress and live 50% longer than the wild type mice. Genetically engineered p66shc$^{-/-}$ mouse have enhanced resistance to apoptosis, oxidative stress and superoxide anions and live 30% longer than the wild type controls [34].

Genetic lesions responsible for early aging associated phenotypes such as human progeroid syndromes might be considered ideal candidates for modeling in mice [35, 36]. Werner's (WRN) syndrome patients after a normal childhood but exhibit a failed growth sprut at puberty, premature graying, loss of hair and skin atrophy and other severe age-pathologies as early as 40 years of age. However, mouse model for WRN showed only a few such pathologies and no obvious reduction in lifespan [37]. Other senescence accelerated mouse models include: (i) Klotho mutant mice that is mutated in gene encoding a membrane protein and share sequence similarity with beta-glucosidase enzymes. These mutant mice live for 60 days and display hunchbacked spine, thymic atrophy, hair loss and osteoporesis early in their life [38–41]. (ii) senescence accelerated mouse (SAM-P) is raised by propagation of recombinant inbred lines [42–44]. These mice live for 9–10 months in comparison to 18–19 months for matched strain controls and experience early appearance of number of age-related disorders, including degenerative changes in multiple tissues, osteoporesis, lordokyphosis, senile amyloidosis, loss of hair, degenerative joint disease, memory and learning dysfunction and enhanced oxidative damage [44]. (iii) Ku80 null mice that are deficient in non homologous end joining following DNA double stranded breaks exhibit early age specific change including tissue degeneration, osteopenia and reduced longevity [45]. (iv) Telomerase deficient mice although were phenotypically normal and had normal lifespan in the first generation of telomerase (mTR$^{-/-}$) knockout mice, the later generations have shortened lifespan accompanied by age-associated phenotypes including loss of body weight, decreased stress responses, increased incidence of skin lesions [46, 47]. Interestingly, some of the aging phenotypes of mTR$^{-/-}$ mouse were rescued by crossing it with the null p53 allele [48].

These studies along with others, as discussed below, have supported that aging phenotype is unlikely to be controlled by any signal gene or pathway [49]. The major pathways recognized to-date to determine cellular senescence are tumor suppressor pathways including p53, pRB and telomere shortening. Each of these pathways has multiple components, expression and activity of which is influenced by intrinsic and extrinsic cellular factors imposing a high level of complexity. Mouse and human system diverge considerably on operation of these pathways and thus display differences in implementation of cellular senescence.

Genes different in mouse and human

Human and mouse genomes have been sequenced; mouse has 99% of the human genes (lacks only 300 genes out of the predicted 30 000). This information is going to be extremely helpful to understand how genes work at most basic level. Major differences appeared in the genes involved in the immunity, reproduction, detoxification and smelling. Female mice have a large repertoire of reproduction genes and may produce about 20 pregnancy specific hormones as compared to human who make only one, prolactin. Similarly, liver detoxification enzymes cytochrome P450s are estimated to be 84 in mice as compared to 63 in human. Although hard to imagine now, it may be possible to have "humanize" lab mice through genetic engineering in future. These mice would metabolize like humans and would offer a system to study drug metabolism and resistance for application in human medicine. In brief, the hard work in last many years has converged to the understanding that it is not the gene composition but the gene regulation that make us different from mice.

Tumor suppressors (p53 and pRB) driven cellular senescence in mouse and human

Progression of cell cycle requires activity of E2F family of transcriptional factors that are captured and inactivated by hypo-phosphorylated pRB proteins (pRB). Phosphorylation of pRB by Cyclin-CDK complexes releases E2F and causes entry of cells to S-phase. Cyclin-CDK complexes are inhibited by two classes of inhibitors, $p21^{WAF1}$ (a down stream effector of p53) and $p16^{INK4A}$. $p19^{ARF}$, an alternative reading frame protein of INK4A locus (encodes $p16^{INK4A}$) acts upstream to p53 to strengthen its negative regulatory effect of cell cycle [50, 51]. Both human and mouse cells undergo replicative growth arrest with their own peculiarities that have been documented for their tendency towards immortalization and role of tumor suppressor and cell cycle regulators, p53 and pRB [5, 52, 53]. Mouse cells show a period of rapid growth, followed by slowing down and cessation of cell proliferation that is accompanied by elevated expression of negative cell cycle regulators including CDK inhibitors, $p16^{INK4A}$, $p21^{WAF1}$, $p19^{ARF}$ and p53. Rare variants arise from the late phase slow cultures and switch to fast proliferation leading to establishment of immortal lines [54, 55] (Figure 1). Genetic analysis identified loss of p53 function and chromosomal instability as prime changes in these cells [56, 57]. Indeed, cells from p53-null or ARF-null mouse were shown to maintain high proliferation potential and immortalize in culture without showing any signs of senescence. Therefore in mouse cells, disruption of p53 function either by mutation or by its functional inactivation by perturbance of its regulators (ARF and Mdm-2) is sufficient to make them immortalize [58]. Both $p19^{ARF}$ and $p16^{INK4A}$ are coded by the same gene locus, genetic manipulation of individual exons of the CDKN2A locus to obtain silencing of one transcript at a time has further documented contributions of $p16^{INK4A}$ and $p19^{ARF}$ in mouse cell senescence. Abrogation of any one of these facilitates immortalization and subsequent carcinogenesis of cells [59–61]. Primary mouse embryonic fibroblasts (MEFs) when exposed to oncogenic Ras, express senescence specific cell cycle inhibitors and undergo premature senescence [17]. In contrast, the immortalized cells lacking p53 or ARF were efficiently transformed [59]. These studies have suggested that senescence operates as a tumor suppressor mechanism in mouse cells wherein $p19^{ARF}$-p53-$p21^{WAF1}$ pathway acts as a main player.

Human cells have more extended proliferation potential than mouse cells but rarely become adapted to continuous proliferation and establishment into permanent cell lines. They are genetically more stable and more resistant than mouse cells to oncogene mediated transformation. Experiments designed to dissect the role of p53 and pRB have suggested that in contrast to the mouse system in which p53 plays a major role with its inactivation sufficient to escape from mortality, the human cells have more rigid constrains that involve participation of p53 and pRB tumor suppressor pathways. Inactivation of either or both of these proteins (e.g., by SV40 LT antigen) is sufficient to confer escape from senescence; inactivation of both proteins has a co-operative effect on lifespan extension [62–64]. However, lifespan extended human cells can only establish into permanent cell lines with additional genetic events culminating in activation of telomerase or ALT (alternative lengthening of telomeres) that lead to maintenance of otherwise eroding telomeres [5, 65–67] (Figure 1). Furthermore, a second product of the early region, the small t antigen (stAg), is essential for its transforming potential for human, but not mouse cells. stAg perturbs one or more isoforms of the widely acting serine/threonine phosphatases (PP2A); the PP2A pathways perturbed by stAg that are essential for human cell transformation remain to be defined [68]. Studies on molecular regulation

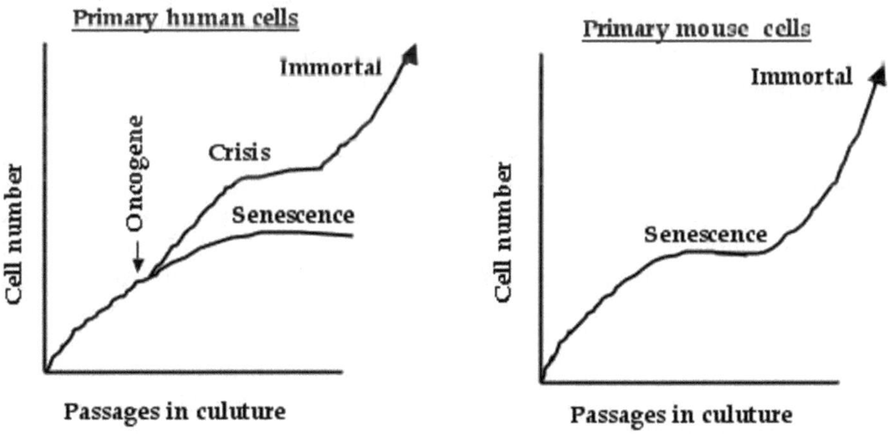

Figure 1. Growth of primary mouse and human cells in culture. Growth of human cells increases and then plateaus at senescence. Senescence can be bypassed by prior introduction of SV40 T-antigen or other oncogenes. If senescence is bypassed, cells continue to divide until crisis. At crisis, rare immortal variant emerge that can grow indefinitely and have active telomere maintaining mechanism(s). Mouse cells lack a distinct senescence checkpoint and frequently immortalize in culture.

of ARF-p53 pathway have also revealed differences in their activity in mouse and human cells. In mouse cells $p19^{ARF}$ was shown to bind to Pex19p in the cytoplasm. Such binding results in cytoplasmic halting of $p19^{ARF}$ before it makes its journey to the nucleus and performs its function of activation of p53 by capturing p53-antagonist mdm-2 [69]. ARF in human cells ($p14^{ARF}$) does not bind to Pex19p due to lack of carboxy-terminus Pex19p binding region, gets translocated to its functional site (the nucleus) soon after its synthesis in the cytoplasm and results in stronger p53 function or mortality check point [70]. These data argue that although human and mice share common set of protein ingredients, there are several important differences in their regulation. Additional complexity is imposed by the way these respond to environment stress including *ex vivo* culture [55, 71].

Premature senescence in mouse and human cells

A senescent-like state can also be induced prematurely (i.e., before the Hayflick limit has been reached) by various stimuli including oxidative stress, γ-irradiation, DNA demethylating agents, accumulation of double-stranded DNA breaks, histone deacetylase inhibitors and DNA topoisomerase inhibitors [5]. Conversely, reduction of oxidative stress can increase lifespan. These observations suggest that DNA and other macromolecular damage can trigger senescence. Inappropriate mitogenic signaling also induces premature senescence. Inhibition of phosphatidylinositol 3-kinase or constitutive mitogenic-activated protein kinase (MAPK) signaling via overexpression of oncogenic Ras, Raf or MEK induces premature senescence in human diploid fibroblasts [17, 72, 73]. Thus, in addition to preventing tumor formation by limiting replicative potential [14], senescence also functions as an alternative mechanism to apoptosis to prevent proliferation of damaged, potentially cancer-forming cells. This dual function of senescence was recently demonstrated by the finding that cells in which limited replicative potential was bypassed by ectopic expression of the human telomerase catalytic subunit gene, hTERT, still undergo senescence following overexpression of oncogenic Ras [74].

Similar to normal replicative senescence, the induction of premature senescence was linked to inhibition of CDK activities by activation of $p16^{INK4A}$ and $p21^{WAF1}$ in pRB and p53 pathways, respectively [17, 75]. In normal MEFs, where ARF-p53-p21 circuit is intact, expression of oncogenic RAS-triggers cell cycle arrest with the up-regulation of $p16^{INK4a}$, $p19^{ARF}$, p53 and $p21^{WAF1}$ [17, 73, 76]. Mutation in either p53 or ARF with retention of intact $p16^{INK4A}$, was enough to render mouse cells resistant to Ras-induced premature senescence [59. 77]. These data suggested that mouse cells depend more on the p53 pathway to limit their replicative potential in culture. In contrast to mouse cells, inactivation of both pRb and p53 pathways is crucial for human cells to escape RAS-induced senescence [75, 78]. Human fibroblasts deficient in $p16^{INK4a}$ but express physiologically relevant level of $p14^{ARF}$ were resistant to arrest by oncogenic RAS. Ectopic expression of RAS and telomerase (hTERT) results in malignantly transformed cells that retain normal p53 function. These experiments have suggested that RAS signaling leading to premature senescence in human cells involves more of $p16^{INK4A}$ than $p14^{ARF}$ activity pointing to differences between the proliferative barriers implemented by the CDKN2A locus in these species [79].

In agreement with the major role of p53 pathway in mouse senescence, MEFs with compromised p53 function are efficiently transformed into tumorigenic cells by co-expression of co-operating oncogenes such as Ras and Myc; the ones with normal p53 function undergo accelerated senescence [59, 77]. In contrast, in primary human cells the settings such as inactivation of p53 by co-expression of SV40 LT Ag and RAS or RAS and Myc could induce transformation only after long periods of selection or further treatment with chemical or physical mutagens suggesting the requirement of additional changes for human cells transformation [80]. This missing component required for human cell transformation was subsequently defined as telomerase (discussed below). Further complexity is added by the observations that the spectrum of tumors associated with oncogenic Ras in humans often differs from those in mice suggesting that development of Ras mediated transformation is mechanistically different in mouse and human cells [81, 82].

Telomere driven senescence in mouse and human

Telomeres are the nucleoprotein structures consisting of GT-rich tandem repeats of sequence (TTAGGG) and their binding proteins. Telomeres extend from 5–15 kb at each chromosome end in human and wild derived mouse strains. Inbred laboratory mouse strains, however, have telomere lengths of approximately 40 kb. These specialized structures protect the ends of linear chromosomes and maintain the chromosomal integrity by preventing their end to end fusion. Due to the semi-conservative mode of replication telomeres face an end replication problem. RNA primed DNA synthesis of the lagging strand results in a terminal gap after degradation of the most distal primer. The single stranded GT-rich 3′ overhang at each telomere is degraded by exonunclease activity leading to progressive shortening of telomeric DNA template with each round of DNA replication. It has been proposed to act as a mitotic clock to indicate the number of times cell has divided and can be expected to divide. In contrast to the normal cells, telomere length is maintained in cancer cells by telomerase, a specialized ribonucleoprotein with reverse transcriptase activity. It is comprised of a catalytic subunit (telomerase reverse transcriptase [83]) and a RNA subunit (telomerase RNA [84]), provides the template for telomere repeat synthesis by hTERT. Telomere hypothesis of aging originally had two components (1) telomeres shorten with a constant rate so that telomere length is a faithful indicator of replicative history and (2) shortening of telomeres to some threshold value triggers senescence [85–87]. This has been further refined on the basis of recent experimental data as follows [88–93].

- Telomeres are not mere cell division counters. These are active complexes and are responsive to internal and external stress signals.

- These are dynamic nucleoprotein complexes that can switch stochastically between two states: capped (functionally defined as preserving the integrity of the telomere, allowing the cell division to proceed) and uncapped (elicits cell cycle arrest)
- The rate of telomere shortening may differ from cell to cell and as a function of macromolecular metabolism and genetic damage.
- Telomere shortening counts not only cell divisions but also the cumulative probability of mutations and cellular oxidative damage; both closely related to the induction of senescence.
- Telomere state rather then length determines the replicative senescence in human cells.
- Telomere shortening acts as a tumor suppressor mechanism in human cells by limiting the number of times cells can divide. However, excessive short telomeres lead to increased genetic instability favoring tumorigenesis.

Mouse and human show major differences in involvement of telomeres in senescence. In the laboratory mouse, inbred strain of *Mus musculus*, telomeres are much longer (~60 kb) than those in human cells (~12 kb). Consistent with their long telomeres, telomerase activity and expression of mTERT is less stringently regulated in mouse somatic tissues; with detectable level of expression in most somatic tissues of inbred mice [94, 95]. Mouse cells although begin their replication *ex vivo* with extremely long telomeres, they undergo replicative arrest after only 15–20 populations doublings, well before the detectable telomere shortening. Although mouse and human senescent cells share many morphological and biochemical features, the very long telomeres in mouse makes it unlikely that telomere shortening provides a signal to activate the senescence program. Telomerase deficient mice have been generated either by elimination of catalytic component, Terc or RNA component, Tert [46, 96]. These mouse showed progressive telomere shortening at a rate of 3–5 kb per generation and increased chromosomal end to end fusions. Late generation of Terc$^{-/-}$ mice show severe proliferation defects and no survivor mouse lines were obtained. Telomere dysfunction resulted in phenotypes such as partial embryonic mortality due to defective closure of the neural tube, small size and severe intestinal atrophy, spleen atrophy and reduced proliferation of B and T lymphocytes upon mitogenic stimulation, impaired germinal center function upon immunization suggesting an essential role of telomerase in highly proliferative organs. Interestingly, from the strain of mice that have longer telomere length (40 kb) at the beginning, six generations were obtained as compared to four generations from the strain that had comparatively shorter (25 kb) telomeres along with the severity of phenotypes. These studies implied that the number of generations that can be derived in the absence of telomerase is directly proportional to the initial length, and telomere length below a threshold is required for this dysfunction. Short telomeres in late generation Terc$^{-/-}$ mice suppressed chemical-induced skin tumors. It was associated with upregulation of p53 suggesting that telomere shortening is recognized as a DNA damage response

[97]. Elimination of p53 in late generation Terc$^{-/-}$ mice accentuates their chromosomal instability caused by excessively short telomeres and increased the incidence of spontaneous tumors [47, 98].

ES cells from telomerase null MEFs showed proliferative defects after 300 populations doublings (PD) and cessation of growth at approximately 450 PDs implicating that telomere maintenance by telomerase dependent mechanisms is necessary in continued proliferation of cultured mouse cells. Reintroduction of mTR into these cells significantly restored the oncogenic potential indicating that telomerase activation is a cooperating event in the malignant transformation of cells containing critically short telomeres [95]. In contrast to mouse, their cells (mTR$^{-/-}$) were able to stabilize their telomeres in the absence of telomerase, were immortalized and transformed in culture [99]. Thus, although telomerase activity is an important component of most, if not all, of human cancers, its elimination in mouse has no effect on tumor formation for many generations. This was in addition to the basic differences in the immortalization frequencies in mouse and human. There is an accumulating evidence that structure of telomere is more important than their length, and manipulation of different components of active telomere complexes influence the senescence or immortal state of cells. Telomere binding protein, TRF2 was shown to be essential for protection of chromosomal ends and an inability of short telomeres to recruit TRF2 may cause replicative senescence. Dominant negative TRF2 induced premature senescence indistinguishable from replicative senescence in human cells. It was mediated by independent activation of p53 and p16^{INK4A}/pRb pathways. Strongest response was observed in primary cells with intact p53 and p16^{INK4A}/pRB pathways; cells deficient in either p53 or p16^{INK4A}/RB were also growth arrested by dominant negative TRF2 suggesting that human cells have two independent pathways that can respond to telomere status and induce senescence when telomere damage is extensive. However, mouse cells lacking p53, but with normal p16^{INK4A}/RB pathway, were not responsive to telomere dysfunction caused by dominant negative TRF2. p16$^{INK4A-/-}$ mouse cells responded to TRF2 inhibition similar to the wild type cells suggesting that p16^{INK4A} is not crucial for response to telomere damage in mouse cells and p53 is the primary pathway in mouse cells [100].

As discussed above, mouse cells when explanted undergo growth arrest long before their telomeres error implying that this growth arrest is independent to telomere shortening pathways. Since these cells have increased p53 and p21^{WAF1} activities (also typical of stress response), one possible interpretation is that replicative senescence in mouse cells may represent a culture stress response due to inappropriate culture conditions. MEFs deficient in DNA damage responsive genes such as p53, Ku, ATM and BRCA2 do not undergo such growth arrest suggesting that elimination of p53 pathway is sufficient for indefinite proliferation of mouse cells and thus explains their easy immortalization in culture [55, 101]. Aging can be viewed as an anti-cancer mechanism(s) that operates at two levels, (i) control of genetic, macromolecular and cellular damage by accuracy of DNA replication and repair, and oxidative defense mechanisms, respectively, and (ii) control by tumor-protective mechanisms including activity of tumor suppressor genes and repression of telomerase that limit the number of cell divisions and thus reducing the chances of accumulations of genetic damage with inevitable restriction in the repair processes. In short lived organisms such as mouse the first level of control may be enough as an anti-cancer mechanism. Long lifespan of humans puts them in the risk of increase genetic damage and instability and thus have evolved a large number of tumor suppressor mechanisms and their stringent controls as discussed above.

Conclusions

Mouse models have provided essential insights into our present day understanding of the biology of aging and cancer and continue to be invaluable tool for future studies using transgenes, knock-out and programmable molecular technologies. At the same time, the approach is complicated by the fact that they show basic differences in ways their proliferation and growth control is executed. Whereas mouse cells are easily transformed *in vitro*, human cells are extremely difficult to transform. These differences in transformation parallel their differences to escape aging and undergo spontaneous immortalization, a prerequisite for transformation. As discussed above some of these differences originate from ways in which the controls on replicative lifespan of mouse and human cells are executed. In human cells three independent check points, the $p14^{ARF}$/p53-pathway, the $p16^{INK4A}$/pRB-pathway and telomere shortening pathway have to be altered to permit indefinite division. Inactivation of p53 or pRB check-point allows escape from mortality stage 1 (M1) leading to extended lifespan. Subsequent activation of telomerase or alternative telomere maintaining mechanism(s) result in escape from mortality satge2 (M2) culminating in unlimited proliferation. In mouse cells, p53 pathway seems to be the major determinant of replicative senescence; pRB and telomeres do not appear to be largely involved.

References

1. Campisi J (2001). Cellular senescence as a tumor-suppressor mechanism. *Trends Cell Biol.* 11: S27–31.
2. Hayflick L, Moorhead PS (1961). The serial cultivation of human diploid cell strains. *Exp Cell Res.* 25: 585–621.
3. Hayflick L (1999). How and why we age. *Exp Gerontol.* 33: 639–53.
4. Smith JR, Pereira-Smith OM (1996). Replicative senescence: implications for *in vivo* aging and tumor suppression. *Science* 273: 63–7.
5. Duncan EL, Wadhwa R, Kaul SC (2000). Senescence and immortalization of human cells. *Biogerontology* 1: 103–21.
6. Campisi J (2001). From cells to organisms: can we learn about aging from cells in culture? *Exp Gerontol.* 36: 607–18.
7. Pawelec G, Barnett Y, Forsey R, *et al.* (2002). T cells and aging, January 2002 update. Front Biosci, 1: d1056–83.
8. Hornsby PJ (2002). Aging of the human adrenal cortex. *Ageing Res Rev.* 1: 229–42.
9. Geiger H, Van Zant G (2002). The aging of lympho-hematopoietic stem cells. *Nat Immunol.* 3: 329–33.

10. Effros RB, Globerson A (2002). Hematopoietic cells and replicative senescence. *Exp Gerontol.* 37: 191–6.
11. Reddel RR (2000). The role of senescence and immortalization in carcinogenesis. *Carcinogenesis* 21: 477–84.
12. Hornsby PJ (2002). Cellular senescence and tissue aging *in vivo. J Gerontol A Biol Sci Med Sci.* 57: B251–6.
13. Allsopp RC, Weissman IL (2002). Replicative senescence of hematopoietic stem cells during serial transplantation: does telomere shortening play a role? *Oncogene* 21: 3270–3.
14. Sager R (1991). Senescence as a mode of tumor suppression. *Environ Health Perspect.* 93: 59–62.
15. Pereira-Smith OM (1997). Genetic theories on aging. *Aging (Milano)* 9: 429–30.
16. Linskens MH, Harley CB, West MD, Campisi J, Hayflick L (1995). Replicative senescence and cell death [letter; comment]. *Science* 267: 17.
17. Serrano M, Lin AW, McCurrach ME, Beach D, Lowe SW (1997). Oncogenic ras provokes premature cell senescence associated with accumulation of p53 and p16INK4a. *Cell* 88: 593–602.
18. Bree RT, Stenson-Cox C, Grealy M, Byrnes L, Gorman AM, Samali A (2002). Cellular longevity: role of apoptosis and replicative senescence. *Biogerontology* 3: 195–206.
19. Zhang Y, Herman B (2002). Ageing and apoptosis. *Mech Ageing Dev.* 123: 245–60.
20. Skulachev VP (2002). Programmed death phenomena: from organelle to organism. *Ann NY Acad Sci.* 959: 214–37.
21. Dierick JF, Eliaers F, Remacle J, *et al.* (2002). Stress-induced premature senescence and replicative senescence are different phenotypes, proteomic evidence. *Biochem Pharmacol.* 64: 1011–17.
22. Toussaint O, Royer V, Salmon M, Remacle J (2002). Stress-induced premature senescence and tissue ageing. *Biochem Pharmacol.* 64: 1007–9.
23. Marcotte R, Wang E (2002). Replicative senescence revisited. *J Gerontol A Biol Sci Med Sci.* 57: B257–69.
24. Biesalski HK (2002). Free radical theory of aging. *Curr Opin Clin Nutr Metab Care* 5: 5–10.
25. Xu D, Finkel T (2002). A role for mitochondria as potential regulators of cellular lifespan. *Biochem Biophys Res Commun.* 294: 245–8.
26. Saretzki G, Von Zglinicki T (2002). Replicative aging, telomeres, and oxidative stress. *Ann NY Acad Sci.* 959: 24–9.
27. te Poele RH, Okorokov AL, Jardine L, Cummings J, Joel SP (2002). DNA damage is able to induce senescence in tumor cells *in vitro* and *in vivo. Cancer Res.* 62: 1876–83.
28. Tominaga K, Olgun A, Smith JR, Pereira-Smith OM (2002). Genetics of cellular senescence. *Mech Ageing Dev.* 123: 927–36.
29. Kirkwood TB (2002). New science for an old problem. *Trends Genet.* 18: 441–2.
30. Mathon NF, Lloyd AC (2001). Cell senescence and cancer. *Nature Rev Cancer* 1: 203–13.
31. Anisimov VN (2001). Mutant and genetically modified mice as models for studying the relationship between aging and carcinogenesis. *Mech Ageing Dev.* 122: 1221–55.
32. Mahler JF (2001). Mutant models of prolonged lifespan. *Toxicol Pathol.* 29: 673–6.
33. Bartke A (2000). Delayed aging in Ames dwarf mice. Relationships to endocrine function and body size. *Results Probl Cell Differ.* 29: 181–202.
34. Migliaccio E, Giorgio M, Mele S, *et al.* (1999). The p66shc adaptor protein controls oxidative stress response and lifespan in mammals. *Nature* 402: 309–13.

35. Ostler EL, Wallis CV, Sheerin AN, Faragher RG (2002). A model for the phenotypic presentation of Werner's syndrome. *Exp Gerontol.* 37: 285–92.
36. Lebel M, Cardiff RD, Leder P (2001). Tumorigenic effect of nonfunctional p53 or p21 in mice mutant in the Werner syndrome helicase. *Cancer Res.* 61: 1816–19.
37. Lebel M (2001). Werner syndrome: genetic and molecular basis of a premature aging disorder. *Cell Mol Life Sci.* 58: 857–67.
38. Kuro-o M, Matsumura Y, Aizawa H, *et al.* (1997). Mutation of the mouse klotho gene leads to a syndrome resembling ageing. *Nature* 390: 45–51.
39. Roush W (1997). Fast-forward aging in a mutant mouse? *Science* 278: 1013.
40. Matsumura Y, Aizawa H, Shiraki-Iida T, Nagai R, Kuro-o M, Nabeshima Y (1998). Identification of the human klotho gene and its two transcripts encoding membrane and secreted klotho protein. *Biochem Biophys Res Commun.* 242: 626–30.
41. Kuro-o M (2001). Disease model: human aging. *Trends Mol Med.* 7: 179–81.
42. Takeda T (1999). Senescence-accelerated mouse (SAM): a biogerontological resource in aging research. *Neurobiol Aging* 20: 105–10.
43. Okuma Y, Nomura Y (1998). Senescence-accelerated mouse (SAM) as an animal model of senile dementia: pharmacological, neurochemical and molecular biological approach. *Jpn J Pharmacol.* 78: 399–404.
44. Mori A, Utsumi K, Liu J, Hosokawa M (1998). Oxidative damage in the senescence-accelerated mouse. *Ann NY Acad Sci.* 854: 239–50.
45. Lim DS, Vogel H, Willerford DM, Sands AT, Platt KA, Hasty P (2000). Analysis of ku80-mutant mice and cells with deficient levels of p53. *Mol Cell Biol.* 20: 3772–80.
46. Blasco MA, Lee HW, Hande MP, *et al.* (1997). Telomere shortening and tumor formation by mouse cells lacking telomerase RNA. *Cell* 91: 25–34.
47. Rudolph KL, Chang S, Lee HW, *et al.* (1999). Longevity, stress response, and cancer in aging telomerase-deficient mice. *Cell* 96: 701–12.
48. Chin L, Artandi SE, Shen Q, *et al.* (1999). p53 deficiency rescues the adverse effects of telomere loss and cooperates with telomere dysfunction to accelerate carcinogenesis. *Cell* 97: 527–38.
49. Martin GM (2002). Keynote: mechanisms of senescence – complificationists versus simplificationists. *Mech Ageing Dev.* 123: 65–73.
50. Sherr CJ (2000). The Pezcoller lecture: cancer cell cycles revisited. Cancer Res, 60: 3689–95.
51. Campisi J (2002). Between Scylla, Charybdis: p53 links tumor suppression and aging. *Mech Ageing Dev.* 123: 567–73.
52. Levine AJ (1997). p53, the cellular gatekeeper for growth and division. *Cell* 88: 323–31.
53. Giaccia AJ, Kastan MB (1998). The complexity of p53 modulation: emerging patterns from divergent signals. *Genes Dev.* 12: 2973–83.
54. Wright WE, Shay JW (2002). Historical claims and current interpretations of replicative aging. *Nat Biotechnol.* 20: 682–8.
55. Sherr CJ, DePinho RA (2000). Cellular senescence: mitotic clock or culture shock? *Cell* 102: 407–10.
56. Todaro GJ, Habel K, Green H (1965). Antigenic and cultural properties of cells doubly transformed by polyoma virus and SV40. *Virology* 27: 179–85.
57. Harvey DM, Levine AJ (1991). p53 alteration is a common event in the spontaneous immortalization of primary BALB/c murine embryo fibroblasts. *Genes Dev.* 5: 2375–85.
58. Donehower LA (2002). Does p53 affect organismal aging? *J Cell Physiol.* 192: 23–33.
59. Kamijo T, Zindy F, Roussel MF, *et al.* (1997). Tumor suppression at the mouse INK4a locus mediated by the alternative reading frame product p19ARF. *Cell* 91: 649–59.

60. Sharpless NE, Bardeesy N, Lee KH, *et al.* (2001). Loss of p16Ink4a with retention of p19Arf predisposes mice to tumorigenesis. *Nature* 413: 86–91.
61. Krimpenfort P, Quon KC, Mooi WJ, Loonstra A, Berns A (2001). Loss of p16Ink4a confers susceptibility to metastatic melanoma in mice. *Nature* 413: 83–6.
62. Shay JW, Wright WE, Werbin H (1993). Toward a molecular understanding of human breast cancer: a hypothesis. *Breast Cancer Res Treat.* 25: 83–94.
63. Bond JA, Webley K, Wyllie FS, *et al.* (1999). p53-Dependent growth arrest and altered p53-immunoreactivity following metabolic labelling with 32P ortho-phosphate in human fibroblasts. *Oncogene* 18: 3788–92.
64. Rogan EM, Bryan TM, Hukku B, *et al.* (1995). Alterations in p53 and p16INK4 expression and telomere length during spontaneous immortalization of Li-Fraumeni syndrome fibroblasts. *Mol Cell Biol.* 15: 4745–53.
65. Duncan EL, Reddel RR (1997). Genetic changes associated with immortalization. A review. *Biochemistry (Mosc).* 62: 1263–74.
66. Stewart SA, Weinberg RA (2002). Senescence: does it all happen at the ends? *Oncogene* 21: 627–30.
67. Maser RS, DePinho RA (2002). Connecting chromosomes, crisis, and cancer. *Science* 297: 565–9.
68. de Ronde A, Sol CJ, van Strien A, ter Schegget J, van der Noordaa J (1989). The SV40 small t antigen is essential for the morphological transformation of human fibroblasts. *Virology* 171: 260–3.
69. Sugihara T, Kaul SC, Kato J, Reddel RR, Nomura H, Wadhwa R (2001). Pex19p dampens the p19ARF-p53-p21WAF1 tumor suppressor pathway. *J Biol Chem.* 276: 18649–52.
70. Wadhwa R, Sugihara T, Hasan MK, Taira K, Reddel RR, Kaul SC (2002). A major functional difference between the mouse and human ARF tumor suppressor proteins. *J Biol Chem.* 277: 36665–70.
71. Hahn WC, Weinberg RA (2002). Modelling the molecular circuitry of cancer. *Nat Rev Cancer* 2: 331–41.
72. Lin AW, Barradas M, Stone JC, van Aelst L, Serrano M, Lowe SW (1998). Premature senescence involving p53 and p16 is activated in response to constitutive MEK/MAPK mitogenic signaling. *Genes Dev.* 12: 3008–19.
73. Zhu J, Woods D, McMahon M, Bishop JM (1998). Senescence of human fibroblasts induced by oncogenic Raf. *Genes Dev.* 12: 2997–3007.
74. Wei W, Hemmer RM, Sedivy JM (2001). Role of p14(ARF). in replicative and induced senescence of human fibroblasts. *Mol Cell Biol.* 21 6748–57.
75. Serrano M, Blasco MA (2001). Putting the stress on senescence. *Curr Opin Cell Biol.* 13: 748–53.
76. Lloyd AC, Obermuller F, Staddon S, Barth CF, McMahon M, Land H (1997). Cooperating oncogenes converge to regulate cyclin/cdk complexes. *Genes Dev.* 11: 663–77.
77. Lin AW, Lowe SW (2001). Oncogenic ras activates the ARF-p53 pathway to suppress epithelial cell transformation. *Proc Natl Acad Sci USA* 98: 5025–30.
78. Peeper DS, Dannenberg JH, Douma S, te Riele H, Bernards R (2001). Escape from premature senescence is not sufficient for oncogenic transformation by Ras. *Nat Cell Biol.* 3: 198–203.
79. Brookes S, Rowe J, Ruas M, *et al.* (2002). INK4a-deficient human diploid fibroblasts are resistant to RAS-induced senescence. *EMBO J.* 21: 2936–45.

80. Hahn WC, Weinberg RA (2002). Rules for making human tumor cells. *N Engl J Med.* 347: 1593–603.
81. Johnson L, Mercer K, Greenbaum D, *et al.* (2001). Somatic activation of the K-ras oncogene causes early onset lung cancer in mice. *Nature* 410: 1111–16.
82. Hamad NM, Elconin JH, Karnoub AE, *et al.* (2002). Distinct requirements for Ras oncogenesis in human versus mouse cells. *Genes Dev.* 16: 2045–57.
83. Banik SS, Guo C, Smith AC, *et al.* (2002). C-terminal regions of the human telomerase catalytic subunit essential for *in vivo* enzyme activity. *Mol Cell Biol.* 22: 6234–46.
84. Adelfalk C, Lorenz M, Serra V, von Zglinicki T, Hirsch-Kauffmann M, Schweiger M (2001). Accelerated telomere shortening in Fanconi anemia fibroblasts–a longitudinal study. *FEBS Lett.* 506: 22–6.
85. Harley CB, Vaziri H, Counter CM, Allsopp RC (1992). The telomere hypothesis of cellular aging. *Exp Gerontol.* 27: 375–82.
86. Blackburn EH (1991). Structure and function of telomeres. *Nature* 350: 569–73.
87. Blackburn EH (2001). Switching and signaling at the telomere. *Cell* 106: 661–73.
88. Karlseder J, Smogorzewska A, de Lange T (2002). Senescence induced by altered telomere state, not telomere loss. *Science* 295: 2446–9.
89. Chan SW, Blackburn EH (2002). New ways not to make ends meet: telomerase, DNA damage proteins and heterochromatin. *Oncogene* 21: 553–63.
90. Wright WE, Shay JW (2001). Cellular senescence as a tumor-protection mechanism: the essential role of counting. *Curr Opin Genet Dev.* 11: 98–103.
91. Blackburn EH (2000). Telomere states and cell fates. *Nature* 408: 53–6.
92. von Zglinicki T (2001). Telomeres and replicative senescence: Is it only length that counts? *Cancer Lett.* 168: 111–16.
93. von Zglinicki T (2002). Oxidative stress shortens telomeres. *Trends Biochem Sci.* 27: 339–44.
94. Prowse KR, Avilion AA, Greider CW (1993). Identification of a nonprocessive telomerase activity from mouse cells. *Proc Natl Acad Sci USA* 90: 1493–7.
95. Greenberg RA, Allsopp RC, Chin L, Morin GB, DePinho RA (1998). Expression of mouse telomerase reverse transcriptase during development, differentiation and proliferation. *Oncogene* 16: 1723–30.
96. Liu Y, Kha H, Ungrin M, Robinson MO, Harrington L (2002). Preferential maintenance of critically short telomeres in mammalian cells heterozygous for mTert. *Proc Natl Acad Sci USA* 99: 3597–602.
97. Gonzalez-Suarez E, Samper E, Flores JM, Blasco MA (2000). Telomerase-deficient mice with short telomeres are resistant to skin tumorigenesis. *Nat Genet.* 26: 114–17.
98. Goytisolo FA, Blasco MA (2002). Many ways to telomere dysfunction: *in vivo* studies using mouse models. *Oncogene* 21: 584–91.
99. Hande MP, Samper E, Lansdorp P, Blasco MA (1999). Telomere length dynamics and chromosomal instability in cells derived from telomerase null mice. *J Cell Biol.* 144: 589–601.
100. Smogorzewska A, De Lange T (2002). Different telomere damage signaling pathways in human and mouse cells. *EMBO J.* 21: 4338–48.
101. Wright WE, Shay JW (2000). Telomere dynamics in cancer progression and prevention: fundamental differences in human and mouse telomere biology. *Nat Med.* 6: 849–51.

Therapeutic Induction of Cellular Senescence: Nuclear Matrix in Senescence

Dai Ayusawa
Kihara Institute for Biological Research, Yokohama City University, Maioka-cho 641-12, Totsuka-ku, Yokohama 244-0813, Japan

Introduction

Normal human cells stop dividing in culture after a limited number of cell division, a phenomenon termed "replicative senescence" [1]. The senescence is featured most typically by specific morphological alterations, flat and enlarged cell shape, and induction of senescence markers such as fibronectin, collagenase I, and senescence-associated β-galactosidase [2]. It is also shown that in normally senesced human fibroblasts the cyclin-dependent protein kinase inhibitors $p21^{waf1/sdi\text{-}1}$ and $p16^{ink4a}$, and the tumor suppressor p53 are upregulated or activated, implicating their roles in replicative senescence [3]. Telomere shortening is one of the events in normal human cells [4]. This is thought to be perceived by p53 as a form of DNA damage and result in cell cycle arrest. On the other hand, accumulating data show that normal human cells can enter senescence prematurely upon treatment with various agents or genes, a phenomenon called "premature senescence." They include, for instance, hydrogen peroxide [5], the activated *c-ras* gene [6], histone deacetylase inhibitors [7], DNA topoisomerases inhibitors [8], and others. In sharp contrast, they have different effects on immortal cells, namely, having no effect or inducing apoptosis or cell lysis.

We have screened various agents which induce premature senescence in normal human fibroblasts, and found that 5-halogenated deoxyuridine (HdU) clearly does so [9–10]. Surprisingly, any types of cell lines and lower eukaryotic cells also responded to HdU similarly. Here we describe our recent studies aimed to understand a molecular basis for the senescence-like phenomena induced by HdU.

Sunil C. Kaul and Renu Wadhwa (eds.), Aging of Cells In and Outside the Body, 239–255.

5-Halogenated deoxyuridine induces a senescence-like phenomenon

5-Bromodeoxyuridine (BrdU) does not rapidly inhibit growth, but gradually stops dividing dose-dependently. BrdU itself does not seem to be toxic to cells. In the presence of BrdU, majority of cells accumulated at G_1 phase of the cell cycle, leaving a small fraction of cells at G_2 phase. The cells become flat and enlarged and strongly induce senescence-associated β-galactosidase 1–2 weeks after addition of BrdU. This phenomenon is quite similar to replicative senescence in normal human cells (Figure 1).

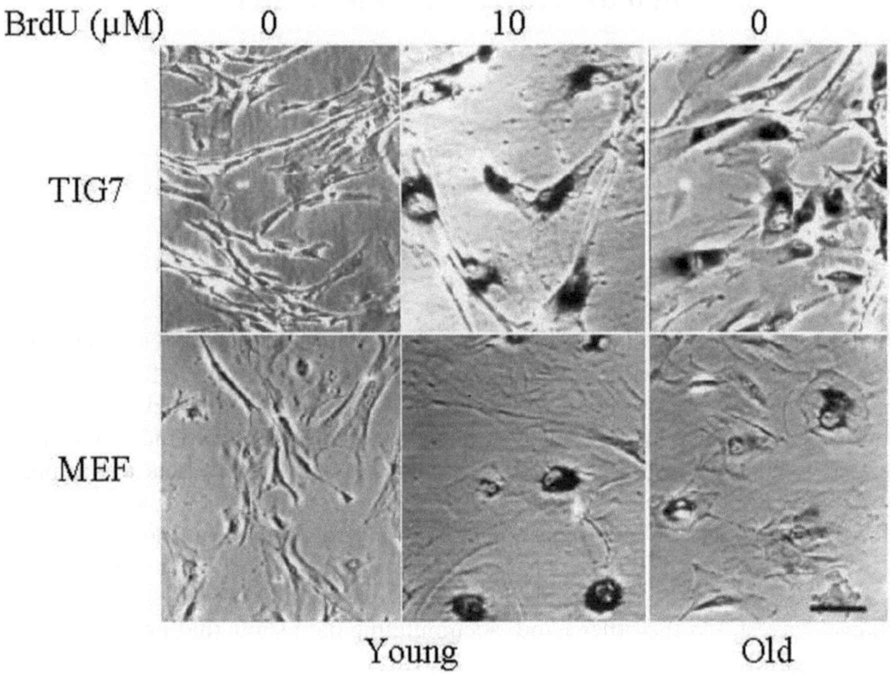

Figure 1. Changes in morphology and induction of senescence-associated β-galactosidase in normal fibroblasts. Cells were cultured for 2 weeks in the absence and presence of 10 μmol/L BrdU. Microphotographs were taken after staining with the β-galactosidase with X-gal. Scale bar, 100 μm. Young and old TIG-7, normal human fibroblasts at 38 and 72 PDLs, respectively; Young and old MEF, mouse embryonic fibroblasts at 5 and 13 PDLs, respectively.

We tested various cell lines including normal type, tumor derived, virus-infected and rodent lines. The human lines included those genetically or functionally defective in $p21^{waf1/sdi-1}$, $p16^{ink4a}$, p53, pRB or others, which are thought to be implicated in cellular senescence. All of the cell lines tested responded to BrdU similarly (Figure 2).

5-Chlorodeoxyuridine (CldU) and 5-Iododeoxyuridine gave similar effects. 5-Bromouracil incorporated into DNA is converted to uracil by irradiation with light

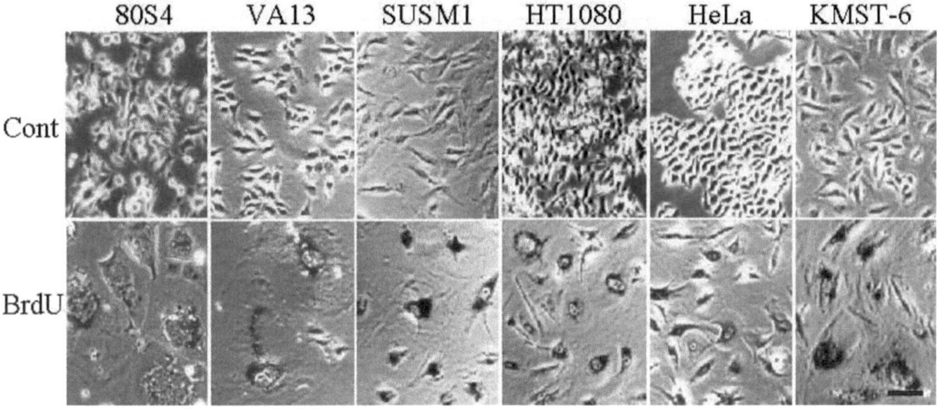

Figure 2. *Changes in morphology and induction of senescence-associated β-galactosidase in immortal human cells. Cells were cultured for 2 weeks in the absence and presence of 10–50 μmol/L BrdU, and processed as in Figure 1. VA13 is SV40-transformed, SUSM-1 is derived from normal fibroblasts, and HCT116/80S14 is p21$^{waf1/sdi-1}$-defective.*

and this causes a nick or gap on DNA during the process of excision repair [11]. However, such DNA-damage, if any, does not contribute to the phenomenon described here because DNA damage-inducible genes, except for the p21$^{waf1/sdi-1}$ gene, were not induced [12]. CldU is effective as BrdU in inducing the senescence-like phenomena, but does not kill cells under the conditions in which BrdU does so extensively. Upon addition of BrdU or CldU to HeLa cells, both p21$^{waf1/sdi-1}$ and fibronectin mRNA levels began to increase one or two days after addition of BrdU and reached levels 100 to 1000 times higher than those of control one week after addition of the analogues (Figure 3).

Their early onset of induction is in agreement with the finding that the cells are destined to stop dividing by 2-3 days after addition of the analogues. Collagenase I mRNA begins to increase slower than the others, but the level increases by 850-fold. Western blot analysis of protein levels for p21$^{waf1/sdi-1}$, p16 ^{ink4a} and p53 revealed no consistent difference upon addition of the analogues among various cell lines examined. Subcellular distribution of mortalin protein was examined by immunostaining with anti-mortalin antibody in senescence induced HT1080, HeLa and SUSM-1 cells because it serves as a reliable marker to distinguish mortal and immortal state of cells [13–14]. When cultured in the absence of BrdU, cells showed the patterns of granular juxtanuclear cap, granular gradient from nuclear membrane to cell membrane, or perinuclear fibres. When cultured with BrdU for 2 weeks, all of them showed a pan-cytoplasmic distribution pattern characteristic of mortal cells. These studies showed that BrdU induced senescence like state in human transformed cells.

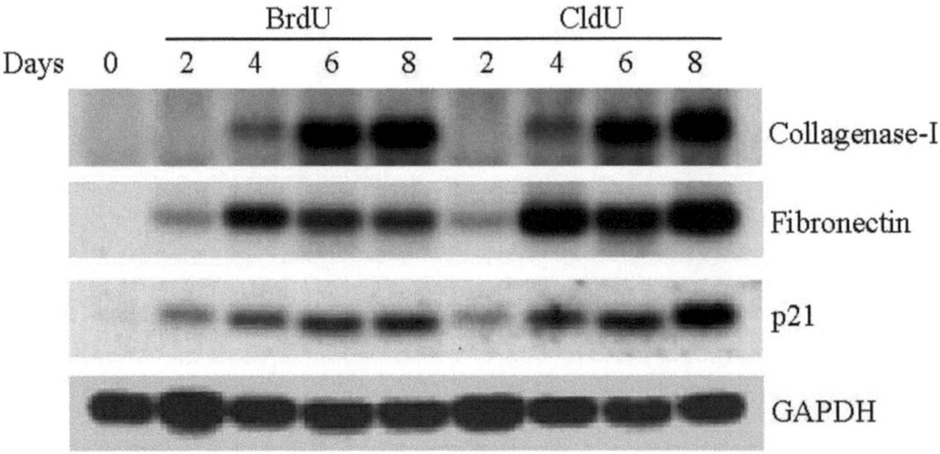

Figure 3. Induction of the senescence-associated genes by HdU in HeLa cells. Cells were cultured in the presence of 50 μmol/L BrdU or CldU for the time indicated. Total mRNA samples were subjected to Northern blot analysis using the radio-labeled probes indicated. GAPDH: glyceraldehyde monophosphate dehydorgenase.

Although telomerase activity in two telomerase-positive cell lines, HeLa and HT1080, was decreased to less than 10% by 2 weeks after addition of BrdU, telomere shortening did not occur in young TIG-7, HeLa, HT1080, SUSM-1, mouse NIH3T3, or rat 3Y1 by addition of BrdU. We do not know whether this result is due to arrest in cell cycle or down-regulation of the genes for telomerase catalytic subunit or its associated proteins.

We examined the effects of BrdU on yeast *Saccharomyces cerevisiae* [15] and nematode *C. elegans* [unpublished data]. Since the yeast cannot use exogenous thymidine, we constructed a derivative strain having herpes virus thymidine kinase and lacking thymidylate synthase. The yeast cells exhibited enlarged cell shape and lost division potential when cultured in the presence of BrdU. The nematode worm lost ability to lay eggs and showed short lifespan when cultured in the presence of BrdU. We isolated mutants resistant to BrdU for their growth from these organisms. Responsible genes in some of the yeast mutants have been identified [unpublished data].

Characterization of genes induced by BrdU

To assess the HdU-induced senescence like phenomenon at the level of gene expression, we constructed a PCR-based subtractive cDNA library enriched for mRNA species that immediately increase by administration of BrdU to HeLa cells [12]. Candidate cDNA clones were isolated by differential colony hybridization, and then positive clones were confirmed by Northern blot analysis. Sequencing analysis revealed that the identified cDNA species were classified into three groups: (1) senescence markers widely used, (2) known species having no relevance to senescence reported, and (3) known or novel ESTs. As expected, the majority of them showed increase in expression in senescent human diploid fibroblasts. These results suggest that similar mechanisms operate in the regulation of BrdU-induced genes and senescence-associated genes.

An ectopic gene integrated in the genome of host cells is occasionally silenced due to a position effect of its adjacent chromatin structure. We have found that BrdU activates a GFP transgene silenced in HeLa cells [16]. The transgene was also activated to various degrees by inhibitors of histone deacetylase, DNA topoisomerases, and DNA methyltransferase. The peptide antibiotic distamycin A (DM) was also found to potentiate the action of BrdU. As described below, transient expression of a synthetic multi AT-hook protein MATH20 similarly activated the transgene. This latter phenomenon resembles the finding that MATH20 suppresses position effect variegation in *Drosophila melanogaster* [17]. Possible mechanisms for these phenomena are described below.

AT-binding ligands potentiate the effects of 5-HdU

We have found that DM, which specifically binds to the minor grooves of AT-tracts [18], dramatically potentiates the effects of HdU. DM alone has no obvious effect on human cells at modest concentrations and only inhibited growth at higher concentrations. However, it significantly induced senescence-like morphology, the senescence-associated β-galactosidase, and mRNA levels for p21$^{waf1/sdi-1}$, fibronectin, and collagenase I when low concentrations of BrdU are present simultaneously (Figure 1).

We then examined the effect of DM on the genes upregulated by BrdU using a very reproducible and reliable cDNA macroarray. DM markedly increased mRNA levels for a set of genes. This group contains well-known typical senescence-associated genes in normal human cells [19]. Their loci are mapped on or nearby AT-rich G-bands of human chromosomes as revealed by computing analysis using the Map Viewer program of NCBI. In contrast, the genes not affected by DM are mapped on both G- and R-bands. DM does not affect incorporation of labeled thymidine or BrdU into cellular DNA. Netropsin and Hoechst 33258, the other types of AT-binding ligands, showed similar effects whereas chromomycin A3, which binds to GC-rich sequences, had no such effects.

HMGI/Y proteins contain multiple copies of a conserved DNA-binding peptide motif called AT hook that preferentially binds to the narrow minor groove of stretches of AT-rich sequence similar to DM [20]. Their intrinsic flexibility allows the HMGI/Y proteins to participate in specific protein-DNA and protein-protein interactions that induce both structural changes in chromatin substrates and the formation of stereospecific complexes called enhanceosomes on the promoter/enhancer regions of the affected genes. We conditionally expressed DsRed-tagged human HMG-I and its derivatives MATH2 and MATH20 in HeLa cells [Satou W, manuscript submitted]. MATH2 is an N-terminal domain of HMG-I containing two

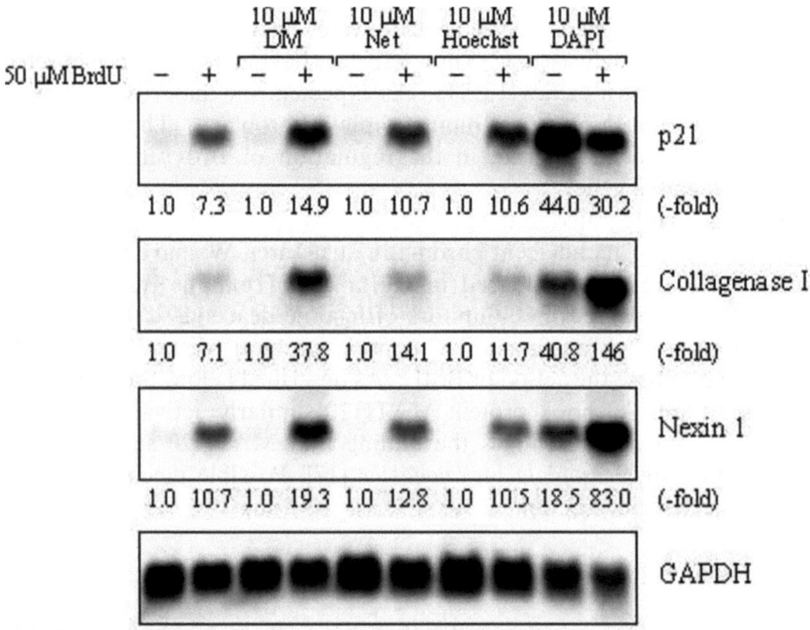

***Figure 4**. Synergistic induction of the senescence-associated genes by simultaneous addition of BrdU and AT-binding ligands in HeLa cells. Cells were cultured in the presence of BrdU and DM, netropsin (Net), Hoechst 33258 (Hoechst), or DAPI for 2 days as indicated, subjected to Northern blot analysis, and processed as in Figure 3.*

AT-hooks but lacking C-terminal or linker domains involved in assembly of enhancesome or protein-protein interaction [21]. MATH 20 is a tandem decamer of MATH2 DNA [17]. These fused proteins were all observed as foci or dots in the nuclei similar to intact HMG-I detected by a specific antibody [22]. Their distribution patterns overlapped with that of Hoechst 33342. Addition of low concentrations of BrdU to the cells overexpressing either of them strongly inhibited growth, induced a flat and enlarged cell shape, and the typical senescence markers (Figure 3). Thus, HMG-I, MATH2 and MATH20 were all shown to dramatically and similarly potentiate the effects of BrdU.

BrdU decondenses centromeric heterochromatin and G-bands in mitotic chromosomes alone [23, 24] or in concert with AT-binding ligands such as DM, netropsin [25], and Hoechst 33258 [26]. With these lines of evidence, we have a novel hypothesis that BrdU might change a chromatin structure, thus inducing or suppressing a number of genes located on or near AT-rich inactive chromatin. One of the possible targets for BrdU is a certain type of scaffold/nuclear matrix attachment region (S/MAR) sequences [27, 28]. S/MARs are very AT-rich regions of several hundred base pairs, preferentially located on G-bands and heterochromatin, and specifically associated with the nuclear scaffold or matrix. They strongly affect expression of genes and chromosome dynamics with DNA-binding proteins such as histone H1 and the high mobility group protein HMG-I/Y [21]. Histone H1, HMG-I/Y and distamycin A all bind selectively to the A tracts of S/MARs, and HMG-I/Y and distamycin A can competitively displace pre-bound histone H1 from an S/MAR template [29, 30]. This may aid opening of condensed chromatin induced by BrdU as discussed below.

5-Bromouracil substitution alters properties of S/MAR DNA

S/MAR sequences bend along the axis of their double-stranded DNA helices [31] and present specific binding sites for DM since they have many A runs or AT tracts. To examine the properties of 5-bromouracil-substituted S/MAR sequences, we used a well-characterized rat S/MAR sequence (Figure 5A). This sequence (740 bp) consists of a tandem dimer of 370 bp fragment and satisfies all of the requirements for S/MAR sequences [32]. We substituted thymine with 5-bromouracil in the DNA by PCR with dTTP or BrdUTP as a substrate. The products were cut with an appropriate enzyme, and run on polyacrylamide gel at 4°C or 52°C. When two thymine-containing fragments having the same size but a bend center at a different position were run at 4°C, one containing a bend center at the central position migrated slower than the other containing it at the terminal position (Figure 5B). Upon substitution with 5-bromouracil, the fragment containing the bend center at the central position migrated faster than the thymine-containing counterpart depending on the degree of 5-bromouracil substitution. The fragment containing the bend center at the terminal position did not show such a change. No such differences were observed at 52°C.

Then, we examined the effect of DM on the mobility of the fragments (Figure 5C). When incubated with DM before electrophoresis, the thymine-containing bent fragment migrated faster as its concentrations increased. In contrast, the 5-bromouracil-containing fragment was less affected by DM than that of the thymine-containing counterpart. Further, we performed DNase I footprinting analysis to determine the affinity of the S/MAR sequence to DM. More than eight binding sites were detected on this sequence. These sites were divided into two groups. Members in one group contained a long A-tract, showed relatively high affinity to DM, and their binding to DM was not affected by 5-bromouracil substitution. Those in the other group showed relatively low affinity to DM, and their binding to DM was weakened by 5-bromouracil substitution. We also examined affinity of the rat S/MAR DNA to nuclear scaffold proteins by filter binding assay. When *E.coli* DNA was used as a competitor, the thymine- and 5-bromouracil-containing DNA bound to the scaffold proteins similarly (Figure 6A). In contrast, when thymine- or 5-bromouracil-containing HeLa DNA was used as a competitor, the 5-bromouracil-containing S/MAR DNA bound more strongly to the scaffold proteins than the thymine-containing counterpart. Further, a S/MAR sequence derived from the human immunoglobulin heavy chain gene increased its binding capacity to the nuclear matrix upon substitution of thymine with 5-bromouracil [33].

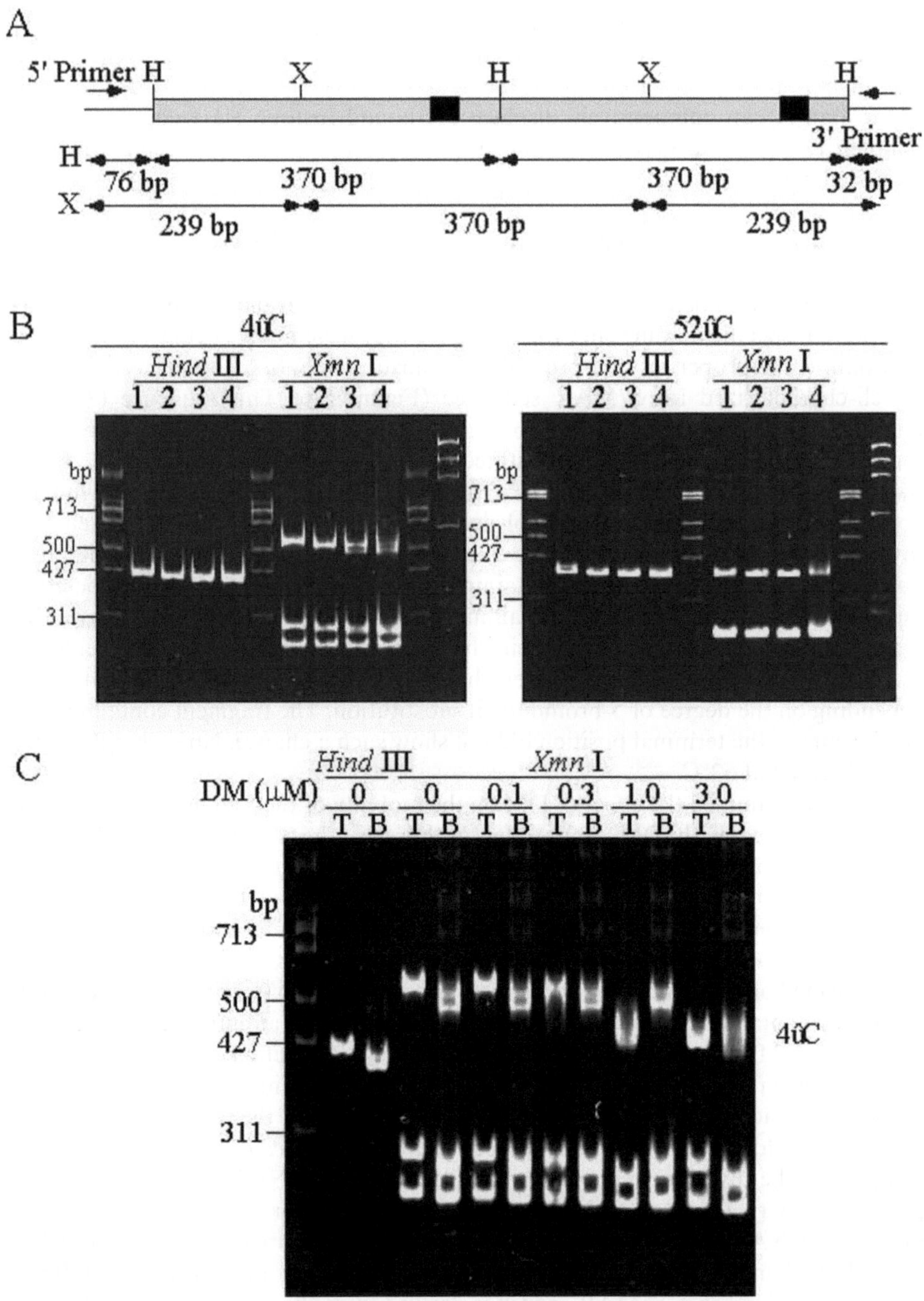
A
5' Primer H
X
H
X
H
3' Primer
H
76 bp
370 bp
370 bp
32 bp
X
239 bp
370 bp
239 bp
B
4ûC
Hind III
Xmn I
1 2 3 4
1 2 3 4
bp
713
500
427
311
52ûC
Hind III
Xmn I
1 2 3 4
1 2 3 4
bp
713
500
427
311
C
Hind III
Xmn I
DM (μM)
0
0
0.1
0.3
1.0
3.0
T B T B T B T B T B T B
bp
713
500
427
311
4ûC

DM inhibited the binding dose-dependently as reported previously [34] when *E. coli* DNA was used as a competitor (Figure 6B). When HeLa DNA was used as a competitor, DM did not affect the binding of the 5-bromouracil-containing S/MAR DNA whereas high concentrations of DM inhibited the binding of the thymine-containing counterpart. When 5-bromouracil-containing HeLa DNA was used as a competitor, DM did not inhibit the binding of the bromouracil-containing S/MAR DNA, but did significantly the thymine-containing counterpart. *E. coli* and HeLa DNA differ greatly in GC contents and/or distribution of AT-tracts. These results suggest that 5-bormouracil-containing S/MAR sequences change their binding to the nuclear scaffolds in the presence of DM.

Changes in nuclear matrix proteins induced by 5-CldU

To study early events in the senescence-like phenomenon, we attempted to search for a nuclear matrix protein that changes in amount or location in the nucleus of HeLa cells upon addition of CldU [35]. Identification of such a protein will facilitate understanding the molecular basis for the action of HdU. The nuclear matrix is a structure that remains when most of the chromatin and loosely bound components are removed from the inter phase nucleus [36]. This structure consists of the nuclear lamina containing the nucleopores and an intranuclear fibrogranular network containing residual nucleoli. The molecular basis for the internal matrix is poorly understood [37, 38]. Many components of the nuclear matrix are involved in DNA replication, RNA transcription, and hnRNA processing [39–44]. Therefore, the nuclear matrix may play an important role in the structural and functional organization of the cell nucleus. Tissue specific nuclear matrix proteins have been found in the rat, mouse, and human, and are shown to have a role in tissue development.

We prepared samples of nuclear matrix and total nuclear proteins from HeLa cells cultured with CldU for 3 days, examined them by pH 3-10 non-linear gradient 2-D PAGE. In the nuclear matrix fraction, several proteins were increased or decreased by addition of CldU (Figure 7A, a–f). However, most of them (a, c, d, e, and f) also

---

Figure 5. *(opposite) Effects of 5-bromouracil substitution on the topology of S/MAR DNA. (A) Structure of S/MAR DNA. A tandem dimer of 370-bp Hind III fragment derived from a rat S/MAR sequence [32] was used. The bend centers were indicated by black boxes. H, Hind III; X, Xmn I. (B) Polyacrylamide gel electrophoresis of S/MAR DNA containing thymine or 5-bromouracil. The DNA was amplified by PCR with dTTP or BrdUTP as a substrate, digests with Hind III or Xmn I, and run on polyacrylamide gel at 4°C (left panel) and 52°C (right panel). Lane 1, DNA amplified by PCR with dTTP; lane 2–4, DNA amplified by PCR with BrdUTP for 1, 2, and 5 cycles, respectively; lane M1 and M2, øX174 DNA digested with Hinf I and Hae III, respectively. (C) Polyacrylamide gel electrophoresis of S/MAR DNA containing thymine or 5-bromouracil after incubation with distamycin A (DM). The same fragments as in (B) were incubated with increasing concentrations of DM and run on polyacrylamide gel at 4°C. Lanes T, DNA amplified by PCR with dTTP; lanes B, DNA amplified by PCR with BrdUTP for 5 cycles.*

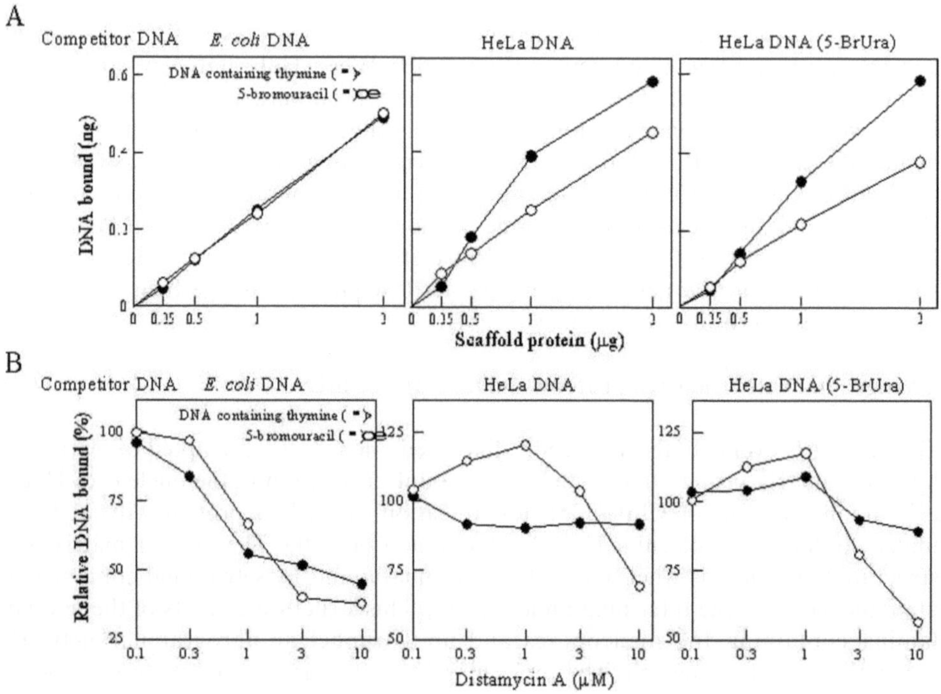

Figure 6*. Binding properties of S/MAR DNA to nuclear scaffold proteins. (A) Effects of 5-bromouracil substitution on the binding of S/MAR DNA to nuclear scaffold proteins. The labeled rat S/MAR DNA (see Figure 5A) containing thymine or 5-bromouracil was incubated with increasing amounts of scaffold proteins in the presence of competitor DNA. HeLa DNA (5-Br Ura) was prepared form the cells cultured with 50 µmol/L BrdU for 1 week. Radioactivity of the DNA bound to nuclear scaffold proteins was determined by an imaging analyzer. The values are averages of triplicate assays. (B) Effects of DM on the binding of S/MAR DNA to nuclear scaffold proteins. The DNA was incubated with a fixed amount of scaffold proteins and competitor DNA in the presence of increasing concentrations of DM, and processed as in (A).*

changed similarly in the total nuclear protein fraction (Figure 7B). One spot (b) showed an increase in the nuclear matrix fraction, but did not in the total nuclear protein fraction. To examine the time-dependent fluctuation of the altered proteins, samples were prepared from HeLa cells cultured for 0, 24, 48 and 72 h in the presence of CldU, and subjected to the same 2-D PAGE. They began to change within 24h and then continued to change up to 72 h after addition of CldU.

We identified the protein spots by MALDI-TOF and ESI-Q-TOF mass spectrometry. The most abundant protein matched to nucleophosmin /B23, a major protein of the internal nuclear matrix of HeLa cells [45]. The protein that moved to the matrix upon addition of CldU matched to heterogeneous nuclear ribonucleoproteins C1 (hnRNP C1) [46], another major protein of the internal nuclear matrix in HeLa

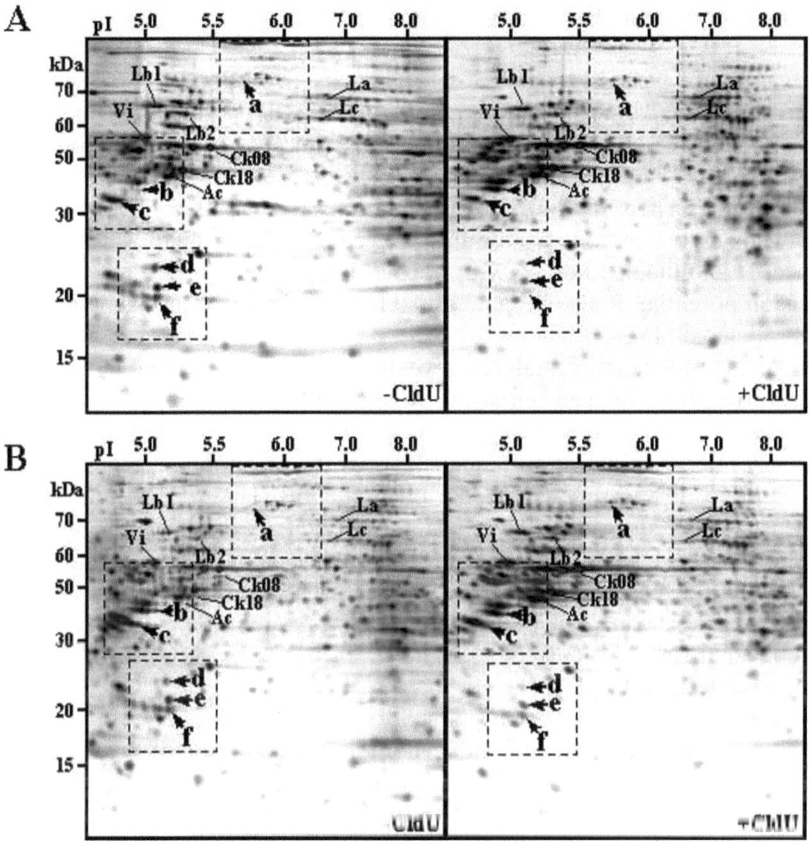

Figure 7. *2-D PAGE images of nuclear matrix and total nuclear proteins in HeLa cells. Cells were cultured in the presence (right panel) and absence (left panel) of 50 μmol/L CldU for 3 days. Nuclear matrix proteins (A) and total nuclear proteins (B) were prepared, and subjected to 2-D PAGE as described [35]. The spots a–f were found to be altered in amount by addition of CldU. Several known spots were indicated: Ac, actin; Ck08, cytokeratins 8; Ck18, cytokeratins 18; La, nuclear lamin A; Lb1, nuclear lamin B1; Lb2, nuclear lamin B2; Lc, nuclear lamin C; and Vi, vimentin.*

cells. hnRNPs contain C1 and C2 splicing variants differing in the presence of an additional 13 amino acids in C2. C1 and C2 proteins form stable heterotetramers ($C1_3C2$) that bind cooperatively to RNA [47, 48]. Biochemical functions of hnRNP C are unknown, but it is recently suggested to be required for mouse postimplantation development although dispensable for cell viability [49].

Concluding remarks and remaining issues

Immortal cells are thought to have lost a capacity to respond to stresses to enter premature senescence. Contrary to this notion, HdU induced a phenomenon indistinguishable form replicative or premature senescence in any types of mammalian cells. The HdU-mediated senescence-like system provides several advantages in the study of senescence. It enables us to obtain a large number of homogeneously senescent cells in a short time from immortal cells, enabling to study early events in senescence. It is difficult to do so with normal human cells due to heterogeneity in their division potential. It also became possible to study senescence genetically using standard rodent cell lines.

Biochemical and genetic analyses revealed that none of the known genes is involved in this HdU-induced senescence-like phenomenon. Therefore, HdU might activate a novel senescence pathway that is common to normal and immortal cells. Alternatively, BrdU might simultaneously activate or suppress particular genes, thereby manifesting the senescence-like phenomenon. These genes may well play roles downstream of stress-response machineries that may be genetically impaired in immortal cells. BrdU is incorporated into DNA in place of thymidine and shown to change interaction between DNA and proteins such as histones and transcription factors [50, 51]. This change is thought to alter expression of a class of genes by an unknown mechanism. This current view, however, has been unsuccessful to disclose the molecular mechanism of BrdU.

As the senescence-associate genes, majority of which are located on G-bands [19], are immediately and independently induced upon incorporation of BrdU into DNA, a change in chromatin structures rather than in specific transcription factors seems to be involved in our senescence-like phenomena. In fact, it is known that BrdU decondenses hetrochromatin in mitotic chromosomes, especially in human chromosome 1, 9, 15, 16 and Y [52, 53]. Consistently, BrdU is found to induce transcripts highly homologous to the AT-rich human satellite III sequences located in the constitutive heterochromatic region 9q12 [12]. Therefore, we hypothesize that S/MAR sequences might be involved in the BrdU-mediated induction of the senescence genes.

The *in vitro* binding assay revealed that S/MAR DNA more tightly binds to the nuclear scaffolds or matrix upon substitution of thymine with 5-bromouracil. Many reports show that ligation of a S/MAR sequence to reporter gene constructs dramatically increase their expression when the constructs are integrated onto the host genome [54]. It is also reported that disruption of the gene encoding a specific S/MAR binding protein SATB1 modulates expression of a set of tissue-specific genes [55]. These findings are in good agreement with the above hypothesis.

On the other hand, the AT- binding ligands are well known to interfere with the association of AT-rich sequences with their specific DNA-binding proteins. The proteins affected by DM include AT-hook proteins [56], S/MAR binding proteins [57, 58], an ARID family (AT-rich interactive domain) of proteins [59], members of SWI-SNF remodeling complexes [60], and others. These proteins act as transcriptional activators or suppressors by recruiting transcription factors to target genes.

DM was also shown to decondense centromeric heterochromatin competitively with histone H1 [29] and HMG-I [61] and to interfere with binding of the nuclear matrix or scaffold to S/MARs *in vivo* and *in vitro* [29, 34]. S/MARs can modulate chromatin structure by competitively binding two types of proteins [29, 30]. One type of proteins, as exemplified by histone H1 and topoisomerase II, cooperatively binds to S/MARs to lead to compaction of the chromatin fibers, chromosome condensation, or heterochromatin formation. The other type of proteins, such as HMG-I and its derivative MATH20, non-cooperatively bind to S/MARs to displace the above compacting proteins by disrupting their cooperative interactions, hence resulting in chromatin accessibility. Apparently, the roles of BrdU and the AT-binding ligands seem to contradict as regards their effects on the binding of S/MAR sequences to the nuclear scaffolds or matrix. This contradiction may be solved by the *in vitro* experiments that substitution of thymine with 5-bromouracil changed the conformation of the rat S/MAR DNA leaving the affinity of some of its AT-tracts to DM unaltered. Notably, the S/MAR DNA containing 5-bromouracil tightly bound to the nuclear scaffold and this binding was not weakened by DM. This increased and persistent interaction between S/MAR DNA and the nuclear scaffolds, and the opening of condensed chromatin in AT-rich regions by simultaneous addition of BrdU and AT-binding ligands are thought to have synergistically effects on the senescence-associated genes.

To date, changes in nuclear matrix proteins have been reported during cellular senescence although none has been identified [62, 63]. In HeLa cells undergoing the senescence-like phenomenon, we have found that at least five nuclear matrix proteins increase or decrease in amount by addition of CldU to the cells. These results seem to be reasonable because nuclephsomin/B23 and hnRNP C1/C2 are thought to be involved in RNA metabolism, and in aged organisms the abilities to synthesize RNA and protein are shown to decrease. The change in the location of hnRNP C1 is particularly interesting since this observation is in agreement with our hypothesis that an interaction between S/MAR DNA and the nuclear matrix should be responsible for the effects of HdU. Although the hnRNP C1/C2 protein is thought to be involved in RNA metabolism, a proteolytic product of the C proteins influences the activity of DNA polymerase, suggesting that hnRNPs C1 and C2 may also influence some aspects of DNA metabolism [64]. As the nuclear matrix play an important role in gene expression, the change in the distribution of hnRNP C1 may affect expression of particular genes.

The several novel findings described here will facilitate to solve the old and new question why BrdU modulates expression of specific genes. This, in turn, may lead to the molecular mechanisms of replicative and premature senescence in normal cells.

References

1. Hayflick L (1965). The limited *in vitro* lifetime of human diploid cell strains. *Exp Cell Res.* 37: 614–36.
2. Smith JR, Pereira-Smith OM (1996). Replicative senescence: implications for *in vivo* aging and tumor suppression. *Science* 273: 63–7.

3. Campisi J (1997). The biology of replicative senescence. *Eur J Cancer* 33: 703–9.
4. Harley CB (1991). Telomere loss: mitotic clock or genetic time bomb? *Mutat. Res.* 256: 271–82.
5. Von Zglinicki T, Saretzki G, Docke W, Lotze C (1995). Mild hyperoxia shortens telomeres and inhibits proliferation of fibroblasts: a model for senescence? *Exp Cell Res.* 220: 186–93.
6. Serrano M, Lin AW, McCurrach ME, Beach D, Lowe SW (1997). Oncogenic ras provokes premature cell senescence associated with accumulation of p53 and $p16^{INK4a}$. *Cell* 88: 593–602.
7. Ogryzko VV, Hirai TH, Russanova VR, Barbie DA, Howard BH (1996). Human fibroblast commitment to a senescence-like state in response to histone deacetylase inhibitors is cell cycle dependent. *Mol Cell Biol.* 16: 5210–18.
8. Michishita E, Nakabayashi K, Ogino H, Suzuki T, Fujii M, Ayusawa D (1998). DNA Topoisomerase Inhibitors Induce Reversible Senescence in Normal Human Fibroblasts. *Biophys Biochem Res Commun.* 253: 667–71.
9. Michishita E, Nakabayashi K, Suzuki T, *et al.* (1999). 5-Bromodeoxyuridine induces senescence-like phenomena in mammalian cells regardless of cell type or species. *J Biochem (Tokyo)* 126: 1052–9.
10. Michishita E, Matsumura N, Kurahashi T, *et al.* (2002). 5-Halogenated thymidine analogues induce a senescene-like phenomenon in HeLa cells. *Biosci Biotechnol Biochem.* 66: 877–9.
11. Morris SM (1991). The genetic toxicology of 5-bromodeoxyuridine in mammalian cells. *Mutation Res.* 258: 161–88.
12. Suzuki T, Minagawa S, Michishita E, *et al.* (2001). Induction of senescence-associated genes by 5-bromodeoxyuridine in HeLa cells. *Exp Gerontol.* 36: 465–74.
13. Wadhwa R, Kaul SC, Ikawa Y, Sugimoto Y (1993). Identification of a novel member of mouse hsp 70 family, its association with cellular mortal phenotype. *J Biol Chem.* 268: 6615–21.
14. Wadhwa R, Pereira-Smith OM, Reddel RR, Sugimoto Y, Mitsui Y, Kaul SC (1995). Correlation between complementaion group for immortality and the cellular distribution of mortalin. *Exp Cell Res.* 216: 101–5.
15. Fujii M, Ito H, Hasegawa T, Suzuki T, Adachi N, Ayusawa D (2002). 5-Bromo-2′-deoxyuridine efficiently suppresses division potential of the yeast Saccharomyces cerevisiae. *Biosci Biotechnol Biochem.* 66: 906–9.
16. Suzuki T, Yaginuma M, Oishi T, *et al.* (2001). 5-Bromodeoxyuridine suppresses position effect variegation of transgenes in HeLa cells. *Exp Cell Res.* 266: 53–63.
17. Girard F, Bello B, Laemmli UK, Gehring WJ (1998). *In vivo* analysis of scaffold-associated regions in Drosophila: a synthetic high-affinity SAR binding protein suppresses position effect variegation. *EMBO J.* 17: 2079–85.
18. Melnikova AF, Zasedatelev AS, Kolchinsky AM, *et al.* (1975). Accessibility of the minor groove of DNA in chromatin to the binding of antibiotics netropsin and distamycin A. *Mol Biol Rep.* 2: 135–42.
19. Suzuki T, Michishita E, Ogino H, Fujii M, Ayusawa D. (2002). Synergistic induction of the senescence-associated genes by 5-bromodeoxyuridine and AT-binding ligands in HeLa cells. *Exp Cell Res.* 276: 174–84.
20. Reeves R (2001). Molecular biology of HMGA proteins: hubs of nuclear function. *Gene* 277: 63–81.
21. Strick R, Laemmli UK (1995). SARs are cis DNA elements of chromosome dynamics: synthesis of a SAR repressor protein. *Cell* 83: 1137–48.

22. Amirand C, Viari A, Ballini JP, *et al.* (1998). Three distinct sub-nuclear populations of HMG-I protein of different properties revealed by co-localization image analysis. *J Cell Sci.* 111: 3551–61.
23. Kaback MM, Saksela E, WJM (1964). The effect of 5-bromdeoxyuridine on human chromosomes. *Exp Cell Res.* 34: 182–212.
24. Zakharov AF, Baranovskaya LI, Ibraimov AI, Benjusch VA, Demintseva VS, Oblapenko NG (1974). Differential spiralization along mammalian mitotic chromosomes. II. 5-bromodeoxyuridine and 5-bromodeoxycytidine-revealed differentiation in human chromosomes. *Chromosoma* 44: 343–59.
25. Ronne M, Thust R (1983). Chromosome banding by *in vitro* exposure to dA-dT probes and BUdR. Relationships between DNA base clusters, replication pattern, and banding. *Hereditas* 99: 245–50.
26. Matsukuma S, Utakoji T (1978). Asymmetric decondensation of the L cell heterochromatin by Hoechst 33258. *Exp Cell Res.* 113: 453–5.
27. Gasser SM, Amati BB, Cardenas ME, Hofmann JF (1989). Studies on scaffold attachment sites and their relation to genome function. *Int Rev Cytol.* 119: 57–96.
28. Mirkovitch J, Mirault ME, Laemmli UK (1984). Organization of the higher-order chromatin loop: specific DNA attachment sites on nuclear scaffold. *Cell* 39: 223–32.
29. Kas E, Poljak L, Adachi Y, Laemmli UK (1993). A model for chromatin opening: stimulation of topoisomerase II and restriction enzyme cleavage of chromatin by distamycin. *EMBO J.* 12: 115–26.
30. Zhao K, Kas E, Gonzalez E, Laemmli UK (1993). SAR-dependent mobilization of histone H1 by HMG-I/Y *in vitro*: HMG-I/Y is enriched in H1-depleted chromatin. *EMBO J.* 12: 3237–47.
31. Yamamura J, Nomura K (2001). Analysis of sequence-dependent curvature in matrix attachment regions. *FEBS Lett.* 489: 166–70.
32. Nakamura K, Ikeda Y, Iwakami N, Hibino Y, Sugano N (1991). Bending of a highly repetitive component in rat nuclear DNA. *Biochem Int.* 25: 355–62.
33. Ogino H, Fujii M, Satou W, Suzuki T, Michishita E, Ayusawa D (2002). Binding of 5-bromouracil-containing S/MAR DNA to the nuclear matrix. *DNA Res.* 9: 25–9.
34. Kas E, Izaurralde E, Laemmli UK (1989). Specific inhibition of DNA binding to nuclear scaffolds and histone H1 by distamycin. The role of oligo(dA).oligo(dT). tracts. *J Mol Biol.* 210: 587–99.
35. Michishita E, Kurahashi T, Suzuki T, *et al.* (2002). Changes in nuclear matrix proteins during the senescence-like phenomenon induced by 5-chlorodeoxyuridine in HeLa cells. *Exp Gerontol.* 37: 885–90.
36. Berezney R, Coffey DS (1974). Identification of a nuclear protein matrix. *Biochem Biophys Res Commun.* 60: 1410–17.
37. Cook PR (1988). The nucleoskeleton: artefact, passive framework or active site? *J Cell Sci.* 90: 1–6.
38. Stuurman N, de Jong Land van Driel R (1992). Nuclear frameworks: concepts and operational definitions. *Cell Biol Int Rep.* 16: 837–52.
39. Berezney R, Mortillaro MJ, Ma H, Wei X, Samarabandu J (1995). The nuclear matrix: a structural milieu for genomic function. *Int Rev Cytol.* 162A: 1–65.
40. Getzenberg RH (1994). Nuclear matrix and the regulation of gene expression: tissue specificity. *J Cell Biochem.* 55: 22–31.
41. Jackson DA, Cook PR (1995). The structural basis of nuclear function. *Int Rev Cytol.* 162A: 125–49.

42. Nickerson JA, Blencowe BJand Penman S (1995). The architectural organization of nuclear metabolism. *Int Rev Cytol.* 162A: 67–123.
43. van Driel R, Wansink DG, van Steensel B, Grande MA, Schul W, de Jong L (1995). Nuclear domains and the nuclear matrix. *Int Rev Cytol.* 162A: 151–89.
44. Verheijen R, van Venrooij W, Ramaekers F (1988). The nuclear matrix: structure and composition. *J Cell Sci.* 90: 11–36.
45. Mattern KA, Humbel BM, Muijsers AO, de Jong L, van Driel R (1996). hnRNP proteins and B23 are the major proteins of the internal nuclear matrix of HeLa S3 cells. *J Cell Biochem.* 62: 275–89.
46. Stone JR, Collins T (2002). Rapid phosphorylation of heterogeneous nuclear ribonucleoprotein C1/C2 in response to physiologic levels of hydrogen peroxide in human endothelial cells. *J Biol Chem.* 277: 15621–8.
47. McAfee JG, Soltaninassab SR, Lindsay ME, LeStourgeon WM (1996). Proteins C1 and C2 of heterogeneous nuclear ribonucleoprotein complexes bind RNA in a highly cooperative fashion: support for their contiguous deposition on pre-mRNA during transcription. *Biochemistry* 35: 1212–22.
48. Merrill BM, Barnett SF, LeStourgeon WM, Williams KR (1989). Primary structure differences between proteins C1 and C2 of HeLa 40S nuclear ribonucleoprotein particles. *Nucleic Acids Res.* 17: 8441–9.
49. Williamson DJ, Banik-Maiti S, DeGregori J, Ruley HE (2000). hnRNP C is required for postimplantation mouse development but is dispensable for cell viability. *Mol Cell Biol.* 20: 4094–105.
50. Bick MD, Devine EA (1977). Interaction of chromosomal proteins with BrdU substituted DNA as determined by chromatin-DNA competition. *Nucleic Acids Res.* 4: 3687–700.
51. Lin S, Lin D, Riggs AD (1976). Histones bind more tightly to bromodeoxyuridine-substituted DNA than to normal DNA. *Nucleic Acids Res.* 3: 2183–91.
52. Kaback MM, Saksela E, WJ M (1964). The effect of 5-bromdeoxyuridine on human chromosomes. *Exp Cell Res.* 34: 182–212.
53. Zakharov AF, Baranovskaya LI, Ibraimov AI, Benjusch VA, Demintseva VS, Oblapenko NG (1974). Differential spiralization along mammalian mitotic chromosomes. II. 5-bromodeoxyuridine and 5-bromodeoxycytidine-revealed differentiation in human chromosomes. *Chromosoma* 44: 343–59.
54. Bode J, Benham C, Knopp A, Mielke C (2000). Transcriptional augmentation: modulation of gene expression by scaffold/matrix-attached regions (S/MAR elements). *Crit Rev Eukaryot Gene Expr.* 10: 73–90.
55. Alvarez JD, Yasui DH, Niida H, Joh T, Loh DY, Kohwi-Shigematsu T (2000). The MAR-binding protein SATB1 orchestrates temporal and spatial expression of multiple genes during T-cell development. *Genes Dev.* 14: 521–35.
56. Aravind L, Landsman D (1998). AT-hook motifs identified in a wide variety of DNA-binding proteins. *Nucleic Acids Res.* 26: 4413–21.
57. Hibino Y, Nakamura K, Tsukada S, Sugano N (1993). Purification and characterization of nuclear scaffold proteins which bind to a highly repetitive bent DNA from rat liver. *Biochim Biophys Acta* 1174: 162–70.
58. Kipp M, Gohring F, Ostendorp T, *et al.* (2000). SAF-Box, a conserved protein domain that specifically recognizes scaffold attachment region DNA. *Mol Cell Biol.* 20: 7480–9.
59. Kortschak RD, Tucker PW, Saint R (2000). ARID proteins come in from the desert. *Trends Biochem Sci.* 25: 294–9.
60. Nie Z, Xue Y, Yang D, *et al.* (2000). A specificity and targeting subunit of a human SWI/SNF family-related chromatin-remodeling complex. *Mol Cell Biol.* 20: 8879–88.

61. Radic MZ, Saghbini M, Elton T S, Reeves R, Hamkalo BA (1992). Hoechst 33258, distamycin A, and high mobility group protein I (HMG-I) compete for binding to mouse satellite DNA. *Chromosoma* 101: 602–8.
62. Dell'Orco RT, Whittle WL (1994). Nuclear matrix composition and *in vitro* cellular senescence. *Exp Gerontol.* 29: 139–49.
63. Pienta KJ, Getzenberg RH, Coffey DS (1992). Characterization of nuclear morphology and nuclear matrices in ageing human fibroblasts. *Mech Ageing Dev.* 62: 13–24.
64. Herrick G, Delius H, Alberts B (1976). Single-stranded DNA structure and DNA polymerase activity in the presence of nucleic acid helix-unwinding proteins from calf thymus. *J Biol Chem.* 251: 2142–6.

Use of Hammerhead Ribozymes for Aging and Cancer

Hiroshi Minoshima[1], Eigo Suyama[1,2], Renu Wadhwa[2] and Kazunari Taira[1,2]

[1]Department of Chemistry and Biotechnology, School of Engineering, The University of Tokyo, Hongo, Tokyo 113-8656, Japan; [2]Gene Function Research Center, National Institute of Advanced Industrial Science and Technology (AIST), Central 4, 1-1-1 Higashi, Tsukuba Science City 305-8562, Japan

Introduction

"Ribozymes" are RNA molecules with enzymatic properties [1]. These are naturally occurring moieties that provide fine-tuning of the gene expression cascades (replication, transcription and translation) by catalyzing essential steps such as cleavage and ligation [2]. Hammerhead ribozymes (for resemblance of their two-dimensional structure to a hammerhead) are the smallest ribozymes that are used as molecular scissors in molecular biology and biotechnology to elucidate and eliminate gene functions. The RNA is induced to fold into its active conformation by the binding of metal ions. It forms two domains; the scaffold ("domain 2") on which the ribozyme is built and the active centre (catalytic domain called "domain 1") of the ribozyme [3, 4]. Over the last two decades, mechanism of action of hammerhead ribozymes describing the requirement of divalent metal ion, definition of the catalytic domains and sequence specificity usually referred to as target site have been largely demonstrated [4–6]. Ribozyme activity *in vivo* critically depends on its effective level of expression, specificity, intracellular stability, target colocalization and accessibility to the target sites [7]. These technical issues imposed a major difficulty in the use of ribozymes *in vivo*. Various modifications of the ribozyme expression plasmid have subsequently thus come into play. For example, it was shown that the ribozymes expressed under the control of RNA polymerase III promoter (tRNAVal promoter) were efficiently expressed, highly stable and were exported to the cytoplasm [7]. Such expression improved the ribozyme activity *in vivo* by many folds. However, efficiently expressed and highly stable ribozymes may still lack action due to their inaccessibility to the target sites that in turn imposed a major hurdle because of the

Sunil C. Kaul and Renu Wadhwa (eds.), Aging of Cells In and Outside the Body, 257–267.

unforeseeable secondary and tertiary RNA structures. Recently, this obstacle was elegantly resolved by modifying the ribozyme expression plasmids. A novel hybrid ribozyme that combined the cleavage activity of hammerhead ribozyme with the unwinding activity of the endogenous RNA helicase was developed. Such RNA helicase based hybrid ribozymes expressed from RNA polymerase III promoter ($tRNA^{Val}$) were indeed shown to have substrate-unwinding as well as strong cleavage activity [8, 9]. Such hybrid ribozymes can attack most, if not all, sites within mRNA. Thus, it was possible to prepare libraries of hybrid ribozymes with randomized binding arms and to examine their effects after they have been introduced into cells. This novel system has many potential applications such as, elucidation of molecular mechanism of biological phenotypes including aging and cancer by identification of novel genes and their intervention by gene targeting. This chapter is dedicated to the molecular and technical details on the use of hammerhead ribozymes and their improved versions for identification of novel functional genes and their role in biological phenotypes.

Shorter and smarter ribozymes (minizymes and maxizymes)

Ribozymes first recognized in the satellite RNAs of certain viruses have three duplex stems and a conserved "core" of two non-helical segments that are responsible for the self-cleavage reaction (*cis* reaction). These studies led to the construction of hammerhead ribozymes that are capable of cleaving other RNA molecules (*trans* reaction) in a highly sequence specific way and thus could be used as molecular scissors. These *trans*-acting hammerhead ribozymes consist of antisense sections (stems I and stem III) and a catalytic core with a flanking stem-loop II section. These have been the focus of extensive analysis for their mechanism of action and potential as therapeutic agents (Figure 1) [5].

The catalytic domain captures the Mg^{2+} ion that is indispensable for catalytic activity [10–12]. Several studies have proved the activity of hammerhead ribozymes *in vitro*. We have recently used hammerhead ribozymes to understand the differential regulation of $p19^{ARF}$-p53–$p21^{WAF1}$ senescence pathway in mouse and human cells. We had previously cloned farnesylated protein Pex19p (essential for peroxisomal biogenesis) as a binding partner of p19ARF and found that it interacts with p19ARF in the cell cytoplasm resulting in its cytoplasmic retention and functional repression in mouse cells [13]. Human ARF protein (called $p14^{ARF}$) is truncated at the carboxy-terminus and was shown to lack the Pex19p binding region. Due to lack of interactions with Pex19p, human ARF acts as a stronger tumor suppressor and provides a stronger tumor surveillance. By using hammerhead ribozymes for Pex19p we have demonstrated that its cleavage enhanced the p19ARF, but not p14ARF function [14]. Similar use of hammerhead ribozymes can be employed for understanding of gene functions and regulations involved in cellular senescence and cancer.

Size and sufficient activity were of major concern for *in vivo* and medical applications of hammerhead ribozymes. Subsequently, short ribozymes (minizymes) were generated [15]. Although minizymes were more compatible for *in vivo* use, most

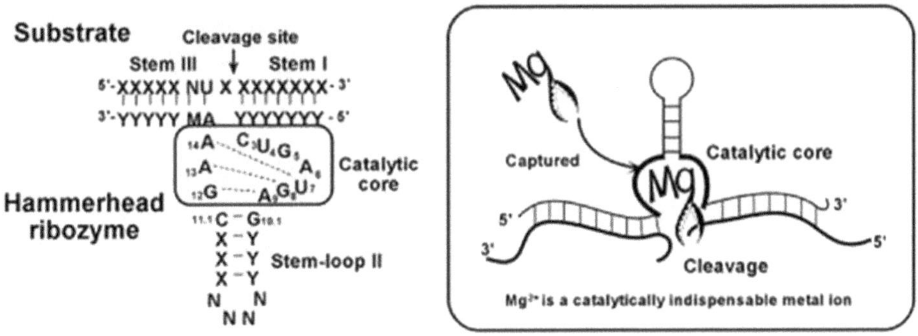

Figure 1. Schematic representation of the secondary structure of the hammerhead ribozyme and its mode of action. The hammerhead ribozyme consists of a substrate-binding region (stems I and III) and a catalytic core with a stem-loop II region. Using a construct in which the substrate-binding region of the ribozyme is complementary to the target RNA (locations marked "X-Y" and "N-M" in the substrate-binding region), we can create molecular scissors that cleave the target RNA in a site-specific manner. When the catalytic core captures the catalytically indispensable Mg^{2+} ions, cleavage occurs only adjacent to the site with the triplet sequence NUX (where N is any base and X is A, C, or U).

of them showed lower activity than their parental full-sized ribozymes. Detailed analysis and subsequent modifications brought in the minizymes that lacked the entire stem-loop II region. These short and smart ribozymes exhibited essentially the same activity as the wild-type parental ribozymes [15]. Studies on their reaction kinetics and NMR spectroscopy elucidated that these modified ribozymes were essentially inactive as a monomer but had extremely high cleavage activity as a dimer. These active ribozyme were named as "dimeric minizyme" and subsequently as "maxizyme" [16–19]. The maxizyme has sensor arms that recognize two target sequences that induce it to form a cavity for capturing Mg^{2+} ions essential for catalysis. The activity of the heterodimeric ribozyme can be allosterically controlled through one of the two substrate-binding regions. A heterodimeric system in which two different monomers, maxizyme left (MzL) and maxizyme right (MzR) were combined to form a dimeric structure that can cleave the target substrate as shown in Figure 2.

Maxizymes, which form highly active dimers, have two substrate-binding regions and have been used to cleave target mRNAs efficiently both *in vitro* and *in vivo* [16–19]. The maxizymes can bind to two different target sites to develop a system that can be used to inactivate gene expression. For example, maxizymes with tumor suppressive effect were demonstrated. Chronic myelogenous leukemia (CML) is a haematopoietic malignant disease associated with the expression of a chimeric BCR-ABL gene. Ribozyme that can exclusively disrupt chimeric *BCR-ABL* mRNA and is neutral to normal *ABL* mRNA was designed. Indeed, it specifically cleaved *BCR-*

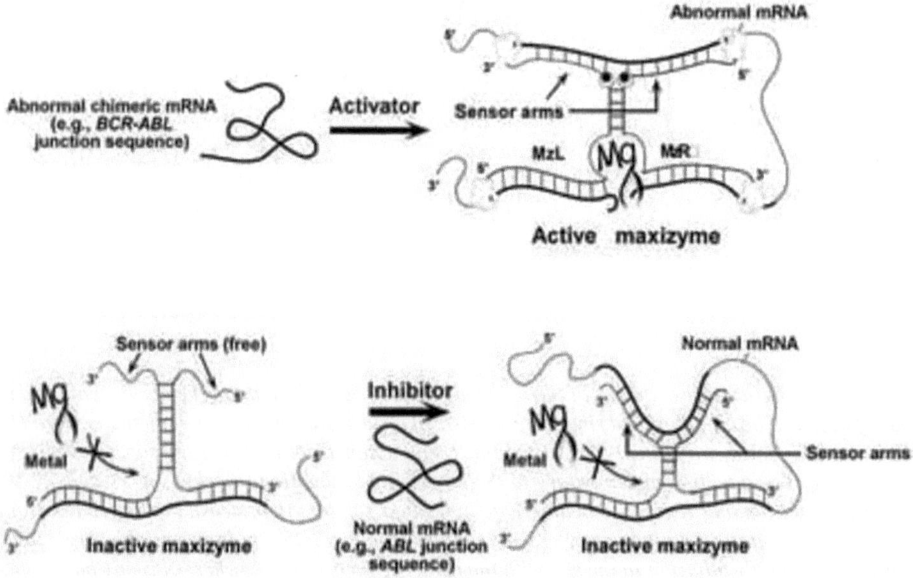

Figure 2. Development of an allosterically controllable maxizyme. Formation of active or inactive maxizymes via dimerization that is regulated allosterically by specific effector sequences. The heterodimer (MzL and MzR) can generate two different binding sites: one is complementary to the sequence of interest (activator or inhibitor), the other is complementary to a cleavable sequence. In order to achieve high substrate-specificity, the maxizyme should be in an active conformation only in the presence of the target sequence, which in this case is abnormal BCR-ABL junction, while the conformation should remain inactive in the presence of normal ABL mRNA or in the absence of the BCR-ABL junction.

ABL mRNA and induced apoptosis in cultured CML cells. Only in the presence of the correct target sequence, i.e., chimeric *BCR-ABL* mRNA, heterodimer form the cavity that can capture catalytically indispensable Mg^{2+} ions. The maxizyme remains in an inactive conformation in the presence of normal *ABL* mRNA or in the absence of the junction sequence. To apply maxizyme technology to the treatment of CML, we used a retroviral system for the expression of the maxizyme in leukemic cells. A line of CML cells (BV173) was either transfected with the control or the maxizyme-encoding vector. We injected the transfected cells into the tail vein of individual NOD-SCID mice. The differences between the two groups of mice were clearly reflected in the mortality rates (Figure 3). All mice that had been injected with the control BV173 cells died of diffuse leukemia, confirmed at necropsy, 6–13 weeks after the intravenous injection (median survival time, 9 weeks). In contrast, mice injected with the maxizyme-transfected BV173 cells remained disease-free. These results demonstrated *in vivo* validity of this technology [19].

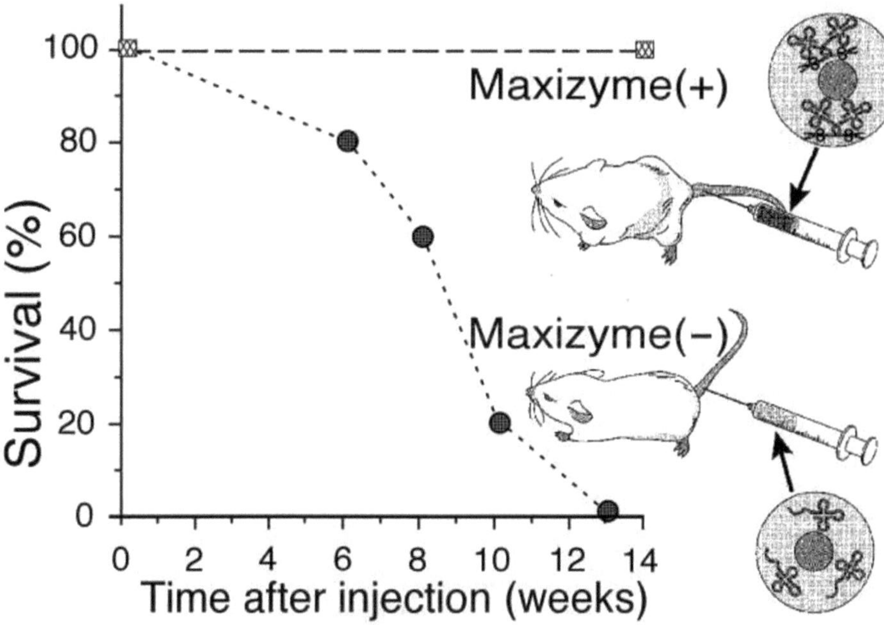

Figure 3. *Reduction of the tumorigenicity of BV173 cells. A total of 2×10^6 BV173 cells that were transfected with either the control or maxizyme vector were selected by incubation with puromycin and injected into the tail vein of NOD-SCID mice. The survival of animals was monitored daily for more than 20 weeks after inoculation. All control mice died within 13 weeks, whereas mice injected with maxizyme-treated cells remained disease-free for the entire period of the investigation.*

Sliding ribozymes (hybrid ribozymes)

The rate-limiting step for the cleavage of phosphodiester bonds *in vivo* is the association and annealing of the ribozyme with its target site [7]. In a long RNA molecule, with its secondary and tertiary folded structure, a significant number of target sites are inaccessible to the ribozyme. This phenomenon is often a serious problem in attempts to exploit ribozyme activity, in particular, *in vivo*. To overcome problems of accessibility, computer-generated predictions of secondary structure can be typically used to identify target sites that are most likely to be in an open and accessible conformation. However, such predictions are often inaccurate because of unpredictable interactions between RNAs and proteins that influence the intracellular structures of RNAs. To circumvent this problem, some researchers have employed an unwieldy systematic approach that involves the use of a large number of candidate antisense molecules [20].

A significant recent modification of ribozymes for its improved target accessibility involved its coupling with a protein with RNA-unwinding activity [8]. Ribozyme was linked to an RNA helicase with nonspecific RNA-binding, sliding and unwinding activities (Figure 4) via an RNA motif, the RNA helicase-binding motif (RBM), that interacts with RNA helicases *in vitro* and *in vivo* [8]. Unwinding and enhanced cleavage activity of these RNA helicase coupled ribozymes was demonstrated in *in vitro* and *in vivo* assays [8, 9, 21, 22].

Such sliding ribozymes were shown to be effective for cleavage of transcripts that were difficult to be cleaved by conventional hammerhead [HH] ribozymes. For example, we have found that mortalin, a heat shock 70 family protein that is involved in cellular aging and immortalization [21] has a secondary structure not accessible to conventional hammerhead ribozymes. These ribozymes were ineffective to target mortalin expression level. Howeever, when sliding hybrid ribozymes were used for the same target sites in the mortalin mRNA, decrease in the level of mortalin was obtained and was accompanied by an induction of senescence phenotype in

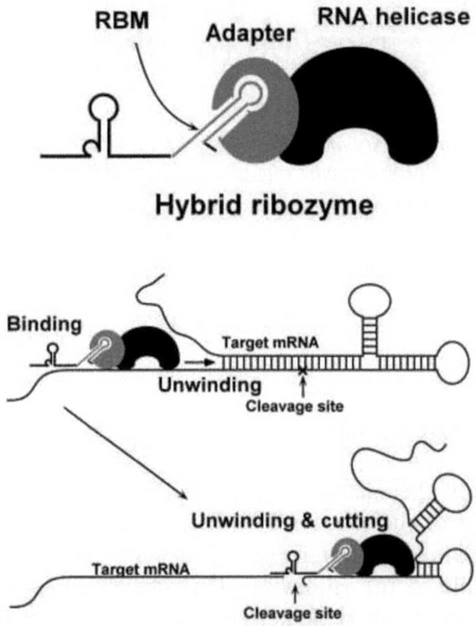

Figure 4. *Schematic representation of cleavage of a normally inaccessible target site by a sliding hybrid ribozyme. The ribozyme is linked to an RNA helicase-binding motif (RBM) that is capable of recruiting a helicase(s) to the target site where it unwinds the substrate structure to expose the cleavage site. The helicase might even be able to slide the ribozyme along the transcript. The attachment of the RBM to the ribozyme allows suppression of the expression of genes whose transcripts were found previously recalcitrant to cleavage by the conventional ribozyme.*

transformed cells [23]. Apparently, such improved ribozymes are extremely powerful tools to define and modify important gene functions.

A novel approach to clone functional genes was based on the strong activity of the sliding hybrid ribozymes. We constructed libraries of hybrid ribozymes with randomized binding arms and introduced into cells. This was combined with the selection of a cellular phenotype, isolation of ribozymes from selected cells, sequence analysis and identification of genes databases. Examples of this ribozyme based novel technique to identify genes functioning in the signaling pathway leading to Fas-induced apoptosis, metastasis and invasion. In each case, selection of a cellular phenotype in conjunction with the activity of ribozyme identified genes that have been implicated in each of these phenomena. Besides many novel genes were identified [9, 22, 24–27].

The Fas protein is a member of the tumor necrosis factor receptors family and it induces apoptosis when associated with Fas ligand or a Fas-specific antibody. Randomized ribozyme (ten nucleotides in each substrate-binding arm of an Rz-RBM) library was introduced into HeLa cells that stably expressed the gene for Fas (HeLa-Fas) followed by treatment with Fas-specific antibodies. The cells that survived were collected, analyzed for the ribozyme sequence by isolation of genomic DNA and sequencing of the randomized region of the Rz-RBM (Figure 5) [9]. Many

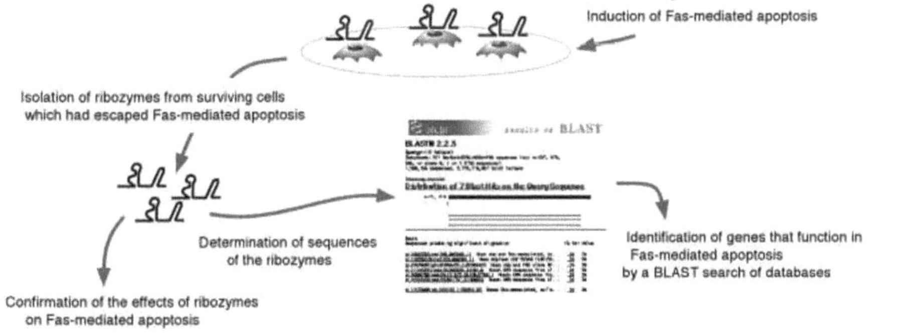

Figure 5. Schematic representation of the system for identification of functional genes in the Fas-mediated pathway to apoptosis. A randomized RBM-connected ribozyme library was introduced into HeLa-Fas cells using a retroviral expression system. Cells expressing the randomized ribozyme library were treated with Fas-specific antibody (α-Fas). After 36 h, surviving clones were picked up and the genomic DNA was purified from each clone. Sequences of RBM-linked ribozymes were determined by direct sequencing and the target genes of ribozymes were identified in databases by a BLAST search.

genes that have been demonstrated in apoptosis pathway were identified demonstrating that the use of randomized hybrid ribozymes is an efficient and reliable tool for isolation of functional genes [9, 21]. Similarly many genes responsible for metastasis were identified [26]. Since cellular motility is a prerequisite for metastasis, we chose to select cells for motility after transfection of randomized ribozyme library. A chemotaxis assay (directional migration of cells toward attractants) was performed in a Boyden chamber (Figure 6).

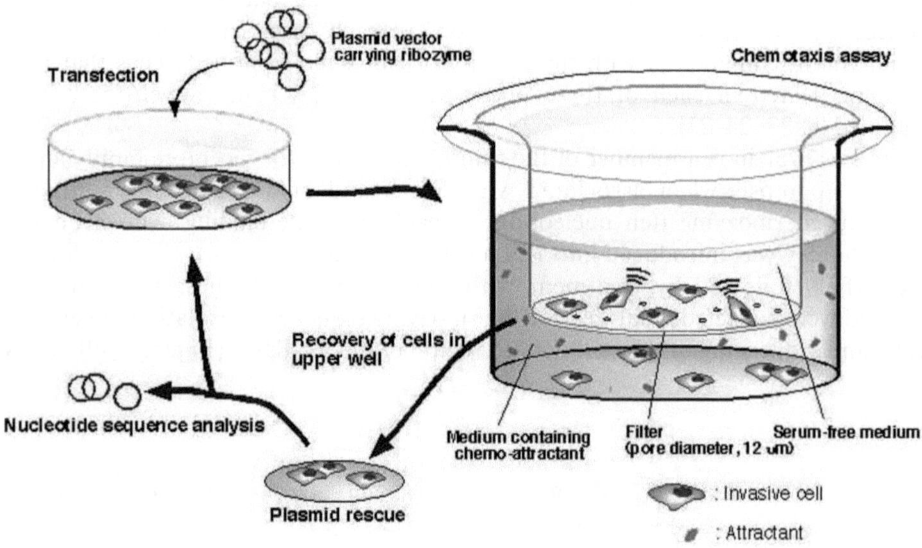

Figure 6. Schematic representation of the chemotaxis assay for gene identification using a randomized hammerhead ribozymes. Invasive cells were transfected with the ribozyme library. Then cells were subjected to the chemotaxis assay. In the chamber, cells migrated towards the chemo-attractant in the lower chamber. Non-migrating cells were recovered. Plasmid DNA for ribozymes was isolated from these cells. The isolated plasmids were used to transform cells in the next round of the assay for concentration of positive clones. Finally, ribozymes encoded by positive plasmids were sequenced and identified by DNA sequence database search.

In this assay, cell culture well is separated into two parts, one contains attractant and is separated from the other by a porous filter that allows the cells to migrate through. Metastatic fibrosarcoma (HT1080) cells essentially are highly motile. When plated in Boyden chamber, cells migrate through the pore to the chemo-attractant. The transfected variants that were affected in their migration were selected. Isolation and identification of ribozyme sequences from these derivatives resulted in identification of genes functionally involved in metastasis. These included ribozyme for ROCK1, confirmed to be involved in metastasis by regulating the actin cytoskeleton [26]. Many novel genes were also identified in the same assay and prove to be an effective novel approach for recognition of biologically important gene function. Besides, the ribozymes can be used for defining important gene functions in carcinogenesis and its therapeutics. One such recent example is the use of ribozymes for telomerase. Telomerase activity is necessary and sufficient for immortality (an essential prerequisite for tumorigenesis) of many cells and hence represents a prime target for antitumor strategies. Hammerhead ribozyme that cleaves human telomerase (hTERT) mRNA inhibited telomerase activity in breast cancer cells, resulted in their shortened telomeres and inhibition of their net growth. The data validated a ribozyme approach for telomerase inhibition therapy [27, 28].

Smarter and sliding ribozymes for aging research

Aging is a complex phenotype, expressed and regulated at all levels of organizations and involves multiple pathways and their integrated functions. It is well accepted that there is no single gene or pathway for aging as discussed in other chapters of this book. Age-related functional deterioration of biological system is the outcome of gene activities that are beneficial for the biological system in the beginning but obligatorily result in aging due to build-in deficiencies of the biological system. This includes inefficient repair and maintenance, integrated of tumor suppressor genes that protect the system from unlimited division, cancer and accumulation of damage. The identification of pathways and genes involved in aging have been difficult in past due to the inherit difficulty of obtaining the aging samples. Induction of aging by drugs in mass cell cultures although may mimic many of the characteristics of aging cells *in vivo*, yet they can be substantially different. Thus an alternate approach "escape from aging" is usually employed. Use of smarter and sliding ribozyme libraries can prove to be an efficient and rapid way of identification of major genes and gene pathways involved in aging similar to one described above for apoptosis and metastasis. Irrespective of what constitutes a major component of aging phenotype, its modulation can be attempted by altered activities of more than one kind of genes such as stress response genes, DNA repair genes and tumor suppressor genes. One possible way to achieve this goal is to use smarter and sliding ribozyme technologies as discussed in this chapter.

References

1. Cech TR (1986). Biologic catalysis by RNA. *Harvey Lect.* 82: 123–44.
2. Doudna JA, Cech TR (2002). The chemical repertoire of natural ribozymes. *Nature* 418: 222–8.
3. Cech TR, Uhlenbeck OC (1994). Ribozymes. Hammerhead nailed down. *Nature* 372: 39–40.
4. Hammann C, Lilley DM (2002). Folding and activity of the hammerhead ribozyme. *Chembiochemistry* 3: 690–700.
5. Takagi Y, Warashina M, Stec WJ, Yoshinari K, Taira K (2001). SURVEY AND SUMMARY: Recent advances in the elucidation of the mechanisms of action of ribozymes. *Nucleic Acids Res.* 29: 1815–34.

6. Pyle M (2002). Metal ions in the structure and function of RNA. *J Biol Inorg Chem.* 7: 679–90.
7. Kato Y, Kuwabara T, Warashina M, Toda H, Taira K (2001). Relationships between the activities *in vitro* and *in vivo* of various kinds of ribozyme and their intracellular localization in mammalian cells. *J Biol Chem.* 276: 15378–85.
8. Warashina M, Kuwabara T, Kato Y, Sano M, Taira K (2001). RNA-protein hybrid ribozymes that efficiently cleave any mRNA independently of the structure of the target RNA. *Proc Natl Acad Sci USA* 98: 5572–7.
9. Sawata S Komiyama M, Taira K (1995). Enhancement of the cleavage rates of DNA-armed hammerhead ribozymes by various divalent metal ions. *J Am Chem Soc.* 117: 2357–8.
10. Takagi Y, Taira K (2002). Detection of a proton-transfer process by kinetic solvent isotope effects in NH_4^+-mediated reactions catalyzed by a hammerhead ribozyme. *J Am Chem Soc.* 124: 3850–2.
11. Zhou JM, Zhou DM, Takagi Y, *et al.* (2002). Existence of efficient divalent metal ion-catalyzed and inefficient divalent metal ion-independent channels in reactions catalyzed by a hammerhead ribozyme. *Nucleic Acids Res.* 30: 2374–82.
12. Amontov SV, Taira K (1996). Hammerhead minizymes with high cleavage activity: a dimeric structure as the active conformation of minizymes. *J Am Chem Soc.* 118: 1624–8.
13. Sugihara T, Kaul SC, Kato J, Reddel RR, Nomura H, Wadhwa R (2001). Pex19p dampens the p19ARF-p53-p21WAF1 tumor suppressor pathway. *J Biol Chem.* 276:18649–52.
14. Wadhwa R, Sugihara T, Hasan MK, Taira K, Reddel RR, Kaul SC (2002). A major functional difference between the mouse and human ARF tumor suppressor proteins. *J Biol Chem.* 277: 36665–70.
15. Kuwabara T, Warashina M, Orita M, Koseki S, Ohkawa J, Taira K (1998). Formation of a catalytically active dimer by $tRNA^{Val}$-driven short ribozymes. *Nature Biotechnol.* 16: 961–5.
16. Kuwabara T, Warashina M, Tanabe T, Tani K, Asano S, Taira K (1998). A novel allosterically trans-activated ribozyme, the maxizyme, with exceptional specificity *in vitro* and *in vivo*. *Mol Cell.* 2: 617–27.
17. Kuwabara T, Warashina M, Nakayama A, Ohkawa J, Taira K (1999). tRNAVal-heterodimeric maxizymes with high potential as gene inactivating agents: simultaneous cleavage at two sites in HIV-1 Tat mRNA in cultured cells. *Proc Natl Acad Sci USA* 96: 1886–91.
18. Tanabe T, Kuwabara T, Warashina M, Tani K, Taira K, Asano S (2000). Oncogene inactivation in a mouse model. *Nature* 406: 473–4.
19. Krupp G, Gaur RK, eds. (2000). *Ribozyme: Biochemistry and Biotechnology.* Natick: Eaton Publishing.
20. Kawasaki H, Onuki R, Suyama E, Taira K (2002). Identification of genes that function in the TNF-alpha-mediated apoptotic pathway using randomized hybrid ribozyme libraries. *Nature Biotechnol.* 20: 376–80.
21. Wadhwa R, Taira K, Kaul SC (2002). Mortalin: a potential candidate for biotechnology and biomedicine. *Histol Histopathol.* 17: 1173–7.
22. Kawasaki H, Taira K (2002). A functional gene discovery in the Fas-mediated pathway to apoptosis by analysis of transiently expressed randomized hybrid-ribozyme libraries. *Nucleic Acids Res.* 30: 3609–14.
23. Wadhwa R, Ando H, Kawasaki H, Taira K, Kaul SC (2003). Conventional and RNA helicase coupled hammerhead ribozymes for mortalin. *EMBO Rep.* 4: 595–601.
24. Taira K, Warashina M, Kuwabara T, Kawasaki H (1999). Japanese Patent Application.

25. Beger C, Pierce LN, Kruger M, *et al.* (2001). Identification of Id4 as a regulator of BRCA1 expression by using a ribozyme-library-based inverse genomics approach. *Proc Natl Acad Sci USA* 98: 130–5.
26. Suyama E, Kawasaki H, Kasaoka T, Taira K (2003). Identification of genes responsible for cell migration by a library of randomized ribozymes. *Cancer Res.* 63: 119–24.
27. Ludwig A, Saretzki G, Holm PS, *et al.* (2001). Ribozyme cleavage of telomerase mRNA sensitizes breast epithelial cells to inhibitors of topoisomerase. *Cancer Res.* 61: 3053–61.
28. Yokoyama Y, Wan X, Takahashi Y, Shinohara A, Liulin T, Tamaya T (2002). Divalent hammerhead ribozyme targeting template region of human telomerase RNA has potent cleavage activity, but less inhibitory activity on telomerase. *Arch Biochem Biophys.* 405: 32–7.

Index

MIX
Papier aus verantwortungsvollen Quellen
Paper from responsible sources
FSC® C105338

If you have any concerns about our products,
you can contact us on
ProductSafety@springernature.com

In case Publisher is established outside the EU,
the EU authorized representative is:
Springer Nature Customer Service Center GmbH
Europaplatz 3, 69115 Heidelberg, Germany

Printed by Libri Plureos GmbH
in Hamburg, Germany